DL

For c

Cooperative Microeconomics

Cooperative Microeconomics:

A GAME-THEORETIC INTRODUCTION

Hervé Moulin

PRINCETON UNIVERSITY PRESS

PRINCETON, NEW JERSEY

Library of Congress Cataloging-in-Publication Data

Moulin, Hervé.
Cooperative microeconomics: a game-theoretic introduction / Hervé Moulin
p.   cm. — (???)
Includes bibliographical references and index.
ISBN 0-691-03481-8
1. Microeconomics. 2. Game theory. 3. Equilibrium (Economics)
4. Competition. I. Title.
HB172.M73   1995
338.5--dc20   95-3606

This book has been composed in 10/12 Times Roman

Princeton University Press books are printed on acid-free paper
and meet the guidelines for permanence and durability of the
Committee on Production Guidelines for Book Longevity of the
Council on Library Resources

Printed in the United States of America
10 9 8 7 6 5 4 3 2 1

This book is dedicated to Victor

Contents

Acknowledgements

THIS BOOK is the product of several years of graduate teaching micro-economics at Duke University and at the University of Aix-Marseille; my gratitude goes first of all to students who had no choice but to endure my lectures: their interested or bored reactions to the material were critically helpful.

During the writing of some of the final chapters I benefitted from the hospitality of Tel Aviv University and Bielefeld University. Special thanks are due to the Groupe de Recherches en Economie Quantitative d'Aix-Marseille and to its Director Louis André Gerard-Varet who hosted me for an entire academic year during which the key components of the manuscript were assembled. I also wish to acknowledge the material and intellectual support of Duke University throughout this entire project.

Comments and critical remarks from Marina Bianchi, William Drake, Marc Fleurbaey, Alison Watts and Jianlin Zhai were especially useful. The excellent typing skills of Barbara Boon and Grassinda das Neves deserve the last but by no means the least of these acknowledgements.

Special thanks to Hans Van Miegroet who proposed the cover of the book.

Overview of the Book

IN THE PAST fifty years, our understanding of the logic of cooperative processes has been revolutionized by the formal tools of microeconomics, in particular the (neo-classical) competitive analysis and the theory of games.

This book gives a partial account of these developments organized in a systematic conceptual framework. I submit that cooperation between selfish economic agents can be conceived in three fundamental "modes," namely, direct agreement, justice, and decentralized behavior. I argue that actual cooperative institutions must accommodate all three modes and postulate that an ideal institution is one where all three modes "converge" on the same outcome.

The first mode of cooperation is by *direct agreements* of the concerned agents. This mode is free of any institutional context, as agents engage voluntarily and freely in face-to-face transactions. Under the efficiency postulate, the formal concept of the core describes the self-enforcing agreements. A configuration where the core is empty (a not infrequent pattern) is one where cooperation by direct agreements is not workable. Another drawback of the direct agreement mode is the transaction cost: when many agents are involved and/or when many decisions must be made over a long period of time, transaction costs may become so high as to wipe out the benefits from cooperation.

Each one of the two other fundamental modes of cooperation is a solution of this difficulty. In the *justice* mode, the community of agents produces a mechanical formula to divide equitably the cooperative surplus among the concerned agents. The justification of the formula is given by a set of ethically meaningful axiomatic properties. Some difficult choices are called for when several desirable axioms turn out to be logically incompatible. The just outcome is enforced by the indivisible collective authority (akin to Rousseau's general will). The justice mode is hard to implement when some of the information necessary to compute the equitable outcome is privately held by the individual participants: they may strategically misrepresent this information, and the eventual outcome will not be just.

In the *decentralized* mode, the community chooses to distribute fully the decision power among individual participants. Cooperation then takes place in a game of strategy, and the role of the collective authority is merely to enforce the rules of the game. The actual cooperative outcome results from the (noncooperative) equilibrium of selfish interests. The decentralized mode may lead to an inefficient strategic equilibrium (as in the Prisoner's Dilemma) or to the absence of any equilibrium.

The design of a cooperative mechanism tries to combine the virtues of all three modes. An ideal mechanism is one such that i) its noncooperative equilibrium should be unique and compelling, ii) the resulting outcome should be just, and iii) this outcome is also stable in the direct agreement mode. Examples of such convergence of the three modes of cooperation are rare; indeed, a recurrent theme of this book is that the existence of an "ideal" mechanism is often a logical impossibility.

Chapter 1 presents the thesis of this book in nontechnical language.

Chapters 2 and 3 look at the exchange and production of commodities in the private property regime. The core (direct agreement mode) and the competitive equilibrium (decentralized mode) are discussed; the celebrated Edgeworth proposition establishing the conditions under which both concepts pick the same outcomes (hence two modes of cooperation converge) is discussed in some detail.

Chapters 4, 5, and 6 look at the distribution and production of (private or public) commodities in the common property regime (when all agents have equal claims on the resources). Chapter 4 discusses the interpretation of fairness as no envy (justice mode), and its relation to competitive equilibrium analysis. Chapter 5 considers the alternative interpretation of fairness as the stand alone test, and its relation to core analysis. Chapter 6 examines decentralized mechanisms for the production of private or public goods, and their relation to the above fairness properties.

Chapter 7 is a concise exposition of cooperative game theory (in the transferable utility context), focusing on core existence results (direct agreement mode) and the Shapley value (justice mode).

The book is elementary and self-contained. No attempt is made at giving complete proofs of the relevant theorems under the weakest assumptions. Instead, the intuition of the general results is developed in the simplest possible numerical examples. Numerous exercises allow the reader to appreciate the capabilities and the limits of the modeling tools.

The book lies at the crossroads of several research fields actively plowed by the literature of the last four decades. Those are competitive equilibrium analysis, cooperative game theory, collective decision (in particular, axiomatic) and fair division, mechanism design, and the implementation problem. Although all of these tightly related topics are discussed at some length, none of them is properly surveyed.

All bibliographical references are gathered at the end of the book, along with a brief list of books of general interest to our topic.

The Three Modes of Cooperation:
Agreements, Decentralization, and Justice

1.1. COOPERATION IN ECONOMIC THEORY

Cooperation in the economic tradition is mutual assistance between egoists. The archetypical example is Rousseau's staghunt: by coordinating their efforts, the two hunters more than double the returns of one isolated hunter; therefore each hunter can receive a positive profit from cooperation. Thus is cooperation justified (from the normative angle of rational pursuit of one's interest) and predicted (a positive statement about the behavior of selfish hunters).

That cooperation can be explained by the rational choice of self-interested parties, rather than by altruism, was most clearly stated in the celebrated passage from The Wealth of Nations: "In almost every other race of animals each individual, when it is grown up to maturity, is entirely independent, and in its natural state has occasion for the assistance of no other living creature. But man has almost constant occasion for the help of his brethren, and it is in vain for him to expect it from their benevolence only. He will be more likely to prevail if he can interest their self-love in his favour, and show them that it is for their own advantage to do for him what he requires of them. Whoever offers to another a bargain of any kind, proposes to do this. Give me that which I want, and you shall have this which you want, is the meaning of every such offer; and it is in this manner that we obtain from one another the far greater part of those good offices which we stand in need of. It is not from the benevolence of the butcher, the brewer, or the baker, that we expect our dinner, but from their regard to their own interest." (Smith [1784]).

The concept of Pareto optimality (also called "economic efficiency," or simply "efficiency") follows logically from this view of cooperation. An outcome x is Pareto optimal if the concerned agents cannot find another outcome y (one that they could achieve by mutual consent) such that every agent is better off in y than in x, or at least, in y no one is worse off and someone is better off (we say that outcome y is Pareto superior to outcome x). Thus if an outcome x is not Pareto optimal, it should not be the eventual outcome inasmuch as our agents can grab a cooperative opportunity by enforcing a Pareto superior outcome. Conversely, if outcome x is Pareto optimal, it cannot be dismissed by unanimous consent: x can only be rejected because it violates certain individual rights or certain principles of justice (see the discussion in Section 1.4).

Pareto optimality is the one and only uncontroversial normative argument in economic theory (the New Welfare Economics, shortly after World War II, made this claim very forcefully; see Samuelson [1947]). One of the main tasks of economic analysis (some would say its only legitimate task) is to look for ways and means to promote a Pareto optimal outcome of the economy.

In some simple cases (involving a few agents who have complete knowledge of the cooperative opportunities, as in the staghunt story, or when I agree with my neighbor that we will keep an eye on each other's homes), it is realistic to postulate that efficiency must prevail among rational agents: they will cooperate if by doing so they both benefit. Coase [1960] provides the classical statement of this *efficiency postulate*. Note that the postulate, if it considerably simplifies the discussion of microeconomic behavior, does not leave the economist unemployed. For her skill is needed to work out the analytical consequences of Pareto optimality (including such facts as the equality of marginal rates of substitution and other efficiency tests; see Chapters 2 and 3). Moreover, under the efficiency postulate the formation of coalitions and alliances cannot be dismissed: if the grand coalition captures all the cooperative opportunities, surely a smaller group of agents has no difficulty doing the same. This, in turn, raises a deep question: which efficient outcomes are stable when each and every coalition can cooperate as well? The corresponding concept of core is first discussed in Section 1.4 and is one of the key concepts of this book.

Most actual markets for the allocation of goods and services involve a large number of agents who cannot directly communicate; the efficiency postulate, then, is not a realistic assumption, because it rests on direct agreement of all the participants in the market. Economists have argued since Adam Smith that a price signal is an efficient decentralization device for the cooperative organization of the market. The advantage of the price system over face-to-face bargaining is that after learning the price, an agent need not further communicate with other participants in the market. Taking the price as given, he acts selfishly (buying or selling as much as pleases him); the results of these uncoordinated selfish decisions is, miraculously enough, a Pareto optimal outcome. Thus everyone in the market is "led by an invisible hand to promote an end (efficiency) which was no part of his intention" (Smith [1784]).

The invisible-hand metaphor captures the gist of the competitive price mechanism (discussed in Chapters 2 and 3). The model of competitive exchange, still the unchallenged masterpiece of economic theory, is the prototype of a cooperative institution which satisfies the economists (and many others). It is discussed in Section 1.6 and also in Chapters 2 and 3.

An important feature of the competitive exchange model is the absence of any reference to justice. The standard interpretation refrains from making an ethical judgment on whether or not the market price yields an equitable outcome. Moreover, a deep skepticism about arguments of justice has become a virtual trademark of the economic profession, under the venerable leadership of Adam Smith. Two sentences after the invisible-hand quotation (above), he writes: "By pursuing his own interest he frequently promotes that of the society more effectually than when he really intends to promote it. I have never known much good done by those who affected to trade for the public good." Two levels of skepticism are mixed here. On one hand, there is the suspicion that behind an argument of justice is hidden a lobbyist ("affecting" to trade for the public good) who speaks for the special interest of a subgroup.[1] On the other hand, and more importantly, there is the concern that a well-intentioned politician who invokes ethical principles to interfere with the market process (e.g., by controlling prices or by ruling out usury) is likely to be countereffective: rent control discourages investment and makes cheap housing even less available to the poor; illegal usury is more expensive because the borrower must pay a premium to insure the lender against the risk of being caught; and so on. The point is twofold: first, in order to assess the impact of a regulatory policy, we must account for its effect on the market equilibrium; second, any policy that reduces the set of feasible outcomes can only be detrimental to efficiency (by the familiar second-best argument: an outcome deemed Pareto optimal when the feasible set is constrained by the policy may become Pareto inferior when the policy is dropped and new outcomes are available). As arguments of justice inevitably reduce the feasible set (by constraints such as equal pay for equal work, no salary below a certain minimal level, and so on), so our zealous economist justifies his stubborn distrust of these arguments.

---

[1] This view, taken to the extreme, leads to a completely cynical view that all such arguments are frauds, and justice should be eliminated altogether from the scientific discourse on society ; this position is untenable, because justice accounts for so much of the political discourse, the regulatory process, and all forms of public intervention. "If normative justice does not exist, what positive theory will account for the ubiquity of the discourse of justice?" (Zajac [1985]).

The first point, about the countereffective impact of some well-mentioned regulations, is well taken; it is one of the most useful contributions of economic thinking to the social debate (going back as far as Mandeville's Tale of the Bees, where "private vices" yield "public virtues"). The second point, however—the second-best argument—has no real bite; when the goals of efficiency and of social justice are partially incompatible, we may choose to give priority to efficiency, but the opposite choice is morally justifiable as well. Postal tariff is an example: charging the same price to deliver a letter anywhere is clearly inefficient; yet a nonuniform price system, that would place an additional burden on rural customers, is widely perceived as unjust and so the uniform tariff prevails. Another example is the use of lotteries for allocating public offices in the Athenian democracy or in certain Italian city-states.[2] Such a radically egalitarian electoral process had a serious impact on the incentive to do a good job while in office (a moral hazard problem, in modern economic terminology; or in the words of a 15th century scholar: "this practice of extraction by lot extinguishes any motivation for prudent conduct"; see Elster [1989], p. 85). Yet the practice survived a long time, confirming Zajac's claim that both in the political and the economic realm, fairness often has preeminence over efficiency.[3]

To a majority of economists today, the ethical choices of distributive justice are alien to economic analysis. No one, of course, will deny that ethical choices influence individual behavior and hence have important economic consequences. The standard view simply incorporates those concerns about justice (distributive and otherwise) in the description of individual characteristics: some of us derive utility from giving to the needy, some of us do not. However, it is not an economist's job to tell if one should care about the needy, any more than to tell if one ought to drink coffee with or without sugar. I will argue below (Section 1.3) that this position is untenable. Before that, a much different approach to social cooperation will be reviewed.

1.2. Cooperation in Political Theory

The object of political theory is to analyze the legitimacy of various forms of government, and to give some formal content to the concept of ideal state.

The notion of a social contract is the key concept of modern political theory. Like the market, it rests on the idea of mutual assistance: "men

---

[2] See Moore [1972], p. 276, and Elster [1989], pp. 80–87.
[3] See Zajac [1985].

being by nature all free, equal and independent, no one can be put out of this estate and subjected to the political power of another without his consent. The only way whereby any one divests himself of his natural liberty, and puts on the bonds of civil society, is by agreeing with other men to join and unite into a community for their comfortable, safe and peaceable living one amongst another, in a secure enjoyment of their properties and a greater security against any that are not of it" (Locke [1690], p. 95). Yet, unlike the market, it is legitimate because it is just. "Men are born free and equal in rights" says the Declaration des Droits de l'Homme, and, "the maintenance of his rights is the only object of political union, and the perfection of the social art consists in preserving them with the most entire equality, and in their fullest extent [. . .]. The will of the majority is the only principle which can be followed by all, without infringing upon the common equity" (Condorcet [1793]).

The argument is beautifully simple. A government must provide certain services indispensable to the proper functioning of social cooperation.[4] To provide those services, the government must be vested with an authority that will infringe upon individual liberties. The only way an individual will agree to "divest himself from his natural liberty" is if he sees everyone else making the same sacrifice: as we are all born equal in rights, we will then remain so after the establishment of the government. In a practical sense, the "will of the majority" is one way to respect the equality of rights and still allow for operational decision-making (as opposed to the unanimity rule whereby the government power to act would be strictly limited to Pareto improving moves). Of course, there are some serious practical difficulties in using the will of the majority as the only benchmark for choice (these are discussed in Section 1.4). Yet the success of the contractarian tradition is unquestionable now that democracy has become, perhaps forever, the only acceptable format of political legitimacy.

The two models of democratic voting and of the competitive market are similar in two important ways. First, both institutions replace direct case-by-case agreement (practical in small-scale exchange, quite impractical for political decision-making, of universal concern by definition) by one decentralized mechanism (majority voting and competitive price, respectively). Second, a collectively optimal outcome results from the

---

[4] As Adam Smith puts it: "The sovereign has only three duties to attend to [. . .]: first the duty of protecting the society from the violence and invasion of other independent societies; secondly [. . .] the duty of establishing an exact administration of justice; and, thirdly, the duty of erecting and maintaining certain public works and certain public institutions, which it can never be for the interest of any individual, or small number of individuals, to erect and maintain" [1784].

sum of uncoordinated selfish actions by individuals (trading at the market price in one case, voting according to one's own preferences in the other case). But there are two important differences as well. In the market, the outcome is collectively optimal inasmuch as it is efficient, whereas the outcome of voting is optimal because it is just (Pareto optimality goes without saying, but it would hold even in a dictatorship!). Moreover, democratic voting is a *visible-hand* process, as the constitution must be posted for everyone to read and can also be changed, because "the right of changing (the constitution) is the security of every other right" Condorcet [1793]. To be fair, it can also be argued that the market mechanism is just (because the price is the same for everyone; see Chapter 2), and in some contexts (e.g., auction with a small number of participants) its hand is fairly visible. However, the two fundamental differences remain, namely, that economic analysis of markets is oblivious to considerations of justice, whereas political theory does not exist without them; and that the rules of the game played by political agents are constantly reasserted and occasionally changed, while in the marketplace the rules of the game are largely self-enforcing and need no monitoring.

When assessing the difference between the economic and the political theories, a crucial factor is the nature of goods and services provided by the market on one hand and the political process on the other hand. The market is well suited to allocate private goods (that can be consumed by only one agent, e.g., food), but usually fails to provide efficiently the *public* goods (those goods which all agents can consume without diminishing each other's consumption, e.g., national defense, the judicial system, and the road network). Failure of the market to provide public goods efficiently is, once again, summarized most clearly by Adam Smith: "it can never be for the interest of any individual, or small number of individuals, to erect and maintain [certain public works]; because the profit could never repay the expense to any individual or small number of individuals, though it may frequently do much more than repay it to a great society" [1784]. (Chapter 6 elaborates on this observation.)

Market failures affect also certain private goods of which the production involves increasing returns to scale (such as telephone services or railroad transportation). We speak of natural monopoly when economic efficiency commands to organize the production of a certain good by operating a single firm; in this case, the government's task is to regulate the exercise of this monopoly power (for it would be both inefficient and unjust to let that firm extract the full monopoly profit).

Thus the political process must interfere with market forces in the allocation of the "goods of the public sector," namely, public goods, and private goods produced by natural monopolies. For such goods, normative arguments of (distributive) justice will naturally complement the positive analysis of efficiency, as the discussions of Chapter 5 will make clear.

On the other hand, voting mechanisms fail when they are used to allocate purely private goods. Their failure is not a lack of efficiency, but rather the irrevocable instability of their outcome when coalitions can directly agree on a joint strategy: voting on the division of a bundle of private goods yields the most extreme instance of the Condorcet paradox (as will be explained in Section 1.4).

The economic model of the competitive market and the political model of voting mechanisms are well suited to handle respectively the allocation of "pure" private goods (produced under decreasing returns to scale) and the provision of pure public goods (the legal system, the staffing of public offices). For most goods of the public sector, however, both models (market and voting) will be useful but neither one will be enough. Education and health are two examples of such goods (they both can be supplied by private entrepreneurs, but the government must regulate their production). These two account for at least one-third of the GNP in developed economies, and the range of government regulation appears to be increasing, reaching, for instance, more and more food products and technologies with an environmental impact.

1.3. THE THEME OF THIS BOOK

I define three idealized "modes" of cooperation and observe that each mode is present in every actual cooperative institution. Any such institution can be examined in the light of the three modes; I submit that their tension and/or compatibility assesses the effectiveness and/or the limits of the institution in question.

The three modes are i) the *direct agreement* mode, where groups of agents reach a consensus after direct, face-to-face bargaining; ii) the mode of *decentralized behavior*, where decision power is fully distributed among the individual agents (whereas it is the indivisible property of all agents signing an agreement), and the collective outcome results from the strategic interplay of selfish actions; and iii) the *justice* mode, where the decision power is vested in an arbitrator (be it a mechanical arbitration formula or a benevolent dictator) whose choices follow from normative principles. As the motto of the French Republic says concisely: *Liberté* (mode of decentralized behavior), *egalité* (mode of justice), *fraternité* ("Brotherhood," i.e., mode of direct agreements)!

Each mode in its pure form is an abstraction. Real face-to-face agreements are guided by fairness principles (the equal-split rule is a shining focal point), as ample experimental evidence demonstrates (see Yaari and Bar-Hillel [1986], Roth [1988]). Real arbitrators must rely on information provided by the interested parties (monitoring perfectly the accuracy of their report is too costly); therefore, the arbitration process shares some aspects of a decentralized game. Real players in a decentralized process can often bypass decentralized behavior and agree directly with one another. A firm, as a cooperative institution, combines justice principles (e.g., regulating promotions and salary increases), decentralized behavior (as required by the division of labor), and direct agreement when unforeseen problems arise.

The numerous connections between the three modes of cooperation are the very subject matter of this book. Looking back at the two previous sections, we say that the mechanism of exchange by competitive price successfully combines the agreement and the decentralized modes of cooperation. The competitive equilibrium outcome results from the aggregation of decentralized price-taking behavior, and at the same time it is stable against direct coalitional agreements. The Edgeworth proposition (Chapters 2 and 3) gives the precise conditions under which price-taking behavior and stable coalitional agreements (the core) produce the same outcome.

Next, consider voting. A voting rule must be just; at the same time it defines a game of strategy, where admissible ballots are the individual strategies. Third, the citizens have a fundamental right to form alliances (political parties) where they directly coordinate their voting strategy; hence, direct coalitional agreements play an important role as well (see Section 1.4). The optimal design of a voting rule attempts to make the modes of justice, of decentralized behavior, and of direct agreement converge to the same outcome. As we shall see (Section 1.4 and Chapter 6), majority voting succeeds only partially in fulfilling these objectives.

The "three modes" methodology will be applied to other cooperative institutions: fair division (Chapter 4), cooperative production of private goods (Chapters 5 and 6), and provision of public goods (Chapters 5 and 6). Nonconvergence of the three modes toward a specific outcome will be the rule rather than the exception; perfect convergence occurs only in some contexts of majority voting (e.g., when preferences are single-peaked; see Sections 1.6 and 6.2) and in some contexts of competitive exchange and fair division (see Chapters 2 and 4 and Section 1.7). However, the presence of externalities in production, whether in the cooperative production of private goods or in the provision of public goods, generally rules out this threefold convergence (see Chapter 6).

To each one of the three modes of cooperation corresponds a specific formal model. The games in characteristic form (also called cooperative games) are suited to the mode of cooperation by direct agreement, as games in strategic (or normal) form (commonly called noncooperative games) are to the mode of decentralized behavior. The axiomatic method is the backbone of the justice mode. Their most important features are reviewed in the next three sections: for the mode of direct agreement, existence or nonexistence of "core" outcomes (Section 1.4; see also Chapter 7); for the justice mode, the difficulty of interpreting the egalitarian program (Section 1.5); and for the mode of decentralized behavior, the frequent inefficiency and multiplicity of equilibrium outcomes (Section 1.6).

My approach is, in Popper's terminology, piecemeal (social) engineering ([1971], p. 156), namely, looking at different problems (such as fair division, or exchange, or cost sharing of a public good) one at a time, without attempting to derive a unifying theory of social organization. The latter grand project, called "utopian (social) engineering" by Popper, has, within the field of political economy, a record of repeated failures, from Bentham's wholistic utilitarianism (a simple "arithmetics of pleasures and pains" to derive the proper course of action in any circumstances) to Nash's program (as defended by Binmore [1987]) attempting to derive universal (that is, context-free) patterns for surplus division through noncooperative bargaining.

Welfare economics traditionally deals with the allocation of goods of the public sector. In many ways, the label of theoretical welfare (micro) economics fits the second part of this book, starting with Chapter 4. The main difference with mainstream welfare economics is that the long-standing dichotomy between economic efficiency and distributive justice[5] is more vigorously rejected than is customary. First of all, I want to stress the well-known fact (see, e.g., Atkinson and Stiglitz [1980] or Starrett [1988]) that redistributive policies and the search for economic efficiency are inextricably interwined because, on one hand, an efficiency-enhancing move that would not hurt anyone's welfare would require compensation patterns of insurmountable complexity as soon as society is large, and on the other hand, any redistributive policy—e.g., the income tax rate–affects the overall output of the economy—e.g., via the incentives to work. Next, I suggest to go one step further, and address jointly the issues of efficiency, of justice and of incentive compatibility as they arise in the context of allocating goods under the common property, as well as the private property, regime. At this point,

---

[5] Resulting in statements such as "for a given income distribution, the economist's job is to recommend policy actions to move the economy toward efficiency without hurting anyone's welfare, while it is someone else's job—the politician? the expert in social justice?—to determine what the income distribution ought to be" (Zajac [1985]).

this tri-pronged methodology has been aimed with some success at only half a dozen problems; of these, this book tries to give a fair account.

1.4. DIRECT AGREEMENTS: THE EFFICIENCY POSTULATE AND THE CORE

The efficiency postulate is an impeccable normative principle (see Section 1.1), but is it a realistic *positive* property as well? In other words, can we expect actual agents to exhaust cooperative opportunities of their own accord, thus reaching a Pareto optimal outcome? The question is both essential and complex, and the only short answer is: it depends.

Under the fairly vague term of "transaction costs," Coase [1960] and others suggested gathering several factors making direct agreement more difficult, more "costly" to the interested parties. For one, those parties must have a way to communicate with each other, and the access to this medium is a primary source of transaction costs: the cost is nil when wife and husband agree on the education of their child (as long as they live together), but for two individuals in separate locations, a link (e.g., by telephone) often involves some nonnegligible costs. Another source of costs is the psychological stress of haggling and hassling (obviously, this cost varies greatly from one individual to another—a few of us actually enjoy the challenge of bargaining—and also from culture to culture, but I am unaware of any systematic analysis). Third, the mutual information of the parties (how much I know about your opportunity costs, about what you know of my opportunity costs, and so on) affects crucially the chances of reaching an efficient agreement. On this third source of transaction costs, some very precise statements have been obtained by the game theoretical literature in the last two decades (more on this in Section 1.7: a general reference is Osborne and Rubinstein [1990]).

In line with the traditional theory of cooperative games, I will assume that the mode of cooperation by direct agreement occurs under the positive efficiency postulate, namely, in a context where transaction costs of all kinds are negligible. The underlying political institution is a minimal state (in the sense of Nozick [1974]) that merely enforces property rights and counts on the initiative of individuals—and coalitions thereof—to bring about collective optimality, which means no more and no less than achieving a Pareto optimal outcome. The question is: what agreements (what Pareto optimal outcomes) are then likely to occur? What agreements are deemed stable?

An agreement reached by the set of all concerned agents (this set is called the "grand coalition") can sometimes be challenged by a (partial) agreement within a smaller set of agents (a "coalition"). Indeed, if the efficiency postulate is a plausible positive statement for the grand coalition, it is at least as plausible for each smaller coalition (within which transaction costs are not larger). An agreement that cannot be challenged by any subcoalition will be called a core outcome, and the set of such outcomes (the core) will be the prime subject of concern while investigating cooperation by direct agreement.

To feel the force of the core concept, the simplest example is a situation of pure exchange, where two sellers each own a car for which their reservation price is, respectively, $3000 and $10,000 (that is, seller 1 would not sell for less than $3000, but would be better off selling for more than $3000 than not selling), whereas two buyers each want a car (and no more than one car) with a respective reservation price of $15,000 and $11,000 (that is, buyer 1 would not pay more than $15,000, but wishes to buy for any price less than $15,000). The two cars are identical in all respects. Here we have a variety of efficient (Pareto optimal) outcomes, among them the following: seller 1 trades with buyer 1 and seller 2 trades with buyer 2; each trading pair splits the difference between their reservation prices, so that buyer 1 pays $9000 to seller 1, while buyer 2 pays $10,500 to seller 2. But this is not a stable outcome if the coalition {seller 1, buyer 2} satisfies the efficiency postulate as well: any trade between these two at a price larger than $9000 and less than $10,500 would make both of them better off, and the postulate tells us that they will not miss this opportunity. In our example, the only core outcomes (the only agreements robust against direct agreements from each subcoalition) are those where a) two trading pairs form where the matching of seller to buyer is arbitrary, b) each pair trades at the same price, and c) this price is between $9000 and $10,000. The proof is given in Section 2.3. Thus the core of this simple exchange economy equals the set of its competitive equilibrium allocations.

We now illustrate some basic features of the core analysis in the particularly transparent context of voting. Voting models will be used similarly in the next two sections to introduce the other two modes of cooperation; a systematic discussion is found in Chapter 6 (Section 6.2 and Appendix to Chapter 6). Voting occurs in an environment favorable to the efficiency postulate inasmuch as the political process encourages ample discussion of the menu of candidates and coordination through the political parties. Yet, in all rigor, the application of the postulate to voting requires the farfetched assumption that a new political party springs up each time a coalition of voters has an opportunity to cooperate.

Consider the popular method of plurality voting. Each voter casts a ballot with the name of one candidate, and a candidate with highest plurality is elected. An outcome (namely a candidate) will be in the core if it is not threatened by an objection, namely, another competing outcome. To see what this means, say that 51 out of 100 voters have formed a coalition to elect candidate a (by jointly casting a vote in his favor). Call those 51 voters white and the 49 others blue. Then an objection to the white coalition may come from 30 white voters and 21 blue voters who realize that a certain candidate b is preferable to candidate a for all of them. Of course, this new outcome may be threatened in turn by another objection, so that electing b may not be the end of the story.

From this emerges the notion of domination. We say that outcome a is dominated by outcome b if there exists a coalition S with the power and the desire (because all members of S prefer b to a) to switch from a to b. The domination relation is the common ingredient in all models of the core; what varies from one model to the next is the allocation of power to coalitions. In plurality voting, the "minimalist" representation of this power says that a coalition can force an outcome of its choice only if it contains a strict majority of voters (clearly, if its opposition is divided, a 49% coalition usually prevails, so to represent it as powerless is an exaggeration).

The core is defined as the set of outcomes that are not dominated by any other outcome. The easy proof of the following observation is left to the meticulous reader. In plurality voting (or in any voting that gives full control over the outcome to a united majority coalition), the core consists of the set of those outcomes a such that if we compare a to *any* other outcome x, at least one-half of the voters view a as at least as good as x (thus they are not ready to enter into a coalition to switch from a to x). Such outcomes are called Condorcet winners.

Note that a Condorcet winner is normally unique. This is formally true if we have an odd number of voters and each voter's preferences are strict (between any two candidates, a voter is never indifferent, and strictly prefers one of the two over the other). With an even number of voters, the chances of two Condorcet winners or more are small (if we have a large number of voters and voters' preferences are independent[6]), and the chances of three Condorcet winners or more are very small.

On the other hand, it is not uncommon to find a preference profile where *no* Condorcet winner exists (hence the core is empty). This observation is known as the *Condorcet paradox* (or paradox of voting) and is easily demonstrated by the familiar configuration where three voters have cyclical preferences over three candidates (voter 1 prefers a over b over c, voter 2 prefers b over c over a, and voter 3 prefers c

---

[6] See Fishburn [1973].

over *a* over *b*). Condorcet's original example is worth recalling here because it has no peculiar symmetry:

Candidates	Voters				
	23	2	17	10	8
Top	Pierre	Paul	Paul	Jacques	Jacques
	Paul	Pierre	Jacques	Pierre	Paul
Bottom	Jacques	Jacques	Pierre	Paul	Pierre

Here Pierre defeats Paul (33 to 27), Paul defeats Jacques (42 to 18), and Jacques defeats Pierre (35 to 25).

Thus the core of the majority voting game is either empty or (essentially) unique. As shown by the exchange example above, we may also have a nonempty core with a continuum of outcomes; this will be the typical situation in exchange (Chapters 2 and 3) and in production with externalities (Chapter 5). In all of these contexts, core analysis revolves around one technical question and one methodological question. The technical question is: how often do we expect the core to be empty and, in particular, in what models can we guarantee a nonempty core? The methodological question is: how do we interpret a situation where there is no core outcome? How, then, do we expect the players to behave, under the efficiency postulate? We start with the second question.

Suppose we are in a situation where the core is empty (to fix ideas, think of majority voting when the preference profile exhibits the Condorcet paradox). If our players stick to cooperation by direct agreement as described earlier, we may call the situation "unstable," with potential coalitions rapidly forming (to agree on a certain candidate) and disbanding once a coalition capable of objecting becomes aware of its own opportunity. In this interpretation (which, I believe, reflects the current mainstream view among political scientists, e.g., Riker [1982], Schofield [1985]), instability can only be revealed in a dynamic context (as when we see the government change "too often" in a multiparty parliamentary system), but always implies (even in the static context) a lack of predictability, a genuine indeterminacy of the final outcome.[7] In the context of majority voting, such instability/indeterminacy has been almost universally regarded as a "problem."[8]

---

[7] At least within the region covered by the possible sequences of successive objections, which is normally a large subset of the original set of outcomes, as shown by McKelvey [1976] in the context of spatial voting.

[8] In the words of Miller [1983], preference profiles exhibiting the Condorcet paradox have been called "discordant," "anarchic," lacking "inner harmony"; they are said to result in "arbitrary" political decisions, political "incoherence," "inconsistency," to threaten political "viability"—see the references in Miller [1983], p. 738.

Thus an empty core is (normatively) bad in the context of voting rules[9] because it leads to a prediction of instability (under the efficiency postulate). The claim is that the emptiness of the core does make cooperation by direct agreement impractical and implausible (who cares about agreeing if we expect the agreement to be overturned anyway?), a positive statement that can ultimately only be proved or disproved empirically. Indeed, the only way in which our players can stick to a particular agreement (say, to elect a particular candidate a) is by taking a more sophisticated view of objections. This means that upon considering electing b (as an objection against a), the players in the coalition anticipate further objections against b (that could elect c, say) and refrain from agreeing to the election of b (despite a shortsighted gain from a to b) because they still prefer a over c (so in the "long run," the initial objection will prove harmful). This line of thinking leads to the second-order stability concepts (not covered in Chapter 7); it rests on the assumption that our players internalize a fairly sophisticated sort of strategic thinking, an assumption that strikes me as too optimistic. Moreover, second-order stability concepts yield typically large sets of stable outcomes and have rarely shown much bite in specialized models such as those reviewed in the following chapters. Thus, I shall interpret the emptiness of the core as a strong signal that cooperation by direct agreements is unlikely to work, and that our players are more likely to cooperate in the justice mode and/or in the mode of decentralized behavior. In the voting example, this means that the Condorcet paradox can be "ignored" by reverting to decentralized voting (where each voter refuses to be seduced by coalitional temptations and casts his vote independently, though strategically) or by calling upon an arbitrator.

An example involving production externalities (see Section 2.8) can be invoked here. Suppose three isolated households consider sharing the cost of a water system. The cost of this equipment for a single house (independent of the other two) is 7 (in thousands of dollars), it is 9 for (any) two houses, and 15 for all three houses. Efficiency commands buying a single system to equip all three houses,[10] but any division of these costs (e.g., each pay 5) is threatened by an objection of at least one pair of players (e.g., equal split is threatened by any two houses:

---

[9] However, note that arguments to the effect that an empty core is good have been put forward as well: the idea is that a Condorcet winner materializes the "tyranny of the majority," since the opposing minority—perhaps as large as 49%—has only an indirect influence on the outcome, via its contribution to the objections against the other outcomes; by contrast, in the case of no Condorcet winner, temporary majorities will form and disband forever, giving to more agents a chance to directly influence the outcome—by being part of the winning coalition. See, again, Miller [1983].

[10] The cost function is subadditive; see Section 2.6.

they can save 1 by building their own system and ignoring the third house). The core of this cost-sharing game is empty. We speak of *destructive competition* in this example, because instability may lead to the inefficient outcome where only two houses are served. The recourse of the justice mode is very plausible here; equal costs is, after all, an efficient outcome and the only one respecting the symmetry of the cost function. Even when the situation is not symmetrical (the cost of serving homes 1 and 2 differs from that of serving homes 2 and 3), cost-sharing formulas such as the Shapley value (see Chapter 7) are the common empirical answer. Cooperation by direct agreements simply does not work.

Much of the formal work on cooperation by direct agreements addresses the following question: how often and when is the core nonempty? The question will be thoroughly discussed in Chapter 7 for the general model of cooperative games. Before that, we present the two main achievements of core analysis in microeconomics. First, in exchange economies (with convex individual preferences), the core captures essentially the same outcomes as the competitive equilibrium (the Edgeworth proposition; see Chapters 2 and 3). Second, when production externalities prevent the existence of a competitive equilibrium (e.g., under increasing returns, or when a public good is produced), the core offers a reasonable, if not universal, substitute to competitive analysis (Chapters 2 and 5). Finally, the core of the majority voting game is discussed in Section 6.2.

1.5. THE JUSTICE MODE: END-STATE JUSTICE

The simplest way to eliminate the transaction costs associated with direct agreements is to vest the decision power into the hands of an arbitrator. The arbitrator's decision will leave none of the concerned agents unhappy only if it is perceived as fair. Moreover, they will accept her decision only if they understand the decision-making process. These two requirements leave not much room to the arbitrator: she must select her decision by means of an equitable mechanical formula. How difficult it usually is to come up with such a formula is one of the themes of this book.

In first approximation, equity means equality: "the passion of mankind for equality is burning, unsatiable, eternal, invincible" (Tocqueville [1860], and more recently Kolm [1985]). Instances abound where the principle of equality yields the unambiguously and uniquely just solution. But one comes up just as easily with examples where, on the contrary, the principle of equality allows several interpretations of the just outcome.

Example 1.1. Estate Division

Ann and Bob are siblings. They inherit a painting from their deceased uncle. For sentimental reasons, Ann cares very much for the painting and would be ready to pay up to $10,000 to keep it. Bob, on the other hand, has no interest whatsoever for the object and, if awarded the painting, he would immediately sell it; the resale value is estimated at $4000. Both Ann and Bob have ample reserves of cash (far beyond $10,000). Efficiency commands giving the painting to Ann; what cash payment to her brother do we call a fair compensation?

Equality here refers to profits or losses measured in dollars (the only available rod), but the question is; equality from where? One answer is to equalize benefits from the initial position where the uncle is still alive and has not bequeathed the painting (note that this could hardly be called the status quo, for it is a hypothetical outcome). This would make a benefit of $5000 for each, i.e., Ann pays $5000 to Bob. Here Bob receives more than the actual value of the painting to him, undoubtedly a problematic feature (on which more in Section 4.2). Another solution uses another (equally hypothetical) initial position where each sibling receives the painting without paying out any compensation. In this case, we equalize the losses due to the necessity of sharing: each must lose $2000 so that the individual benefits ($8000 and $2000) sum up to $10,000. In this solution, Ann ends up paying $2000 to Bob. By construction, Bob will never receive more than his benefit from the full value of the painting. There is yet a third way to apply the egalitarian idea to this problem. This time the "initial position" consists of giving the painting with probability $\frac{1}{2}$ to each of the heirs (a plausible outcome if they cannot reach an agreement), thus yielding a profit of $5000 and $2000, respectively (if our wealthy agents are risk neutral). From this initial position (of which a deeper, nonprobabilistic interpretation is offered in Section 4.3), the surplus $10 - (5 + 2) = \$3000$ is split equally between our two agents, so Ann ends up paying $3500 to Bob.

The three above egalitarian solutions are the most sensible solutions to the estate division problem, as casual experimentation in the classroom reveals. They are depicted in Figure 1.1. Note that the third outcome u_3, corresponding to the initial position where the painting is randomly allocated, is always the middle point between the other two. Later (in Section 4.3) we shall derive two independent arguments in favor of solution u_2.

We turn to May's theorem, a simple and important statement from the theory of voting that follows from a more sophisticated egalitarian principle. Consider voting by n voters among two candidates $\{a, b\}$. Say that a voting rule allows each voter to cast a vote for either a or b

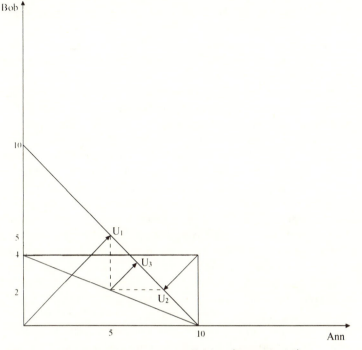

Fig. 1.1. Three solutions to estate division (Example 1.1).

(abstention is not permitted for simplicity). Surely the prime rule of democratic justice is the egalitarian maxim "one man, one vote." It means that from a given profile of ballots $(x_1, \ldots, x_n)$ (where agent i's vote x_i is either a or b), the only information relevant to the election is the number of a votes versus the number of b votes (the urn achieves this by shuffling the ballots). Next, we must implement equal treatment of (no discrimination among) candidates. This means that if a is elected while receiving 60% of the votes (hence b receiving 40%), then b must be elected if b receives 60% of the votes. Intuitively, "one man, one vote" and "equal treatment of candidates" appear to drive us toward the majority rule (a is elected if it receives at least as many votes as b, with ties being resolved by electing both candidates). Yet these two axioms are formally not enough to characterize majority voting, for they are both satisfied by a few crazy rules such as "a is elected if it receives no *more* votes than b," or even worse perhaps, "a is elected if it receives *either* 80% of the votes or more, *or* more than 20% and no more than 50%." To rule out such pathological rules, we add a fairly natural monotonicity axiom saying: "if a is elected when receiving $x\%$

of the votes, it should still be elected when it receives more than $x\%$ (on this last axiom, see Section 1.6).[11]

Aristotle's formal principle of justice ("equals ought to be treated equally and unequals unequally, in proportion to their relevant similarities and differences"; see, e.g., Isaac, Mathieu and Zajac [1991]) is commonly interpreted as the "proportionality rule" saying that "distributive justice involves a relationship between...two persons P_1 and P_2...and their two shares of rewards R_1 and R_2. The condition of distributive justice is satisfied when... $P_1/P_2 = R_1/R_2$" (Homans [1961], as quoted by Bar-Hillel and Yaari [1993]). This relative egalitarianism gives more flexibility to the arbitrator than absolute egalitarianism ($R_1 = R_2$). For instance, some of the inefficiencies implied by absolute egalitarianism (already touched upon in Section 1.1) can be alleviated.

Think of a simplistic model of production where all agents contribute some labor input and the total output (equal to the sum of the labor inputs—the technology transforms one unit of input into one unit of output) must be divided among the workers. (This is a simple version of the model of production cooperatives as discussed by Israelsen [1980], and in Chapter 6.) Compare equal division of total output among all workers (absolute egalitarianism) and proportional division (relative egalitarianism), where each worker receives exactly what he contributed (remember the technology has constant returns to scale). Intuitively, the equal-division rule destroys the incentives to work hard, whereas proportional division stimulates hard work, so that we expect the latter rule to result in a more efficient outcome; perhaps, in fact, *everyone* is better off with the proportional-division rule. The formal analysis confirms that this intuition is by and large correct. (See Section 6.4, in particular Example 6.8, and Exercise 6.14.)

The tension between equality and efficiency has far-reaching implications. It is the driving force behind the analysis of fair division in Chapter 4, for the following reason. Suppose we must divide a pile of commodities between several agents who have equal rights over the goods, but different preferences. Equal split is the unsurpassably just outcome, and yet we do not expect it to be Pareto optimal as well (because preferences differ[12]). The literature on fair division therefore proposes several ways to weaken the fully egalitarian requirement into a

---

[11] Moulin [1988], chapter 9 provides a detailed discussion of May's theorem.

[12] Think of dividing season tickets covering 15 musical performances between three agents—the tickets are neither resalable nor refundable; say that agent 1 is a music fan who wants to get as many as 15 tickets, agent 2 would be fully satisfied with six tickets, and agent 3 with three tickets. Here equal split—five tickets to each—is inefficient because agent 3 giving away one ticket to each one of the others is a Pareto improvement.

property compatible with Pareto optimality. The two main proposals—the no envy property and the egalitarian equivalence property—are the subject of Chapter 4.[13]

The interface between the three modes of cooperation is the unifying theme of this book (see above, Section 1.3). Incompatibility of equality and efficiency reveals a tension between the justice mode and the direct agreement mode (because efficiency is the main force driving direct agreements). Inconsistencies between equality and an individual rationality constraint are another manifestation of the same tension.

Example 1.2. Sharing the Cost of an Indivisible Public Good

The condominium owners in a building wish to hire a gardener to embellish the common areas of their residence. The floral decorations will be a public good for the enjoyment of all the residents. It is only just that all of them share the cost equally. However, if some of the residents are not willing to pay their "fair" share of the costs (because they are broke, or do not enjoy flowers, or for whatever reason), they are likely to veto the project, even though the other residents want the flowers so badly that efficiency commands hiring the gardener. At this point, our residents face a dilemma. Either they stick to the "equal split of costs" rule, but then efficiency may imply charging some agents (perhaps a majority of them) more than the benefit they derive from the flowers (a normatively problematic outcome as well as a positively implausible one); or they go to a system of personalized prices where different agents are billed differently for the same flowers. This makes it possible to achieve efficiency and to make no one unhappy (that is, no one pays more than whatever he or she benefits from the flowers), but it also raises the thorny question of assessing those individual benefits (naively asking each agent to report his or her benefit gives them a strong incentive to underreport).

A simple numerical example is helpful at this point. Say that the yearly cost of the gardener is $2000 and that the five owners are willing to pay, respectively, (800, 600, 450, 350, 300). Total benefit is $2500, so the project will bring a surplus of $500. With equal split of the costs, however, the last two owners end up with a deficit of $50 and $100, respectively. Of course, we may try to make them swallow the pill by

---

[13] In the tickets example, no envy picks the allocation $(6, 6, 3)$ where agents 2 and 3 are fully satisfied, whereas egalitarian-equivalence picks $(9\frac{3}{8}, 3\frac{3}{4}, 1\frac{7}{8})$, where $\frac{5}{8}$ of each agent's demand is satisfied. These numbers are explained in Example 1.7.

arguing that the project is supported by a majority of owners (3 out of 5). Yet for some other benefit profiles, the majority will be supporting the wrong (inefficient) outcome: two instances are (800, 600, 350, 350, 300), where the majority opposes a surplus-generating project, and (600, 500, 450, 150, 100), where the majority supports a surplus-losing project.

We can adapt the formal principle of justice in several ways so as to make it compatible with cooperation by direct agreement. Consider our original benefit profile (800, 600, 450, 350, 300). An efficient outcome consists of hiring the gardner, and charging x_i to owner i, where $\Sigma_i x_i = 2000$. A core outcome is an efficient outcome such that i) no one pays more than his or her benefit, or $x_1 \leq 800$, $x_2 \leq 600$, and so on, and ii) no one receives a subsidy from the others, or $x_i \geq 0$, $i = 1, \ldots, 5$. To see why $x_i \geq 0$ is necessary, notice that if $x_i < 0$, agent i receives money to enjoy the flowers, so the other agents would be better off ignoring agent i altogether and paying for the flowers themselves (further discussion of the core in this and related examples is in Section 5.7). A just formula that picks a core outcome is the proportional division: cost shares are proportional to benefits, or in the example, (640, 480, 360, 280, 240). Another one simply adjusts the equal split of the costs to take into account the constraint $x_i \leq b_i$; here an agent pays either the common cost or her own benefit, whichever is less. In the example, the common cost must be taken at $450, charged to the first three agents, while the last two pay $350 and $300, respectively. Both the proportional division and the constrained equal cost just defined are difficult to implement when the mechanism must elicit the numbers b_i from the agents themselves (the latter difficulty is addressed in Section 1.7 and further discussed in Sections 6.2 and 6.3).

Example 1.3. Sharing a Joint Cost

Four home owners live along a private unpaved path connecting them to the highway. Ann lives 0.2 mile away from the highway, Bob is 0.9 mile away, Cathy is 1.0 mile away, and David is 2.9 miles away. They are willing to share the costs of paving the path; it costs $1000 to pave 0.1 mile of path. What is an equitable division of the total cost of $29,000?

The proportionality rule is the easiest answer: split the cost in proportion to distances from the highway. The argument is that the cost of building the road until Ann's house is $2000, it costs $9000 to build the road until Bob's house, and so on. By splitting costs in proportion to these "stand alone" costs, we make sure that everyone gets the same *rate* of saving. Of course, we could consider giving everyone the same *absolute* saving, but then Ann would end up paying nothing (because total saving is $(2 + 9 + 10 + 29 - 29 = 21,000)$, which seems unfairly low. With cost shares proportional to stand alone costs, we get

$$x_A = 1160, \qquad x_B = 5220, \qquad x_C = 5800, \qquad x_D = 16{,}820,$$

but this is not a core outcome, because Bob and Cathy together have an objection: it would cost them $10,000 to pave the road just for themselves, but the proposed method charges them $11,020! In other words, the proportional rule charges to some coalition of agents more than their stand alone cost, and thus yields an unstable agreement.

A "correct" method is one that conveys a plausible interpretation of justice and always selects a cost allocation in the core. Such a method was proposed nine centuries ago by Rabbi Ibn-Ezra in a formally equivalent context.[14] Consider the segment of the road from the highway until Ann's house. Everybody is using it so its cost $2000 should be equally divided among all; Ann should pay $500 and not a cent more. The next segment from Ann's house to Bob's house costs $7000, equally divided between Bob, Cathy, and David; thus Bob pays $500 + $2333 = $2833; and so on, with Cathy paying $2833 + $500 = $3333, and David $22,333. Of course, this rule is yet another interpretation of egalitarianism, where we equalize the cost shares per mile actually used. The point of the example is that it always picks a core outcome, and therefore is consistent with cooperation by direct agreement. (The proof of this assertion and the discussion of Ibn-Ezra's formula are Example 7.10 in Section 7.6.)

To conclude Section 1.5, define *end-state justice* as the search for general arbitration formulas (selected according to appropriate normative criteria) when we conceive the arbitrator as all-powerful and omniscient, a benevolent dictator. This eliminates the possibility of direct agreements between the agents (that the dictator will see and block) and of selfish manipulation by individual agents (for instance, any attempt to misrepresent one's own characteristics, such as preferences, will be foiled). By contrast, procedural justice (Section 1.7) presupposes an arbitrator ignorant of individual characteristics (although he is aware of their range of variation), who designs a mechanism to elicit this information from individual agents.

In this volume, we look at only one type of end-state justice models (the most popular kind among economists), where individual preferences are only meaningful in an ordinal sense. No meaning is attached to the cardinal measure of welfare;[15] in particular, no interpersonal

---

[14] See O'Neill [1982].

[15] Even though we systematically use the convenient representation of preferences by cardinal utilities.

comparison of absolute levels of welfare is allowed. We also hold every agent responsible for his or her preferences (differences in preferences reflect different tastes, not different needs). These assumptions are discussed in more detail in Chapters 4 and 5.

The main alternative conception of end-state justice is the welfarist model. Its main postulate is that welfare is cardinally measurable and that comparing the welfare level of different agents is ethically relevant. It suits those problems where agents are not held responsible for their personal characteristics (in particular, their preferences). See Moulin [1988] or Roemer [1994] for a systematic account of the welfarist model.

Within our ordinalist model, we devote two chapters (Chapters 4 and 5) to a fundamental puzzle of distributive justice, namely, the interpretation of common property. As illustrated in the examples above, to simply divide equally the resources owned in common by a community is bound to be inefficient (Pareto inferior) when agents hold different preferences. The test of no envy (Chapter 4) is a convenient weakening of the blind equal split that often proves compatible with the efficient allocation of resources. It is also normatively appealing, yet alternative equity tests deserve our attention, too. The most important one is the stand alone test (Chapter 5), especially relevant when the common property resources consist of a technology (as in Examples 1.2 and 1.3). This leads to a normative interpretation of the core stability property and enriches the set of applications of the general cooperative game model developed in Chapter 7.

1.6. DECENTRALIZED BEHAVIOR

Decentralized behavior is the other way of achieving a cooperative outcome at little or no transaction costs. The agents (also called players) are engaged in a game of strategy, the rule of which specifies the set of legal moves, of possible decisions open to each individual player. An agent acts by choosing one among these moves (also called strategies), and a profile of actions (one for each agent) determines the outcome. The only transaction cost consists of reading the rules of the game and enforcing them; in the standard interpretation of strategic games, this service (publication and enforcement of the rules) is provided free of charge to the players, who show no concern whatsoever for the collective implications of the games they are playing. In reality, of course, there is a cost attached to the service: some governmental agency must protect markets against thieves, and elections against frauds. As this cost, ultimately, is borne by the players themselves, the notion that they take the game for granted can only be an approximation. This is, however, the assumption we make in the decentralized behavior mode.

The second crucial assumption is that the players do not communicate directly with one another, that they do not seek to agree on some (coordinated) choice of strategies. In particular, they do not know that the outcome of the game is efficient or not, or if they know, they do not care because they have no way of agreeing with the rest of the players on a Pareto superior outcome. However, from our birdlike point of view assessing the merits of different modes of cooperation, efficiency or inefficiency of the decentralized equilibrium outcome (namely, the outcome that we predict will result from the decentralized behavior of the players) is an issue of utmost importance.

If the equilibrium outcome is efficient, and even robust against coalitional agreements within the strategic game (i.e., agreeing on a coordinated choice of strategies), we see the two modes of direct agreements and of decentralized behavior converging in this very outcome. There will be no need to encourage nor to discourage direct communication between the players themselves; such communication, being worthless to the players, will disappear by itself and the decentralized players will fulfill the invisible-hand prophecy: "promoting an end (efficiency) which was no (longer) part of their intention". The paramount example here, as mentioned earlier, is competitive exchange decentralized by the price signal, because a competitive equilibrium is also robust against direct coalitional exchanges (the competitive equilibrium is a core allocation; see Chapters 2 and 3).

On the other hand, if the decentralized equilibrium outcome is inefficient (or simply threatened by the coordinated move of some subcoalition), then the feasibility of communication between the players is crucial. The inefficient (or simply core unstable) outcome that results from decentralized behavior may disappear if our players (or coalition thereof) can effectively coordinate their strategies. Moreover, the agency publishing and enforcing the rules of the game may also see as part of its mission to promote communication and direct agreements for the sake of its constituents' welfare. This is a policy issue that greatly influences our approach to the game at hand. The realization that there are many patterns of strategic interaction where the decentralized equilibrium is inefficient is one of the major findings of the theory of games. This feature is common to Hardin's tragedy of the commons, to Olson's "logic of collective action," as well as to Cournot's model of oligopolistic competition.

Example 1.4. A Simple Model of the Tragedy of the Commons

Hardin [1968] formulates the tragedy of the commons to explain how the common property of a scarce input may result in inefficient overuse of the input.[16] Consider a pasture "open to all" where herdsmen send their cows. Suppose the pasture produces 100 units of grass every season and that a cow can eat up to 1 unit of grass per season. Each cow transforms grass into meat at a one-to-one rate, and the price of meat is 1. Finally, the opportunity cost of sending a cow to the pasture is 0.5; that is to say, a cow kept at the farm will produce 0.5 units of meat at the end of the season (so the pasture technology is twice as efficient when the pasture is not crowded). What is the equilibrium situation, when each herdsman's decision of how many cows are sent to the pasture is taken independently? Consider a situation where herdsman i has n_i cows on the pasture where there are n cows altogether. If $n < 100$, each cow is eating at capacity, the herdsman benefit is $(1/2)n_i$, and he wants to send at least one more cow. When $n \geq 100$, each cow eats $(100/n)$ units of grass (each cow imposes a production externality on all other cows), and herdsman i's benefit becomes

$$100\frac{n_i}{n} - \frac{n_i}{2}.$$

Thus the herdsman keeps sending more cows until the point where sending one more cow brings a loss, namely, when

$$100\frac{n_i}{n} - \frac{n_i}{2} \geq 100\frac{n_i + 1}{n + 1} - \frac{n_i + 1}{2}. \tag{1}$$

A little algebra shows that with a large number of herdsmen, these inequalities imply that n is nearly 200 (but not quite), namely, the pasture ends up with twice as many cows as would be efficient. Indeed, the cows sent to the pasture will produce no more than 100 units of meat, so efficiency commands sending exactly 100 cows and leaving the rest at the farms. Moreover, the total benefit of the herdsmen in the decentralized equilibrium is $100\text{-}(n/2)$, which means that the overgrazing wipes out most of the 100 units of surplus offered by the pasture technology.[17]

---

[16] A difficulty already noted by Aristotle : "what is common to the greatest number has the least care bestowed upon it. Everyone thinks chiefly of his own, hardly at all of the common interest" (quoted in Ostrom [1991]).

[17] Summing up inequality (1) over all herdsmen and rearranging gives $n + 1 \geq 200(K - 1)/K$, where K is the number of herdsmen. An inequality similar to (1) expresses the fact that herdsman i does not wish to reduce n_i by 1 (assuming n_i is positive). Summing up that inequality over all i, we find that n is worth $200(K - 1)/K$, with an error of ± 1. More generally, suppose that the opportunity cost is c, $0 < c < 1$. Then the same argument yields $n = (100/c) \cdot (K - 1)/K$ accurate within ± 1. The total benefit is $100/K$ (accurate within ± 1). Thus doubling the herdsmen population halves the surplus they extract from the pasture in decentralized equilibrium.

The key to the tragedy is that a herdsman bears only a fraction of the cost imposed on the whole group by the presence of his cows: when maximizing his benefits, he fails to "internalize" the negative externalities in production that he causes. The generality of this phenomenon is well known. For instance, the economist reader will recognize in Example 1.4 a formal model of oligopoly à la Cournot, with free entry and demand function $D(p) = 100/p$; the inefficiency of Cournot's oligopoly equilibrium, is almost universal (see, e.g., Olson [1965]). Other examples include the exploitation of exhaustible resources such as fisheries, oil and minerals, firewood in the third world, as well as overpopulation, the problem of acid rain, and so on (see the numerous references in Chapter 1 of Ostrom [1991]).

The second classic example of inefficient decentralized behavior is the private provision of public goods, again a long-recognized problem (recall Smith's quotation at the beginning of Section 1.2). Its first systematic exposition appears in Olson's influential "logic of collective action": "unless the number of individuals is quite small, or unless there is coercion or some other special device to make individuals act in their common interest, rational, self-interested individuals will not act to achieve their common or group interest" (Olson [1965], 2). Olson points out that most practical achievements serving the common interest of a group can be conceived as a public good. That is to say, no individual in the group can be excluded from its benefit, just as the flowers in Example 1.2 above benefit all the condominium owners; but each owner at the same time wishes to pay as little as possible for the flowers. In team competition such as rowing, each rower has a common interest to win the race with his teammates and will share with them all the credit for victory; at the same time, he would rather take it easy and let them pull the boat to victory. Similarly, if legislation is passed to protect a certain group of professionals (say the dentists), all of them will enjoy the full benefit of the law, but they would rather avoid the trouble of lobbying for it.

Within this perfect identity of interests (all enjoy the same public good, all prefer to contribute less of their own pocket), the inefficiency of the decentralized equilibrium is a strikingly simple observation.

Example 1.5. Voluntary Contribution to a Public Good

A cartel of 16 firms is lobbying for a bill that will bring a tax break of $90,000 to each one of these firms. They are in competition with another interest group lobbying against the proposed bill. This other group has invested $40,000 in its lobbying campaign. It is estimated that the probability of adoption of the bill is proportional to the budgets of the two competing campaigns. Thus, if our industry gathers x thousands of dollars, the probability of adoption of the bill is $x/(x + 40)$ (this assumption is fairly common; see Tullock [1980]). The expected profit of firm i contributing x_i (thousands of dollars) to the campaign is

$$u_i = 90 \cdot \frac{x}{x + 40} - x_i, \quad \text{where } x = \sum_i x_i. \tag{2}$$

The decentralized behavior in the sense of Nash equilibrium means that each firm chooses a contribution x_i maximizing its profit under the assumption that other firms' contributions are fixed. Every outcome where $x = 20$ is such a Nash equilibrium.[18] Thus, total lobbying campaign effort will be $20,000, and this can be divided in an arbitrary fashion between the firms (notice the genuine indeterminacy in the equilibrium behavior, on which more below). Total profit to the cartel is only $\sum_i u_i = 460$ (in thousands of dollars), or an average per-capita profit of $28,750. Contrast this with the profit to a cartel efficiently coordinating its contributions (and enforcing payment by all firms). This efficient cartel maximizes

$$\sum_i u_i = 16 \cdot 90 \left(\frac{x}{x + 40} \right) - x,$$

by choosing $x = 200$ (a tenfold increase in the lobbying effort), generating $\sum_i u_i = 1000$, or an average per-capita profit of $62,500 per firm (more than doubling the "decentralized" profit).[19]

Inefficiency of the equilibrium outcome in decentralized behavior will be a central theme throughout Chapter 6 (generalizing, among other things, the tragedy of the commons example and the voluntary contribution model). Yet another difficulty of the equilibrium analysis is already apparent in Example 1.5, namely, a severe multiplicity of possible equilibrium outcomes. Plurality voting offers a striking example of this problem.

Suppose the voters cast their vote in a decentralized mode. Then *any* candidate is the outcome of some Nash equilibrium, except perhaps a Condorcet loser candidate, namely, one who is defeated by *every other* candidate in binary duels. To see this, assume candidate a is not a Condorcet loser, so there is (at least) one other candidate b such that a

---

[18] The function u_i is concave in x_i and its derivative with respect to x_i, $(60/(x + 40))^2 - 1$, is zero when $x = 20$.

[19] If we have n firms in the cartel, the average per capita profit in a Nash equilibrium is easily computed as $30 - (20/n)$, whereas it is worth $10(3 - (2/\sqrt{n}))^2$ in the efficient cartel agreement. Thus, for n large, the per capita profit is 30 under voluntary contribution, or a third of the efficient per capita profit 90.

majority of the voters prefer *a* over *b*. Now suppose that everybody believes that the only serious contenders are *a* and *b*. Then a rational voter will vote for whichever of *a* and *b* he or she prefers; thus the votes split between *a* and *b*, confirming the expectations, and *a* wins.[20]

The key equilibrium concept for the mode of decentralized behavior is the Nash equilibrium: a profile of strategies is a Nash equilibrium if no player can improve upon her utility by unilaterally switching to another strategy (where unilateral switch means that no other player's strategy changes). The (positive) decentralization postulate is that only a Nash equilibrium outcome can emerge in this mode. The postulate is meaningful only if there is at least one Nash equilibrium outcome in any given game. Although one can construct easily games in strategic form with no Nash equilibrium, several meaningful "extensions" of the initial game (such as the introduction of mixed strategies, or the repetition over time) are known to restore the existence of a Nash equilibrium. This stands in sharp contrast with the direct agreement mode, where the nonexistence of a core outcome is a very robust phenomenon.[21]

The multiplicity of Nash equilibrium outcomes in a given game is the issue of real concern. In such a game, one would like to select a single equilibrium outcome (or a small subset of those) by means of rationality arguments consistent with the mode of decentralized behavior (direct coordination is, of course, ruled out). A great deal of game theoretical research in the last two decades has been devoted to this very problem, with considerable success only in the context of the so-called dynamic games,[22] namely, those games played over several periods where a player takes a strategic decision in each period. For games in strategic form where each player chooses his strategy once and for all, and all choose at the same time (those games are called "one-shot" games, to distinguish them from dynamic games), the problem of equilibrium selection remains almost intact. Two elementary and very well-known games will make the point. In both games, two players choose one of

---

[20] Note that one could make an argument to show that even a Condorcet loser could be elected in equilibrium: if I believe that *a* will receive all the votes, I cannot threaten its election by virtue of my single vote, so I might as well vote for *a* ; hence the prediction that everyone votes for *a* is fulfilled in equilibrium. The problem here is that "to vote for *a*" might be a bad strategy for some voters; for instance, if *a* is my worst choice, I have nothing to lose by supporting a candidate I prefer to *a* (technically speaking, voting for *a* is a dominated strategy, see below). Thus the postulate that I could vote for *a* because I have nothing to gain is not realistic.

[21] On the issue of existence of a Nash equilibrium, and further interpretations of the decentralization postulate, several game theory textbooks can be consulted, e.g., Shubik [1984], Moulin [1986], Friedman [1986], or Myerson [1991].

[22] See the textbooks by Friedman [1986], Fudenburg and Tirole [1991], or Kreps [1990].

two strategies. In the first one, known as the Battle of the Sexes,

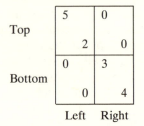

the two Nash equilibria (at (Top, Left) and (Bottom, Right), respectively) are also efficient, but we have no way of selecting an equilibrium within the decentralized behavior mode. The second example, sometime called stag-hunt,[23]

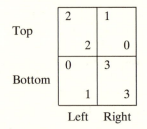

has two Nash equilibrium outcomes, one of which ((B, R)) is Pareto superior to the other ((T, L)). Thus we may select (B, R) on the grounds that each player, aware of their mutual interest in playing (B, R), will count on the same reasoning by the other player. Yet an argument of prudence suggests playing (T, L) instead: T (resp. L) guarantees a payoff of 1 to Row (resp. Column), whereas B (resp. R) guarantees only 0. Prudence is a strong argument if a player is not absolutely sure of the other player's choice.[24]

---

[23] See Crawford [1991].

[24] Check that some degree of uncertainty about other's choice is very likely. Assume that the Row player knows the full payoff matrix but does not know if the Column player knows it too. Row should consider the possibility of using his prudent strategy Top (guaranteeing a pay-off of at least 1). Hence, if Column does not know if Row knows that she (Column) knows, she should view Row playing Top as a serious possibility, hence consider the possibility of playing L herself. This, in turn, implies that Row will consider playing T, not only if he does not know that Column knows, but also if he does not know if Column knows that he knows that she knows. By induction, it follows that the (T, L) equilibrium is a real possibility unless all statements of any length of the form "Row knows that Column knows that...," and "Column knows that Row knows that...," are true (when all such statements are true, we say that the information about the pay-off matrix is common knowledge—see Aumann [1976]). And if (T, L) is a possibility, it can surely be selected by the prudence argument.

Multiplicity of the Nash equilibrium outcomes is not the only difficulty in the interpretation of decentralized behavior. Even if a game has a unique Nash equilibrium outcome, we must explain how the players come up with the corresponding equilibrium strategy without communicating among themselves (or with anyone else, for no one is around to help them). The easy answer is that the profile of preferences (or the payoff matrix, as the case may be) is common knowledge (see above and note 25), that our players have learned about Nash equilibrium at school, that they believe that it is the proper behavior in a decentralized world, and that this way of approaching the game is common knowledge, too. In this story, the least plausible assumption is the common knowledge of preferences.

Under the more realistic assumption that a player knows his or her preferences very well but knows little (or perhaps nothing) of the others' preferences, a number of tatonnement processes and/or learning algorithms have been proposed[25] to explain how players might progressively adjust their behavior over time (while the same one-shot game is played over and over). The simplest such object is the Cournot tatonnement, where players play at period $t + 1$ an optimal (i.e., utility-maximizing) strategy against the strategies used by other players in period t. When a Nash equilibrium is the common limit of a large class of such algorithms, it is a justifiable behavior even under the limited-information assumption. Examples 1.4 and 1.5 above are two instances.

We have two stories to explain how the players implicitly coordinate their strategies on the unique Nash equilibrium, by common reasoning under common knowledge of preferences, or by dynamic adjustment if the game is played many times. Still, in a one-shot game where each player knows only her own preferences (and not other players' preferences), the postulate that decentralized players will play the Nash equilibrium is not convincing. We need the more demanding property of the dominant-strategy equilibrium for a realistic prediction of the deterministic decentralized behavior.

A strategy x_i by player i is a *dominant strategy* if no matter how the rest of the players play, strategy x_i maximizes player i's preferences over his strategic choices. Thus x_i is unambiguously the (a) best strategy for player i, even if he does not have the slightest idea of how other players will act. A *dominant-strategy equilibrium* is a profile of strategies where each player uses a dominant strategy. This equilibrium property is (much) more demanding than the Nash equilibrium property; on the other hand, the positive postulate that players will use a dominant strategy in the decentralized mode seems hardly objectionable, for this

---

[25] From Cournot [1838] tatonnement to Milgrom and Roberts [1991] adaptive behavior.

behavior is insensitive to the amount of information they possess on other players' preferences (the postulate is equally plausible under common knowledge of preferences or when information on individual preferences is strictly private).

Voting by majority rule between two candidates provides an example of dominant-strategy equilibrium. If I prefer a over b, casting a ballot for a is my dominant strategy, because no matter how the rest are voting, I would never be better off voting for b (i.e., against a). So from the individual citizen's point of view, voting becomes an extremely straightforward operation: go to the voting booth, take a look at the two candidates, and cast your vote for whomever you prefer; you do *not* need to peek in other booths to see what other voters are doing (but it does not matter if you do peek). This strategic transparence of the voting process is so appealing that all voting rules sharing the property are worthy of our interest. In a vote over two outcomes only, this property is precisely the monotonicity axiom used in May's theorem (see above at the beginning of Section 1.5). In a vote over three or more outcomes, the property turns out to be too demanding in general; see Section 1.7.

Now to a simple example of dominant-strategy equilibrium in the context of exchange.

Example 1.6. Second-Price Auction

Agent 1 wishes to sell an indivisible object (such as the painting in Example 1.1) for which she has no use (her reservation price for the object is zero). Five potential buyers have the following reservation prices for the object:

Ann: 10   Bob: 25,   Charles: 15,   David: 30,   Emily: 12.

The object is sold by second-price auction (also called Vickrey auction after its inventor): each buyer submits a bid; the highest bidder wins the object (possible ties being resolved by flipping a coin) but pays only the second highest bid. The claim is that the strategy "bid your reservation price for the painting" is a dominant strategy for any buyer; thus the dominant-strategy equilibrium outcome is: David gets the painting for $25.

We check the claim for Charles. We must compare the two strategies "bid 15" and "bid x," and show that Charles is never worse off bidding 15 than bidding x, where "never" refers to all conceivable profiles of bids by other buyers (and we must show this property for all x). To fix ideas, take $x = 20$. If the highest bid y of buyers other than Charles is above 20, then it does not matter whether Charles bids 15 or 20, for he will not get the object anyway; similarly, if y is below 15, then Charles will get the object anyway and pay y for it. Now suppose y is between 15 and 20. Charles will not get the object by bidding 15, whereas he will get it and pay y for it by bidding 20; the latter outcome entails a net loss for Charles so he is better off bidding 15, as was to be shown. The argument extends identically to any x above 15, and with a simple twist to the case of x below 15.

Notice that the argument for a dominant-strategy equilibrium does not apply to the more familiar *first*-price auction where the highest bidder wins the object and pays his own bid for it. In this auction, bidding one's reservation price is typically not a dominant strategy, not even a Nash equilibrium outcome: David would rather bid 26 than 30 if no one is bidding above 25! The second-price auction will be generalized in Section 2.9.[26]

The interface between the decentralized mode of cooperation, the direct agreement mode, and the justice mode is the subject of Section 1.7.

Remark 1.1 Repeated Games

The analysis of purely decentralized behavior[27] allows one test of the positive efficiency postulate (on which rests the mode of cooperation by direct agreements; see Section 1.4). If the Nash equilibria of a given game are all Pareto optimal, the postulate will be confirmed for the situations described by this particular game. The paradigm of repeated games has been offered as a general, context-free justification of the postulate.

The repetition of a given (one-shot) game consists of playing this game many times among the same players, who measure their gains (or losses) by the accumulation of their "instantaneous" payoffs. Many more outcomes result of equilibrium behavior in the repeated game than in the initial game. Take a Pareto optimal outcome x of the initial game that is not a Nash equilibrium there. The outcome of the repeated game where x is played in every instantaneous game might be a Nash equilibrium outcome of the repeated game: a deviation by a single player away from x in one particular occurrence of the (initial) game, although profitable in that particular period, may be retaliated against

---

[26] See Moulin [1986], Chapter 2 for a systematic discussion of dominant strategy equilibrium in normal form games.

[27] Which some authors take to be the only legitimate object of game theory, e.g., Fudenberg and Tirole [1991].

in the following periods, thus making it on balance unprofitable. It turns out that many reasonable Pareto optimal outcomes of the initial game can result in this fashion from a Nash equilibrium of the repeated game. This Folk Theorem[28] is an extremely general statement; it requires *no* assumption on the initial game in strategic form. On the other hand, common knowledge of (cardinal) utilities must be assumed.

As a justification of the efficiency postulate, the Folk Theorem is disappointing because a) it does not eliminate other, Pareto inferior outcomes from the equilibrium set, and b) the choice of a particular equilibrium is left unresolved, and presumably results from a direct agreement between the players prior to the play of the repeated game. So we are back to square one: the direct agreement mode of behavior is still indispensable to the positive explanation of efficiency.[29] The real success of the repeated games model lies elsewhere: when the assumption of complete mutual information is dropped, the model often allows one to measure how much efficiency is lost, whence to make precise evaluations of the impact of informational transaction costs.[30]

1.7. PROCEDURAL JUSTICE

End-state justice, the "pure" mode of cooperation by justice, assumes an omniscient arbitrator. His formula for a just compromise measures all "relevant similarities and differences" (cf. Aristotle's formal principle of justice). As discussed in Section 1.5, such relevant differences must include fine-grained information about personal characteristics, in order to guarantee an efficient outcome. In fair division, equal split is generally inefficient, and computing an efficient and equitable allocation (as defined, for instance, by the no envy test; see Chapter 4) requires detailed information about preferences, and so on.

The information about relevant parameters such as preferences, effort, and the like, is necessarily dispersed across individual agents in a society where all production activities, including education and leisure, are governed by specialization and division of labor. In such a world, the model of end-state justice becomes an impractical abstraction.

---

[28] Of which a good account is in Friedman [1986].

[29] Note that the criticism does not apply to certain "evolutionary" models of cooperation where players' behavior is fully decentralized, although less plausible than the traditional utility-maximizing behavior of noncooperative theory; see, e.g., Axelrod [1984], Young [1993]. It is too early to tell whether the evolutionary models can provide systematic foundations for the efficiency postulate, beyond the few simple cases discussed so far.

[30] See Fudenberg and Tirole [1991].

Procedural justice assigns to the arbitrator the task of designing a mechanism that i) elicits from individual agents a report about their relevant characteristics, and ii) picks the final outcome on the basis of the profile of reports. The "founding fathers" fulfill precisely this assignment when designing the constitution. They choose the rule of the political process that will determine the future of their country, unaware of the particular characteristics of the actual players (most of whom are not even born), and yet aware that these characteristics critically influence the outcome of this process. Of course, the letter of the constitution is not in the format "elicit individual reports, then take a decision," but as we will see below, this format brings no real loss of generality to the discussion (this is the message of the "revelation principle").

From the point of view of the three modes of cooperation, procedural justice combines the mode of justice with that of decentralized behavior. Indeed, with the format "elicit reports, then take decision" each individual agent sends whatever report pleases her, and strategic considerations may well dictate that she does not report truthfully, that she "misreports." The division of power is clear: to the arbitrator, the choice and enforcement of the rules of the game; to the individual agents, the selection of a particular outcome of the game. The possibility of misreport is not only a practical option (when characteristics are private, who will know that I cheated?), it is often a legal right (freedom of opinion in voting implies the right of using my vote strategically—much to Jean Charles de Borda's dismay, who exclaimed: "my method is for honest men!"), and finally it is in general a strategic necessity. The last statement, unlike the previous two, is not obvious; indeed, it was not formally demonstrated before 1973 (see below the discussion of the Gibbard–Satterthwaite theorem). The elaboration and qualification of this statement in a variety of different contexts is a major research theme of the last two decades.

The choice of a mechanism in the spirit of procedural justice is guided by two kinds of considerations: normative principles to assess the justice of the mechanism, and positive analysis of the decentralized behavior of the agents playing the mechanism. There are two degrees in interpreting the justice of the mechanism. The minimal interpretation is justice "*a priori.*" It only requires one to endow equal agents with equal strategic opportunities (namely, equal set of messages and equal influence of everyone's message): what happens next—which outcome emerges from the decentralized behavior of the agents—is automatically just. On the other hand, justice "*a posteriori*" applies the normative criteria of end-state justice (as discussed in Section 1.5) to the decentralized equilibrium outcome of the game.

Justice a priori with the exclusion of justice a posteriori is the criterion we apply to sports competition. Its application to political economy is often called the libertarian doctrine (see, e.g., Hayek [1976] and Nozick [1974]). I will argue that justice a priori is a logical prerequisite of justice a posteriori, but that justice a priori without justice a posteriori is unduly reductionist.

Voting, as usual, provides our first example. The "one man, one vote" principle is justice a priori among voters. Formally, a voting rule associates to each profile of preferences a single outcome; "one man, one vote" is the *anonymity* property, requiring that this function is symmetrical in its various components (thus the outcome is not affected if two voters exchange their votes). Of course, all familiar voting rules (such as plurality, the Borda rule, approval voting, Condorcet consistent rules, and so on) are anonymous. Dismissing justice a posteriori amounts to viewing all voting rules as normatively indifferent, a proposition invalidated by the debate surrounding the choice of voting systems in the political arena; see, e.g., Brams [1993].

Our second example is a simple variant of the voluntary contribution to a public good game (Example 1.5). The game is surely just a priori, but its equilibrium turns out to be unjust a posteriori. In the variant, one of our 16 firms, call it firm 1, is different: its tax break is $102,500 if the bill is adopted. All other firms still receive a $90,000 tax break. Thus, if the total lobbying effort is x, of which firm 1 contributes x_1, firm 1's (expected) profit is

$$(102.5)\frac{x}{x + 40} - x_1, \tag{3}$$

while other firms' (expected) profit remains unchanged. What is now the outcome of decentralized voluntary contributions? It turns out that the *only* Nash equilibrium outcome of this game has firm 1 supporting the *whole* cost of the lobbying, or approximately $24,000. No other firm pays anything and they all end up with an (expected) profit of $33,800, whereas firm 1's profit is only $14,400.[31] The Nash equilibrium outcome is plainly unjust in the sense of end-state justice: the "favored" firm 1 (receiving a larger tax break) provides a completely free ride to all 15 other firms (receiving a slightly smaller tax break). Firm 1's profit is

---

[31] If in a Nash equilibrium, x_1 is positive, then the first-order condition for the maximization of firm 1's profit gives $(x + 40)^2 = 4100$ or $x = 24.030$. Similarly, if x_i is positive for any of the other 15 firms, we find $x = 20$ as before. Thus at a Nash equilibrium, either $x_1 = 0$ or $x_i = 0$ for each of the remaining 15 firms. Assume $x_1 = 0$, so that $\sum_{i \geq 2} x_i = 20$. Note that firm 1 is not maximizing its profit (namely, $102.5(20 + x_1)/(60 + x_1) - x_1$ at $x_1 = 0$.

nearly twice smaller than other firms' profit. This is unfair by any standard, as the discussion of Chapter 6 will confirm.

We can easily multiply examples like the above one, where the unique Nash equilibrium of an a priori just game is a posteriori unjust. The reader may look at the variant of our tragedy of the commons model (Example 1.4) with eleven herdsmen, ten of them with a high opportunity cost of 0.5 per cow, and one of them with a low opportunity cost 0.2. In a Nash equilibrium outcome, the low-cost herdsman sends 59 cows to the pasture, whereas every high-cost herdsman sends 13 or 14, for a total of 133 cows from these high-cost agents, and a grant total of 192 cows (all figures accurate within ± 1). Note that high-cost agents simply have a higher productivity at home than the low-cost agent. Thus it seems unfair that the inefficient agent eats about four times more grass from the commons than any of the efficient agents. Formally, a high-cost agent "envies" the low-cost agent: he could benefit from switching roles with him (high-cost agent's profit increases if he sends 59 cows, while the low-cost agent sends only 13; see Section 6.4 for more discussion of envy in this context).

Procedural justice is defined by Rawls as follows: "the idea is to design a scheme such that the resulting distribution, whatever it is, which is brought about by the efforts of those engaged in cooperation and elicited by their legitimate expectations, is just" (Rawls [1971]). In a more formal fashion, I propose to think of procedural justice as an attempt to combine in a single mechanism the virtues of all three modes of cooperation. Taken literally, this means that we look for mechanisms 1) that are anonymous (just a priori), and where the decentralized equilibrium outcome is ii) just a posteriori, and iii) efficient. An additional property of the mechanism should be iv) core stability, in the following sense: no coalition is able to agree on a profitable change in its equilibrium messages.[32]

The key to the formal development of the above program is the interpretation of the decentralized equilibrium outcome. As underlined by the discussion of Section 1.6 (summarizing in very few words four decades of game theoretical literature), the only uncontroversial equilibrium concept is also a very demanding one, namely, the dominant-strategy equilibrium. This yields the first and foremost question in the formal theory of procedural justice (also its seminal question, from a historical point of view): given a certain choice problem (e.g., sharing joint costs or dividing private goods, or voting), what mechanisms will endow every agent with a dominant strategy, no matter what her

---

[32]Assuming that coalitions are not endowed with any other rights than those derived from their strategic behavior in the mechanism itself.

preferences are? If we can determine the entire set of those mechanisms, we will know, with absolute precision, the flexibility awarded to the mechanism designer (under this interpretation of decentralized behavior). The latter will be able, then, to pick one of those mechanisms according to his own interpretation of justice and his concern for efficiency. Unfortunately, the designer (almost always) faces some difficult choices, because there is no decentralized mechanism satisfying requirements i, ii, and iii above.

A technical (yet easy) argument known as the *revelation principle* (see Myerson [1991]) allows us to restrict the above question to the so-called *direct* mechanisms. Those require that each agent "reveals" his characteristics (e.g., preferences, endowment of wealth, and so on), but do not authorize any other kind of message (such as: conditional on what Bob reports, I will report this or that). Based on these reports, the mechanism chooses the final outcome. A direct mechanism is called *strategy-proof* if for every agent, no matter what her characteristics, sending a truthful report about these is a dominant strategy. The revelation principle says that for every mechanism where each agent has a dominant strategy for every conceivable characteristic (of this agent alone), there exists an outcome-equivalent strategy-proof direct mechanism.

Thus the seminal question of mechanism design becomes: for a given problem, what is the set of its strategy-proof direct mechanisms? Not much is known about this very hard question, except in some specific models such as voting rules. In the voting context, the main result is also among the earliest ones, namely, the theorem jointly attributed to Alan Gibbard and Mark Satterthwaite (who discovered it independently and simultaneously; see Appendix 6.2 for a formal statement). A voting rule is a direct mechanism: each agent reports a preference over a given set of outcomes; by freedom of opinion he is not limited in the shape of preferences he may choose to report, nor in the shape of his true preferences. The theorem says that a strategy-proof voting rule must be either a *dictatorial* rule (choosing a favorite outcome of a certain agent called the dictator, always the same dictator at all preference profiles) or a *binary* rule (choosing one of two given outcomes, always the same pair at all preference profiles). Of course, a dictatorial rule is strategy-proof,[33] and we have seen earlier (Section 1.6) that majority voting among two outcomes is strategy-proof. But it is hard to think of a less palatable menu of mechanisms for a constitution designer: forced to choose between the dramatic inefficiency of a binary rule (where the fixed pair of outcomes are at some profiles Pareto inferior to *every other*

---

[33] Except perhaps for the unimportant case where the dictator has several favorite outcomes.

outcome) and the unsurpassable injustice of a dictatorial rule. That a binary rule by majority is just and a dictatorial rule is efficient is not likely to restore his optimism.

In contexts of resource allocation (such as fair division or cost sharing), the quest for the class of all strategy-proof mechanisms is still a long way from its goal. The most interesting characterization results generally combine strategy-proofness with the anonymity property (justice a priori) and/or with efficiency of the outcome, thus dropping requirements ii and iv from the wish list—although in some miraculous cases, these two properties follow automatically. Here is an example in a particular kind of division problem.

Example 1.7. Fair Division with Single-Peaked Preferences

Ann, Bob, and Charles share a 15-hour baby-sitting job. Charles is able to do his homework while baby-sitting; he needs 3 hours to complete the homework, after which he wants out as soon as possible. Thus Charles's preferences over his time share (in hours) are single peaked with a peak at 3: increasing from 0 to 3 and decreasing from 3 to 15 (note that we do not make any assumption about his preferences across his peak; e.g., whether he prefers to get 2 hours or 4 hours). Ann genuinely likes baby-sitting and would like to do as much of the job as possible; hence her preferences are increasing from 0 to 15, another case of single-peaked preferences with the peak at 15. Finally, Bob wants the money but would rather not stay too long: his ideal share is 6 hours, with the same single-peaked pattern (increasing before the peak, decreasing after the peak). Notice the (only) difference with the tickets-division problem in note 13 of this chapter, where agents are satiated with the good and can dispose of it freely (so preferences increase up to the satiation level, after which they are flat).

The problem when dividing the 15 hours of the item "baby-sitting" (we do not call it a good, as agents can be made unhappier from consuming more of it) is excess demand: it would require $15 + 6 + 3 = 24$ hours of the item to satisfy completely all three agents. The two natural interpretations of a "just" rationing are proportional rationing (each agent receives a fraction of his or her demand: in the example, each gets $\frac{5}{8}$ of his ideal consumption, or $9\frac{3}{8}$, $3\frac{3}{4}$, and $1\frac{7}{8}$, respectively) or uniform rationing (each agent faces the same cap λ but chooses freely his consumption between 0 and λ: in the example, $\lambda = 6$ and so Ann and Bob get 6 hours while Charles get 3 hours). Both versions of rationing are normatively plausible, but from the procedural justice angle, they fare very differently. Look first at proportional rationing as a direct mechanism (where agents report their peak): it is *not* a strategy-proof mechanism. Indeed, if Ann and Bob report truthfully, Charles can get his *true* peak by *pretending* that his peak is at $5\frac{1}{4}$. Of course, Bob will counter that move by raising his own (alleged) peak to $13\frac{1}{2}$, and so on. The unique equilibrium of this game has the following messages: Ann and Bob announce 15 each (note that Ann's truthful report does not mean that she is honest or naive), Charles announces $7\frac{1}{2}$; hence, the uniform rationing outcome (6, 6, 3)!

By contrast, the direct uniform rationing mechanism *is* strategy-proof. Charles and Bob certainly would not want to misreport as they get their peak allocation; Ann, on the other hand, can only lower her report, which is either ineffective (her allocation does not change as long as she reports a peak above 6) or countereffective (by reporting below 6, she will get less than 6).[34] The uniform rationing outcome is efficient; moreover, the uniform rationing mechanism has the core stability property iv. As the mechanism is also anonymous (property i) and its outcome is just a posteriori (in the sense of the no-envy property), it completely fulfills the goals of procedural justice. Moreover, in this fair-division problem with single-peaked preferences, uniform rationing is the *only* mechanism combining the properties of strategy-proofness, anonymity (justice a priori), and efficiency (see Exercise 4.11).

This is the best of all worlds: we have a satisfactory mechanism, and we have only one. Alas, we are not always so lucky, for the theory of strategy-proof mechanisms has been consistently more generous with "negative" results such as the Gibbard–Satterthwaite theorem, than with "positive" results such as the characterization of uniform rationing. The standard negative result is that there is *no* mechanism satisfying the three requirements of strategy-proofness, anonymity, and efficiency. This result holds true in fair division of at least two divisible goods among as few as two agents (Section 4.5); ditto in the cooperative production of a (single) private good (from a single input) (Sections 6.4 to 6.6); ditto in the provision of a (single) public good (Sections 6.2 and 6.3).

As we cannot weaken anonymity, we have to content ourselves with anonymous mechanisms that are either strategy-proof and inefficient or efficient but not strategy-proof. The former class of mechanisms (strategy-proof and anonymous) contains at least the "deaf" mechanisms that always pick the same outcome no matter what the preference

---

[34] The reader may suspect that strategy-proofness comes cheap in a configuration where two out of three agents receive their peak ; it will suffice to look at a couple of other peak profiles such as (10, 8, 3) or (2, 7, 12) to be convinced that the property is perfectly general. For a detailed discussion, see Section 3.8, where it is also shown that the equilibrium outcome of proportional rationing is always the uniform rationing outcome.

profile. The interesting question is to find in that class the subset of the least inefficient mechanisms. The question has not been completely solved for any of the three problems mentioned above (fair division and production of a private or public good), although we have some elements of the answer; see Barbera and Jackson [1993].

Most of the efforts of the mechanism design literature[35] have focused on the second class, namely, anonymous, efficient mechanisms, in which the decentralized equilibrium concept is weaker than dominant strategy. For many alternative concepts, the results remain negative; for instance, if the concept is unique Nash equilibrium outcome (that is, we want a mechanism such that at all preference profiles there is a unique Nash equilibrium profile of messages), both the conclusion of the Gibbard–Satterthwaite theorem and all the negative statements two paragraphs above remain valid. Positive results appear only when we use certain equilibrium concepts requiring the common knowledge of preferences among agents.[36] A simple example will give the flavor of such mechanisms.

Example 1.8. Estate Division (Continued from Example 1.1)

Ann and Bob go to an arbitrator who recommends the following (a priori just) mechanism. Each of them will simultaneously submit a *sealed* bid for the object to the arbitrator, who will then open the bids. Whomever submitted the highest bid (say x submitted by Bob) will receive the painting and pay *half* of the bid to the other party (so Bob would get the painting and pay $\frac{x}{2}$ to Ann). Ties will be resolved by flipping a coin.

Consider first a situation where neither Ann nor Bob knows the worth of the object to the other (so Bob's valuation of $4000 would have to be something else than the market value of the object). Then Ann's prudent bid must be $10,000. For this bid guarantees her a profit of at least $5000 whether her bid is the highest or not, and no other bid guarantees her that much (e.g., a bid of $9000 could leave her with only $4500 and 50¢ *if* Bob were to bid $9001). Thus prudent, uninformed players will bid truthfully, implying that Ann will get the object and give $5000 to Bob. Notice that this outcome is equal split of the efficient surplus, namely, one of the a posteriori just outcomes in Example 1.1.

---

[35] Good surveys of which are Maskin [1985], Groves and Ledyard [1987], and Moore [1993].

[36] The main examples are Nash equilibrium in undominated strategies, equilibrium by iterated elimination of weakly dominated strategies, and in the context of games in extensive form, subgame perfect equilibrium. See Moore [1993].

Now consider the common knowledge situation (each knows the other's valuation, knows that the other knows, and so on). Of course, truthful bidding by both agents is not a Nash equilibrium because Ann's bid of $10,000 is not her best reply to Bob's bid of $4000. It is easy to see that this game has a whole range of Nash equilibrium outcomes where Ann gets the object and pays anywhere between $2000 and $5000.50.[37] But there are good strategic reasons to find one of these Nash equilibria more plausible, namely, this equilibrium where Ann bids $4000 and Bob bids $3999. The reason is that any bid of Bob above $3999 is a dominated strategy.[38] We obtain yet another one of the efficient and a posteriori just outcomes of Example 1.1, namely, the one where Bob receives only $2000.

---

[37] Say Ann bids x and Bob $x - 1$, and suppose the smallest incremental bid is $1: this is a Nash equilibrium profile of strategies if $10,000 - (x/2) \geq (x - 1)/2$ and $(x/2) \geq 4000 - (x/2)$.

[38] Compare Bob's bid 3999 with any bid x, $x \geq 4000$. Check that for all possible values y of Ann's bid, Bob is better off bidding 3999 than x. For instance, if $3999 < y \leq x$, check that $(y/2) \geq 4000 - (x/2)$.

CHAPTER 2

Core and Competitive Equilibrium: One Good and Money

2.1. INTRODUCTION

Markets describe the exchange and production of commodities in the private property regime. The ownership structure assigns each unit of good and each technology to exactly one participant.[1] Under specialization and division of labor, the initial allocation of resources is not expected to be Pareto optimal: farmers produce more vegetables than they care to eat, landowners have more land than they can plow, workers gladly give up some of their leisure for a meal ticket, and so on. Markets are meant to exploit these cooperative opportunities by eliciting Pareto improving trades.

Two concepts yield a unified treatment of the trading behaviour implied by private property: one is the core, corresponding to cooperation by direct agreements (see Chapter 1); the other is the competitive equilibrium, namely the single most important concept of economic analysis. The central theme of this chapter and the next one is the deep structural link of the core and competitive equilibrium in exchange economies. Chapter 2 introduces the main findings in the particularly simple context (often called the *partial equilibrium* model) where a single good is traded for money (this good may come in indivisible units —e.g., cars—or in divisible quantities—e.g., gasoline), and where the market participants are divided into *buyers* who initially have money and are willing to buy some good if the price is right (we may think of buyers of initially owning no good, or not willing to sell what they own) and *sellers* who own initially some of the traded good and are willing to give it up for money (they do not wish to buy more of the traded good). Equivalently, buyers are called *consumers* and sellers are called *firms* (and the cost function of a firm determines its reservation price qua seller). In partial equilibrium analysis, we also assume that the individual preferences of the participants take a very simple form.[2]

---

[1] In subsequent chapters (in particular, Chapters 4 and 5), we relax this assumption and allow "fractional" ownership, where several agents are shareholders of the resources.

[2] They can be represented by utility functions linear in money and separable between money and the traded good; see Section 2.2.

Recall the definition of the core in an exchange situation. Agents agree directly on specific trades (barter), namely a reallocation of their property among themselves. An allocation is stable in the sense of the core (it is a core allocation) if (i) trade is voluntary (every agent ends up at least at the utility level he initially enjoyed by consuming the goods he owned), (ii) allocation is efficient (the allocation is Pareto optimal; this is the efficiency postulate discussed in Chapter 1), and (iii) no coalition of agents can find a better (that is, from the point of view of the coalition members) trade of its own (that is, a trade compatible with the initial structure of private ownership).

A competitive price on the other hand, coordinates the decentralized trade choices of the participants. Each agent reads the posted price and determines, based on the sole price signal and independently from everyone else, his preferred transaction (how much to buy or sell of each good). After all such individual choices are made, the aggregate demand and supply coincide on each good (the equilibrium property of the competitive price); thus it is feasible to meet each and every preferred transaction. The resulting outcome is called a competitive allocation.

A fundamental difference between the two concepts hinges on the anonymity of trade. In the core, any barter is possible; for instance, a farmer may exchange 1 kilo of oranges for \$1 with one buyer, while exchanging 1 kilo of identical oranges for \$2 with another buyer. By contrast, in a competitive allocation, all buyers (resp. sellers) must pay (resp. be paid) the same price for the goods. This property is at the heart of the equity property of the competitive equilibrium discussed in Chapter 4.

In most of this chapter and the next one, we take on face value the (traditional) interpretation of the competitive equilibrium as a decentralization device. This interpretation should be questioned, for competitive allocations are not mechanically computed from a set of individual messages independently selected by the market participants. The familiar story of a Walrasian auctioneer (adjusting prices until demand equals supply), despite its realism, is not in the correct format of a game form. In Section 2.9 (as well as in Section 3.8), we briefly explain how the competitive allocations can be obtained as the strategic equilibrium outcomes of certain legitimate game forms (that is, where the decision power is fully distributed among the market participants, whereas a disinterested outsider enforces the rules of the game).

The three main findings of partial equilibrium analysis (to be confirmed in the more general model of Chapter 3) are as follows. First, we note that the competitive equilibrium is a more demanding concept than the core: a competitive allocation is always core stable.[3] In particular, a competitive equilibrium is a Pareto optimum, a modern formulation of Adam Smith's invisible-hand metaphor (Section 1.1) often called the first fundamental theorem of welfare economics.

In order to state the next result, we need the crucial assumption of convexity of preferences for all agents present in the market. In a partial equilibrium context, this means that buyers and sellers have decreasing marginal utility for the traded good (the more good I am consuming, the less I am willing to pay for one additional unit). Equivalently, firms have increasing marginal costs (a convex cost function). Under these convexity assumptions, the second important finding emerges: the existence of at least one competitive equilibrium (hence, the nonemptiness of the core) is guaranteed. Moreover, in many markets (for instance, those containing many agents who each own a small fraction of the total endowment of the economy), the private ownership core and the competitive price mechanism select approximately the same outcomes. This equivalence is known as the Edgeworth proposition. In our terminology, this means that the modes of cooperation by direct agreement and by decentralized price-taking behavior make approximately the same predictions about exchange. This remarkable proposition and its limits are illustrated in Sections 2.3, 2.4, and 2.5; a more general formulation will be in Section 3.7.

How realistic is the convexity assumption? If it is rather plausible for buyers, as long as they do consume the good here and now, the assumption is problematic for firms. Indeed, several industries (called natural monopolies), for instance, the production of telecommunication services and of many other public utilities,[4] have typical decreasing marginal costs (as well as decreasing average costs). In fact, most production processes require at least some start-up fixed cost, and are accurately modeled by a U-shaped average cost curve (initially decreasing, and eventually increasing). Similarly, a buyer using the traded good as an input in some downstream production process will exhibit increasing marginal utility for a while.[5]

What becomes of our two concepts when preferences of buyers and/or sellers are not convex? The third finding is that under these circumstances the competitive equilibrium collapses (its existence be-

---

[3] This elementary observation uses only the assumption that preferences are monotonic: agents like money and they are never made worse off by consuming more of the good—both money and the good can be disposed of at no cost.

[4] See Baumol, Panzar, and Willig [1982].

[5] Corresponding to the decreasing marginal cost portion of the downstream technology.

comes exceptional, because price-taking behavior does not make sense anymore; see Section 2.6), whereas core stability remains a powerful, if somewhat diminished, analytical tool. Indeed, the core is often nonempty,[6] although it tends to be "big."[7] However, there are market configurations where the core is empty as well. Thus the competition described by the core may be excessive, destructive. This is a paramount argument for the regulation of certain industries; many instances are given in Section 2.8. Still, in markets with nonconvex preferences, the core is generally a more robust concept than the competitive equilibrium.

Even in the simple equilibrium model, the cooperative analysis of the allocation of resources under decreasing marginal costs explains the need for the regulation of production. In Chapter 5, we pursue the normative discussion of such regulated monopolies in the justice mode of cooperation, and in Chapter 6 we discuss the decentralized mode; the convergence of these two modes will be the exception, unlike in the case of increasing marginal costs of production.

2.2. The Partial Equilibrium Model

We assume that there are only two types of goods, one of them perfectly divisible and called money, the other one (that could be divisible or not) being the specific good traded in the market (and interpreted, in various examples, as horses, cars, houses, or tasks if the good comes in indivisible units, or as labor, corn or oil if the good is divisible).

Throughout Chapter 2, we maintain the assumption of transferable utility (also called the assumption of quasi-linear preferences) requiring that every individual preference be separable between money and nonmonetary goods and linear in money. Thus, a convenient representation of preferences is by a "cardinal utility"[8] denoted u_i for agent i, and measuring, in monetary terms, the worth to this agent of consuming the traded good. Thus, if agent i consumes q_i units of this good and $\$t_i$, his overall utility index is $u_i(q_i) + t_i$, and we interpret $u_i(q_i) - u_i(0)$ as his willingness to pay for q_i units of the traded good (similarly $u_i(q_i +$

---

[6] This holds true, for instance, if *all* firms have decreasing marginal costs (if all sellers have increasing marginal utilities); see Section 2.7.

[7] In the case mentioned in the previous note, it allows for a broad range of surplus distribution among buyers.

[8] Of course, this utility can be interpreted as a cardinal index—it can be used to make interpersonal comparisons of utility increments, or even of absolute utilities—but we will scrupulously avoid doing so in this chapter and the following ones: all the concepts discussed use only the underlying preferences.

3) $- u_i(q_i)$ is his willingness to pay for three more units of the good, when he already owns q_i units). This quantity is independent of agent i's "income" (his consumption of money). As the choice of a particular number $u_i(0)$ is irrelevant, we always assume $u_i(0) = 0$.

A further assumption of the transferable utility context is that all agents have unbounded reserves of cash. This assumption, combined with the linearity of the utility function with respect to money, allows us to ignore entirely individual wealths, namely, initial endowments of money.

In the transferable utility context, the structure of Pareto optimal (efficient) allocations is extremely simple. A feasible allocation is efficient if and only if it maximizes the joint utility $\sum_{i \in N} u_i(q_i)$ over all feasible allocations of the nonmonetary goods among the participants in the market (the symbol N represents the set of such participants).[9] The important point is that Pareto optimality is logically independent of the redistribution of money among agents; therefore, the allocation of nonmonetary goods is (almost) completely determined by the efficiency criterion, whereas the distributional constraints derived from the existing property rights (or from any argument of fairness, as in Chapters 4 and 5) determine (or at least limit the set of) monetary transfers. This is the key simplification afforded by the transferable utility assumption, and the very source of its popularity.

2.3. BÖHM–BAWERK'S HORSE MARKET

In this celebrated model, the traded goods are identical (indivisible) horses and (i) each participant wants to consume at most one horse, (ii) each participant owns either one horse or no horse. Thus, the participants are partitioned into (potential) *sellers* who own a horse (hence are not interested in buying one) and (potential) *buyers* who own none. Moreover, agent i's preferences are described by a single nonnegative number, representing agent i's willingness to pay for a horse (if she is buyer), or equivalently her reservation price for selling her horse (if she is seller).

---

[9] Consider a feasible allocation $((q_i, t_i), i \in N)$ that is Pareto inferior to another feasible allocation $((q_i', t_i'), i \in N)$: $u_i(q_i) + t_i \leq u_i(q_i') + t_i'$ for all i with at least one strict inequality. Summing up and taking $\sum_N t_i = \sum_N t_i'$ into account (a consequence of feasibility), we get $\sum_N u_i(q_i) < \sum_N u_i(q_i')$. Conversely, if $Z = ((q_i, t_i, i \in N)$ is a feasible allocation and $\sum_N u_i(q_i) < \sum_N u_i(q_i')$ for some feasible allocation $(q_i', i \in N)$ of the nonmonetary goods, then Z is Pareto inferior to Z', given by $t_i' = t_i + (u_i(q_i) - u_i(q_i')) + \delta/n$, where $\delta = \sum_N (u_i(q_i') - u_i(q_i))$.

A useful computational device is to order the buyers by decreasing utilities (for a horse) and the sellers by increasing utilities:

$$u_1 \geq u_2 \geq \cdots \geq u_m, \qquad v_i \leq v_2 \leq \cdots \leq v_n, \tag{1}$$

where $u_j = u_j(h)$ is buyer j's utility for a horse, and similarly, $v_i = u_i(h)$ for seller i. We also denote by b_j the buyer with valuation u_j and by s_i the seller with valuation v_i. Consider the following example:

$$\text{5 buyers:} \quad 10 \geq 8 \geq 7 \geq 4 \geq 3, \qquad \text{6 sellers:} \quad 1 \leq 3 \leq 3 \leq 5 \leq 6 \leq 8. \tag{2}$$

Compute first the efficient allocations, namely, the optimal assignment of the 6 horses among the 11 agents. An assignment is described by the set of active buyers (who end up with a horse) and a set (of the same cardinality) of active sellers (who end up with no horse). Of course, an inactive buyer ends up with no horse and an inactive seller stays with her horse. In an efficient assignment, if buyer b_{j_1} is active while buyer b_{j_2} is not, we must have $u_{j_1} \geq u_{j_2}$ (otherwise a transfer of one horse from j_1 to j_2 enhances the joint utility); thus the q^* active buyers in an efficient assignment must have the q^* highest utilities among buyers, and similarly, the q^* active sellers must have the q^* lowest utilities among sellers. To determine the efficient number q^* in our example, we compute the best feasible surplus, namely, the largest difference in joint utility between a feasible allocation and the initial allocation (this is enough in view of the general characterization of efficiency under transferable utility):

best surplus	with one active buyer:	$10 - 1 = 9,$
	with two active buyers:	$9 + (8 - 3) = 14,$
	with three active buyers:	$14 + (7 - 3) = 18,$
	with four active buyers:	$16 + (4 - 5) = 15,$
	with five active buyers:	$17 + (3 - 6) = 14.$

Clearly, $q^* = 3$ and the general formula for q^* is as follows (given the inequalities (1)). The optimal number of "trades" is the largest number q^* such that $v_{q^*} < u_{q^*}$ and $u_{q^*+1} < v_{q^*+1}$ if such a number exists (with the convention $u_{m+1} = v_0 = -\infty$ and $v_{m+1} = u_0 = +\infty$; for instance, if $u_1 < v_1$, then $q^* = 0$). If such a number does not exist, there exists at least one index such that $v_q = u_q$; in this case, the optimal number of trades can be anywhere between the largest q such that $v_q < u_q$ and the largest q such that $v_q = u_q$ (whether or not a seller and a buyer with the same utility for a horse do trade is a matter of social indifference).

Among Pareto optimal allocations, we may restrict our attention to those where trade is voluntary (this is a simple consequence of private ownership: no one can be forced to trade), namely, an active buyer pays no more than his utility and an active seller receives at least her utility. This may entail a nonuniform transaction price: in our example, the buyers may pay respectively 6, 2, and 5 and the sellers may be paid respectively 4, 5, and 4. The competitive equilibrium price rules out nonuniform transactions by definition.

Define a *price p* to be *competitive* if at that price, total demand equals total supply. In our example, any price p between 4 and 5 is competitive because (the first) three buyers and (the first) three sellers want to trade. At price $p = 4$, exactly three sellers want to buy, the first three buyers definitely want to trade, while b_4 is indifferent between buying or not; we still call $p = 4$ a competitive price because one of the possible demands equals the unique supply.

The general formula is as follows (see Exercise 2.2). Let q^* be the largest index such that $v_{q^*} \leq u_{q^*}$ (remember our convention $u_{m+1} = v_0 = -\infty$ and $v_{n+1} = u_0 = +\infty$). Then the competitive prices cover the interval $[\sup(u_{q^*+1}, v_{q^*}), \inf(u_{q^*}, v_{q^*+1})]$.

Thus the competitive price is unique when there is an index q such that $v_q = u_q$ or when $u_{q^*} = u_{q^*+1}$ or when $v_{q^*} = v_{q^*+1}$. A justly famous example is the gloves market[10]: we have m identical buyers with common utility u and n identical sellers with common utility v, and $u > v$ (interpretation: each buyer wants a left-hand glove to match the right-hand glove he owns; each seller owns a left-hand glove *and* does not wear gloves). When there are fewer buyers than sellers ($m < n$), then $q^* = m$ and $v_{q^*+1} = v_{q^*} = v$, so the competitive price is v and buyers reap all the net surplus; symmetrically, with a shortage of sellers with respect to buyers ($n < m$), the competitive price is u ($q^* = n$, $u_{q^*} = u_{q^*+1}$), and the buyers get no surplus whatsoever. We come back to the gloves market in Section 7.7 (see Example 7.12) to compare this competitive allocation with the less extreme allocation resulting from the Shapley formula.

To each competitive price corresponds a set of *competitive allocations*, characterized by two properties:

(a) Each transfer of money t_i is the price of whatever horse agent i

---

[10] See Shapley and Shubik [1969a].

owns initially minus the price of the horse she consumes in the allocation in question (so in a Böhm–Bawerk market, $t_i = 0$ for an inactive agent, $t_i = p$ for an active seller, and $t_i = -p$ for an active buyer).

(b) An agent's own allocation maximizes her preferences over the set of all consumption vectors she can afford by selling her initial endowment at the competitive price (in this example, this means $u_j \leq p$ for an inactive buyer, $u_j \geq p$ for an active buyer, $v_i \leq p$ for an active seller, and $v_i \geq p$ for an inactive seller). Finally, we shall call *competitive set* the set of all competitive allocations for all possible competitive prices.

Turning to core stability, we find that all transactions must take place at the same price, implying that the core allocations coincide with the competitive allocations. We check this general fact in the numerical example (2). A core allocation must be Pareto optimal; hence, we know that the first three buyers (and only them) and the first three sellers (and only them) are active. Let p_j be the price paid by b_j and r_i that received by s_i ($1 \leq i, j \leq 3$). Suppose $p_j > r_i$ for some b_j and some s_i. Then the pair $\{b_j, s_i\}$ has an objection: they are both made better off by trading at price $(p_j + r_i)/2$ (b_j pays less and s_i earns more). Thus $p_j \leq r_i$ for all $i, j, 1 \leq i, j \leq 3$. However, feasibility implies $\Sigma_j p_j = \Sigma_i r_i$; hence, all six numbers p_i, r_j are equal to a common value p. It remains to check that p is between 4 and 5. If $p > 5$, then s_4 can form an objection with anyone of the active buyers (say b_1). Remember that s_4 is inactive by Pareto optimality. By trading with b_1 at price $(p + 5)/2$, s_4 does make a positive profit *and* gives a better deal to b_1. Thus $p > 5$ is impossible, and $p < 4$ is similarly ruled out. We have shown a particular case of a general fact:

In a Böhm–Bawerk's market, the competitive set and the core

are nonempty and coincide. (3)

See Exercise 2.2 for a systematic proof; see also a generalization of (3) as Lemma 2.1.

2.4. OLIGOPOLY WITH BINARY DEMANDS

In our first generalization of the Böhm–Bawerk market, the traded good still comes in indivisible units and the potential buyers still want at most one unit, with a willingness to pay u_j, $u_j \geq 0$, for buyer b_j. Thus buyer b_j's willingness to pay for the first unit is u_j and is zero for the second and following units: his marginal utility (willingness to pay for one more unit) is indeed nonincreasing. Sellers, on the other hand, own several units of the good and have nonincreasing marginal utility:

seller s_i: owns Q_i units and for all $q = 1, \ldots, Q_i$: $\partial v_i(q) \geq \partial v_i(q + 1)$

(where $\partial v_i(q) = v_i(q) - v_i(q - 1)$ and we set $\partial v_i(Q_i + 1) = 0$).

Equivalently, we can think of s_i as a firm capable of producing up to Q_i units of the good at the following cost:

$$c_i(q) = v_i(Q_i) - v_i(Q_i - q) \quad \text{for} \quad q = 0, 1, \ldots, Q_i,$$

that is, agent i, qua firm, incurs the same cost to produce q units of the good as agent i, qua seller, to give up q units out of Q_i. Nonincreasing marginal utility for the seller s_i corresponds to *nondecreasing* marginal costs for the firm s_i:

$$\partial c_i(q) = \partial v_i(Q_i + 1 - q), \quad q = 0, 1, \ldots, Q_i,$$
$$\partial c_i(q) = +\infty, \quad \text{for } q \geq Q_i + 1.$$

We retain in most of our discussions the more familiar interpretation of s_i as a firm.

Here is a numerical example. On the buyer side, we have 12 agents whose willingness to pay u_j for one unit of the good are arranged in decreasing order to form the following demand graph:

q	1	2	3	4	5	6	7	8	9	10	11	12
∂u	20	19	18	16	16	16	15	15	13	12	12	11

. (4)

This list is the *demand graph*, and can be used to compute the *demand function* δ (at price p, the number of units that would be sold is at most $\delta(p)$) and the inverse demand function δ^{-1} (to sell q units or more, the price must be at most $\delta^{-1}(q)$), e.g.,

$$\delta(p) = 3 \quad \text{if } 16 < p \leq 18, \qquad \delta(p) = 6 \quad \text{if } 15 < p \leq 16,$$

$$\delta(p) = 8 \quad \text{if } 13 < p \leq 15, \qquad \delta^{-1}(3) = 18,$$

$$\delta^{-1}(4) = \delta^{-1}(5) = \delta^{-1}(6) = 16, \qquad \delta^{-1}(7) = 15, \ldots.$$

Notice the convention that a buyer facing a price equal to her willingness to pay (thus indifferent between buying or not) chooses to buy. We could take the opposite convention instead. By construction, the demand curve is nonincreasing.

On the seller side, we have three firms, each with capacity 5 and with the following marginal costs:

q	1	2	3	4	5
δc_1	6	10	14	18	20
δc_2	8	10	13	16	19
δc_3	11	13	16	18	20

The list $(q, \delta c_i(q))$ is the *supply graph* of firm i and gives similarly the *supply curve* ζ_i: at price p, firm i would sell at most $\zeta_i(p)$ units. Because marginal costs are nondecreasing, the solution of the profit-maximization problem

$$\max_{q=1,2,\ldots} \; p \cdot q - c_i(q)$$

consists of the quantities q such that

$$\delta c_i(q) \le p \le \delta c_i(q + 1).$$

This implies in particular that the supply curve is nondecreasing. For instance, the supply curve ζ_1 (with our convention that the firm picks the largest profit-maximizing quantity) is

$$\zeta_1(p) = 0 \quad \text{if } 0 \le p < 6, \qquad \zeta_1(p) = 1 \quad \text{if } 6 \le p < 10,$$

$$\zeta_1(p) = 2 \quad \text{if } 10 \le p < 14 \quad \text{and so on}$$

The (nondecreasing) aggregate supply curve $\zeta_1 + \zeta_2 + \zeta_3 = \zeta$ gives the overall supply (by all three firms together) for different (uniform) prices at which their output can be sold. It is also the supply curve associated with the efficient cost function c combining optimally all three production possibilities:

define $\quad c(q) = \inf\{c_1(q_1) + c_2(q_2) + c_3(q_3) \mid q_1 + q_2 + q_3 = q\}$

then $\quad \delta c(q) \le p < \delta c(q + 1) \Leftrightarrow \zeta(p) = q$ $\qquad\qquad$ (5)

$\quad$ (where $\zeta = \zeta_1 + \zeta_2 + \zeta_3$).

(The proof is the subject of Exercise 2.18.) In our numerical example, we get the following aggregate supply graph:

q	1	2	3	4	5	6	7	8	9	10	...	15
δc	6	8	10	10	11	13	13	14	16	16	...	20

$$(6)$$

We are now ready to compute the efficient trade and the competitive equilibrium allocations. In an efficient allocation with q^* active buyers, these buyers must have the q^* highest willingness to pay (by the same argument as in Section 2.3). On the other hand, the q^* units traded must be produced at minimal overall cost given the three existing technologies, that is to say, they must be produced at cost $c(q^*)$. Hence the efficient trade obtains whenever the (inverse) demand and (inverse) supply curves cross, namely, at a level q such that

$$\partial c(q) \le \partial u(q) \quad \text{and} \quad \partial u(q+1) \le \partial c(q+1). \tag{7}$$

Indeed, the net surplus generated by a trade of q units is $u(q) - c(q)$ (where u, the aggregate utility of buyers, obtains by integrating the inverse demand (4), namely, $u(1) = 20$, $u(2) = 39$, and so on). The left-hand (resp. right-hand) inequality in (7) says that decreasing (resp. increasing) q by one unit does not increase surplus. As $u - c$ is concave in q, these local optimality conditions imply that a global maximum of $u - c$ has been reached.

In the numerical example, the only solution of inequalities (7) is $q^* = 8$; therefore, efficiency commands trading 8 units (3 from firm 1, 3 from firm 2, 2 from firm 3). The corresponding surplus $v(N)$ is

$$v(N) = u(8) - c(8) = (20 + 19 + 18 + 16 + 16 + 16 + 15 + 15)$$

$$-(6 + 8 + 10 + 10 + 11 + 13 + 13 + 14) = 50.$$

A price p is competitive if (and only if)

firms may supply 8 units $\Leftrightarrow$ $14 \le p \le 16$ and
buyers may demand 8 units $\Leftrightarrow$ $13 \le p \le 15$,

whence the competitive prices cover the interval [14, 15]. The $50 of surplus are divided as follows:

$$
\begin{aligned}
\text{firm 1:} \quad & 12 \le \sigma_1 \le 15, \\
\text{firm 2:} \quad & 11 \le \sigma_2 \le 14, \\
\text{firm 3:} \quad & 4 \le \sigma_3 \le 6, \\
\text{buyers:} \quad & 23 \ge \Sigma \, \beta_i \ge 15.
\end{aligned}
\tag{8}
$$

Remark 2.1

In the next section, we give the traditional graphical representation of demand and supply curves in the context of a divisible traded good. Such a representation is also possible in the case of indivisible units, with the supply and demand curves taking "staircase" shapes and, awkwardly, jumping discontinuously. We shall reserve the use of these useful figures for the divisible-good model. See, for instance, Figure 2.1.

Remark 2.2

The general formula for the interval of competitive prices is

$$[\sup(\partial u(q^* + 1), \partial c(q^*)), \inf(\partial u(q^*), \partial c(q^* + 1))],$$

where q^* is the largest solution of system (7). This generalizes the formula for the Böhm–Bawerk market (Section 2.3) and will be generalized again in Section 2.5.

We turn now to the core. The equality of the core and competitive equilibrium (as in the Böhm–Bawerk market) holds true in this more general model provided two firms at least are active.

Lemma 2.1. (*Two is enough for competition*). *If the traded good is produced in indivisible units; if the demand is binary (each buyer wants at most one unit); if at least two firms (sellers) are active in at least one efficient trade; then the core allocations and competitive equilibrium allocations coincide.*

We prove the result for the numerical example. The general proof is much the same; see Appendix 2.1. Denote firms by s_1, s_2, s_3 and buyers by $b_1, b_2, \ldots$, where buyer 1's willingness to pay is $u_1 = 20$, buyer 2's is $u_2 = 19$, and so on. Let $v(N) = 50$ denote the efficient surplus within the set N containing our three firms and twelve potential buyers. Similarly, $v(s_1, b_4, b_5, b_6)$ is the maximal surplus from efficient trade between firm 1 and these three buyers. Thus

$$v(s_1, b_4 b_5, b_6) = u_4 + u_5 + u_6 - c_1(3) = 18.$$

Now pick an allocation in the core. It is efficient; hence, the first eight buyers are active and the firms produce respectively 3, 3, and 2 units. Denote by σ_i the net surplus of seller i (e.g., s_1 receives $c_1(3) + \sigma_1$ in cash) and by β_j that of buyer j (thus if $j \leq 8$, buyer j pays $u_j - \beta_j$, and if $j \geq 9$, buyer j receives β_j). We write down some inequalities following from the core stability property:

$$\beta_j \geq 0 \quad \text{for } j = 9, \ldots, 12 \quad (\text{by voluntary trade}),$$

$$\sigma_1 + \beta_4 + \beta_5 + \beta_6 \geq v(s_1, b_4, b_5, b_6) = 18$$

(or else that coalition can enforce a better deal),

$$\sigma_2 + \beta_3 + \beta_7 + \beta_8 \geq v(s_2, b_3, b_7, b_8) = 17,$$

$$\sigma_3 + \beta_1 + \beta_2 \geq v(s_3, b_1, b_2) = 15.$$

Adding up the seven inequalities above, we find an equality (because $v(N) = 50$); therefore, all inequalities are in fact equalities, for instance,

$$\sigma_1 + \beta_4 + \beta_5 + \beta_6 = v(s_1, b_4, b_5, b_6) = u_4 + u_5 + u_6 - c_1(3).$$

Now the choice of the partition $\{4, 5, 6\}\{3, 7, 8\}\{1, 2\}$ was arbitrary; we could have picked any other 3-3-2 partition and applied the same argument. Therefore,

$$\sigma_i + c_i(3) = (u_{k_1} - \beta_{k_1}) + (u_{k_2} - \beta_{k_2}) + (u_{k_3} - \beta_{k_3})$$

for all $i = 1, 2$, all k_1, k_2, k_3 distinct in $\{1, \ldots, 8\}$.

This, in turn, implies that $u_k - \beta_k$ does not depend on $k = 1, \ldots, 8$; we call it the price p paid by each buyer. Moreover, $\sigma_i = 3p - c_i(3)$, $i = 1, 2$ and $\sigma_3 = 2p - c_3(2)$. To complete the proof that p is competitive, we only need to show $14 \leq p \leq 15$. This follows from $\sigma_1 + \beta_1 + \beta_2 \geq v(s_1, b_1, b_2)$ and $\beta_8 \geq 0$, respectively. We have shown that a core allocation is competitive as well. The converse property (a competitive allocation is in the core) is a very general property that will be stated with full generality in Section 3.6. Q.E.D.

A corollary of Lemma 2.1 is that free entry wipes out the possibility of a positive profit for the firms (at the competitive outcome). Indeed, fix a technology (with nondecreasing marginal costs) and a binary demand. Call q^* the overall demand at price $c(1) = \partial c(1)$ (so $\partial u(q^*) \geq c(1) > \partial u(q^* + 1)$). Assume that $(q^* + 1)$ firms or more have entered the market. Then the industry supply is zero for $p < c(1)$ and is $q^* + 1$ or more for $p > c(1)$. Thus, the only competitive price is $c(1)$, and in a typical competitive equilibrium, q^* firms produce one unit each and make no profit. By Lemma 2.1, this conclusion is also implied by the core stability property. See the comments at the end of Section 2.5 for a general proof of "free entry wipes out the firms' profits" (which requires neither increasing marginal costs nor binary buyers).

Finally, we look at the case of a monopolist, where the conclusion of Lemma 2.1 does not hold anymore. For the sake of illustration, we suppose that the three firms in our numerical example merge in a single supplier, with cost function c (see (5)). Of course, the competitive equilibrium prices do not change (they are still located where ∂u and ∂c cross), and the corresponding surplus division gives at least \$27 and at most \$35 to the monopolist (out of an overall \$50). The core, on the other hand, is much bigger.

In one of the core allocations, the monopolist keeps the entire surplus by perfectly discriminating between the eight efficient buyers (thus charging \$20 to buyer 1, \$19 to buyer 2, and so on). At the other extreme, the monopolist could get as little as \$27, but no less (as in the competitive set). To see this, note that in a core allocation, buyer b_i cannot get more than $v(N) - v(N \setminus \{b_i\})$, where $N \setminus \{b_i\}$ is the coalition containing the monopolist and all buyers but himself; otherwise, the coalition $N \setminus \{b_i\}$ would object. Therefore

$$\beta_1 \leq 6, \qquad \beta_2 \leq 5, \qquad \beta_3 \leq 4, \qquad \beta_4 \leq 2,$$

$$\beta_5 \leq 2, \qquad \beta_6 \leq 2, \qquad \beta_7 \leq 1, \qquad \beta_8 \leq 1.$$

On the other hand, the competitive price $p = 14$ still yields a core allocation, and the corresponding surplus shares for the buyers are precisely these upper bounds. Our numerical example illustrates a general property.

Lemma 2.2. *If the traded good is produced in indivisible units; if the demand is binary, then in the core, a monopolist receives at least his worst competitive equilibrium profit and at most the whole profit. In particular, a set of (nondecreasing marginal costs) firms never sees its joint profit decrease when it colludes.*

Lemma 2.2 (of which the proof is in the Appendix) says that, according to the core logic, collusion always pays (at least it never hurts).[11] By contrast, the competitive equilibrium surplus division is unaffected by collusion. Note, however, that the core does not determine the surplus division between buyers and monopolist with any degree of accuracy (unlike the competitive equilibrium).

2.5. EXISTENCE OF THE COMPETITIVE EQUILIBRIUM UNDER CONVEX PREFERENCES

Denote by S the set of sellers (firms) and by B the set of buyers. Each agent can trade more than one unit of the good:

firm s_i: utility function $t_i - c_i(q_i)$, where t_i is its revenue and $q_i \geq 0$ is the quantity of good it produces,
buyer b_j: utility function $u_j(q'_j) - t_j$, where $q'_j \geq 0$ is the quantity of good he buys and t_j his payment.

We always assume $c_i(0) = 0$ and $u_j(0) = 0$.

In this section, the good traded is perfectly divisible and traded quantities are any nonnegative real numbers. This switch to a continuous model (as opposed to the discrete model of Sections 2.3 and 2.4) is not required by the logic of the core and competitive equilibrium; it is meant to illustrate the analytical tractability of the continuous model (when combined with differentiability assumptions on utility and cost functions), as well as the convenience afforded by the demand/supply diagrams. It should be obvious how to adapt our discussion in this section to a model where goods come in indivisible units; see note 14 below, as well as Exercise 2.5.

We define the competitive equilibrium and the core.

Definition 2.1. An allocation of the traded goods $(q_i, i \in S; q'_j, j \in B)$ is a competitive equilibrium with competitive price p if we have:

$$u_j(q'_j) - pq'_j \geq u_j(q) - p \cdot q \quad \text{for all } q \geq 0, \quad \text{all } j \in B,$$

$$p \cdot q_i - c_i(q_i) \geq p \cdot q - c_i(q) \quad \text{for all } q \geq 0, \quad \text{all } i \in S,$$

$$\sum_S q_i = \sum_B q'_j \quad \text{(feasibility)}.$$

[11] Here, by collusion we mean that the firms pool their technologies as in (5): this does not change the aggregate supply function, hence the competitive equilibrium is unaffected.

The price p determines the transfers of money: firm s_i receives $p \cdot q_i$ and buyer b_j pays $p \cdot q'_j$. Note the slight abuse of notation identifying seller s_i with index i and buyer b_j with j.

In order to define the core, we need one piece of notation. For a coalition $S' \cup B'$, where $S' \subseteq S$ and $B' \subseteq B$, we denote by $v(S', B')$ the efficient surplus generated by trade within the submarket restricted to S', B':

$$v(S', B') = \max\left\{ \sum_{B'} u_j(q'_j) - \sum_{S'} c_i(q_i) \right\},$$

where the maximum bears on all trades feasible within $S' \cup B'$, $\sum_{S'} q_i = \sum_{B'} q'_j$. Any efficient allocation $((q_i, t_i), i \in S; (q'_j, t'_j), j \in B)$ [where t_i (resp. t'_j) is the net payment of seller s_i (resp. buyer b_j)] determines a surplus distribution $(\sigma_i, i \in S; \beta_j, j \in B)$:

$$\sum_{S} \sigma_i + \sum_{B} \beta_j = v(N).$$

Definition 2.2. An efficient allocation is in the core if the corresponding surplus distribution satisfies the following:

$$\text{for all } S' \subseteq S, B' \subseteq B: \quad \sum_{S'} \sigma_i + \sum_{B'} \beta_j \geq v(S', B'). \qquad (9)$$

A very general property (stated as Lemma 3.6 below) is that *any competitive equilibrium allocation is a core allocation as well*. This fact requires no assumption whatsoever on the functions u_j, c_i (they may even decrease at times); see Exercise 3.17. To ensure the existence of a competitive allocation (implying, then, the existence of a core allocation), we will make the following convexity assumptions on utilities and costs:

Increasing marginal costs: for all i, c_i is increasing convex and infinite beyond the finite capacity Q_i.
Decreasing marginal utilities: for all j, u_j is increasing, concave, and flat beyond the satiation quantity Q'_i.

Lemma 2.3. Under the above convexity assumptions, a competitive equilibrium exists.

Proof. The assumptions allow us to define buyer b_j's demand curve δ_j and seller i's offer curve ζ_i: for all $p \geq 0$

$$\delta_j(p) = q_j \quad \text{iff} \quad q_j \text{ maximizes } u_j(q) - p \cdot q \text{ among all } q \geq 0,$$

$$\zeta_i(p) = q_i \quad \text{iff} \quad q_i \text{ maximizes } p \cdot q - c_i(q) \text{ among all } q \geq 0.$$

A fully rigorous definition allows δ_j and ζ_i to be multivalued whenever the price p corresponds to a linear piece of u_j or c_i.[12]

The convexity assumptions guarantee that (i) δ_j is nonincreasing (downward sloping) and continuous, for all buyer b_j, and (ii) ζ_i is nondecreasing (upward sloping) and continuous for all seller s_i.[13] Therefore, the aggregate demand $\delta = \sum_{j=1}^m \delta_j$ and aggregate supply $\zeta = \sum_{i=1}^n \zeta_i$ are both continuous and respectively downward and upward sloping. At $p = 0$, we have $\delta(0) \geq \sum_j Q_j'$, whereas $\zeta(0) = 0$. When p becomes arbitrarily large, ζ goes to $\sum_i Q_i$, whereas δ goes to zero. Therefore, the curves δ and ζ must cross: any price p where $\delta(p) = \zeta(p)$ is a competitive price with corresponding competitive allocation $q_j' = \delta_j(p), j = 1, \ldots, m, q_i = \zeta_i(p), i = 1, \ldots, n.$ \hfill Q.E.D.

Therefore the convexity assumptions guarantee the existence of at least one competitive equilibrium.[14] With a few buyers and a few firms, the core is typically a large set (containing the competitive allocation); but when the number of buyers and of firms increases, the core shrinks around the competitive allocation. The next numerical examples underline this important point.

One Seller, One Buyer

The buyer's demand is $\delta_1(p) = 100 - p$ and the firm supply is $\zeta_1(p) = 4p$ (exercise: what are the utility and cost functions?). Figure 2.1 (with the usual convention setting p on the vertical axis and q on the horizontal axis) shows the competitive allocation $q^* = 80$, $p^* = 20$, and the corresponding surplus division $\beta = 3200$, $\sigma = 800$. By contrast, the core contains all allocations where the efficient quantity q^* is traded and the surplus 4000 is divided in arbitrary nonnegative shares between seller and buyer.

---

[12] That is to say, if for all q, $3 \leq q \leq 5$, we have $u_j(q) = 2q$, then $\delta_j(2)$ contains the whole interval [3, 5]. Note that δ_j is necessarily multivalued at $p = 0$: $\delta_j(0) = [Q_i' + \infty]$.

[13] These properties hold true even at those points where the demand (resp. supply) is multivalued.

[14] In the case of indivisible units of traded goods, the corresponding assumptions are $\partial c_i(q) = c_i(q) - c_i(q-1)$ nondecreasing in q and infinite beyond Q_i and $\partial u_j(q) = u_j(q) - u_j(q-1)$ nonincreasing in q and zero beyond Q_j'.

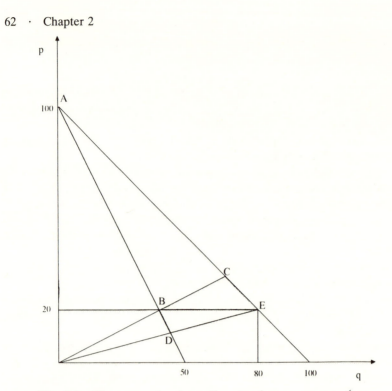

Fig. 2.1. The core of a two-seller, two-buyer economy (area *BCDE* measures the size of the core).

Two Sellers, One Buyer

Assume two identical firms with supply function $\zeta_i(p) = 2p$. The competitive price is unchanged, with the two firms splitting production and profit: $\sigma_1 = \sigma_2 = 400$. In the core, the presence of two competing firms limits their profit to 667 each:

$$\left. \begin{array}{l} \beta + \sigma_2 \geq v(s_2, b) = 3333 \\ \beta + \sigma_1 + \sigma_2 \geq v(s_1, s_2, b) = 4000 \end{array} \right\} \Rightarrow \sigma_1 \leq 4000 - 3333 = 667.$$

Thus the firms will not get more than *one-third* of total surplus (the surplus distribution $\beta = 2667$, $\sigma_1 = \sigma_2 = 667$ is in the core). Of course, they may get as little as zero.

Two Sellers, Two Buyers

Assume two identical buyers with demand $\delta_i(p) = 50 - p/2$. The core now has the equal-treatment property: the two identical buyers get the same surplus share ($\beta_1 = \beta_2$) and so do the two identical firms ($\sigma_1 = \sigma_2$).[15] To check the claim, consider the inequalities

$$\beta_1 + \sigma_1 \geq v(s_1, b_1) = 2000,$$

$$\beta_2 + \sigma_2 \geq v(s_2, b_2) = 2000.$$

Upon summing up, we get an equality; hence, $\sigma_i + \beta_j = 2000$ for all i, j (since the pairing $(s_1, b_1), (s_2, b_2)$ was arbitrary), implying the claim.

Thus the core yields a single line in the space of surplus distributions, namely,

$$1333 \leq \beta_1 = \beta_2 \leq 1777,$$

$$222 \leq \sigma_1 = \sigma_2 = 2000 - \beta \leq 667.$$

Indeed, $\beta_1 + \sigma_1 + \sigma_2 \geq v(s_1, s_2, b_1) = 2222$ implies $\beta_2 \leq 1777$, and similarly, $v(s_1, b_1, b_2) = 3333$ implies $\sigma_2 = 667$ as in the two sellers, one buyer example. Note that the surplus distribution from the competitive allocation ($\beta_j = 1600$, $\sigma_i = 400$) is near the middle of the interval of core distributions.

Figure 2.1 provides a geometrical illustration: in the core allocation must favorable to the buyers, each buyer receives (a surplus measured by the area of) the triangle ADE and each firm receives OBD. In the core allocation most favorable to firms, each buyer receives ABC and each firm OCE.

n Sellers, n Buyers

The equal-treatment property is preserved and the interval of core surplus distributions is computed to be

$$\text{for all } j: \quad \beta_j = \lambda \cdot \left(\frac{4000}{n} \right), \qquad \text{for all } i: \quad \sigma_i = (1 - \lambda) \cdot \left(\frac{4000}{n} \right),$$

$$\text{where} \quad \frac{4n - 4}{5n - 4} \leq \lambda \leq \frac{4n}{5n - 1}. \tag{10}$$

---

[15] The core of a two-sellers, one-buyer economy may *not* treat equals equally, because the monopsonist can discriminate among the two buyers: e.g., $\sigma_1 = 667$, $\sigma_2 = 0$ in the above example.

When n goes to infinity, the ratio β/σ goes to 4, as required at the competitive equilibrium for all n. See Exercise 2.4 for a proof of (10).

The example yields a particular case of the Edgeworth proposition: the core shrinks to the competitive equilibrium set when the economy is replicated. This will be the case for any economy with convex firms and buyers as described above. See Section 3.7 for a general discussion of the Edgeworth proposition.

Our discussion of the model with a perfectly divisible good shows that if the demand curves and supply curves are continuous, there is a unique competitive price, hence a unique competitive equilibrium outcome.[16] By contrast, the core is not a singleton, even utilitywise, except in some degenerate cases. Here is a two sellers, two buyers example. Take $\delta_i(p) = 50 - p/2$ if $p > 20$, $\delta_i(20) = [40, 80]$, $\delta_i(p) = 90 - p/2$ if $p < 20$ (compatible with a decreasing marginal utility), and $\zeta_i(p) = 2p$ if $p < 20$, $\zeta_i(20) = [40, 80]$, $\zeta_i(p) = 2p + 40$ if $p > 20$.

In the model with an indivisible good (as in Section 2.4), the competitive price is normally not unique (it varies in the interval given in remark 2.2, where c is the aggregate cost defined as in (5) and u is the aggregate utility defined as $u(q) = \sup\{\sum_{j \in B} u_j(q_j) \mid \sum_j q_j = q\}$), and the core contains precisely the competitive allocations in some important cases such as a Böhm–Bawerk market, or the economy described in Lemma 2.1.

The Edgeworth proposition states that in an economy where we have many identical copies of each type of agent, the core allows no further distribution of overall surplus than the competitive allocations. It should not be confused with another—much easier and much more general—proposition, namely, *"free entry wipes out the firms' profits."* (See the discussion after Lemma 2.1.) Consider a market where all firms (sellers) have the same technology (the same cost function); we do not make any assumption about marginal costs: this cost function is perfectly general. Suppose that not all firms are active in at least one efficient allocation; then all firms (active or not) make zero profit in all core allocations (hence in every competitive allocation as well). This proposition is easier to prove in the case of an indivisible good where, once the set B of buyers is fixed, the efficient number of active firms is necessarily finite.[17] Say that for a given cost function c, efficiency

---

[16] Continuity of (possibly multivalued) demand curves holds if the utilities u_j are concave, and continuity of (possibly multivalued) supply curves holds if the cost functions c_i are convex. Thus the competitive price is unique for any problem with decreasing marginal utilities and increasing marginal costs.

[17] In the divisible good case, the easy limit argument is left as an exercise to the reader.

requires at most $n - 1$ active firms, whereas the market has n firms all endowed with the cost function c. Denote by $S \setminus \{s_i\}$ the coalition of all firms except firm i. By assumption, we have

$$v(S, B) = v(S \setminus \{s_i\}, B) \quad \text{for all } i.$$

We invoke a general fact about the core in the transferable utility context. Call marginal contribution of agent i the number $v(N) - v(N \setminus \{i\})$. The marginal contribution is an upper bound on agent i's surplus share π_i in the core:

$$\left\{ \sum_N \pi_i = v(N) \text{ and } \sum_{N \setminus i} \pi_j \geq v(N \setminus \{i\}) \right\} \Rightarrow \pi_i \leq v(N) - v(N \setminus \{i\}).$$

$$(11)$$

In particular, if $v(N) - v(N \setminus \{i\}) = 0,$[18] then $\pi_i = 0$ (this remark has been used already twice in the above examples).

Remark 2.3

A useful consequence of the upper bound (11) is the following. If the vector of all marginal contributions is feasible, then it is the only surplus distribution in the core (hence the only competitive surplus vector as well):

$$\sum_{i \in N} (v(N) - v(N \setminus \{i\})) = v(N) \Rightarrow \pi = (v(N) - v(N \setminus \{i\}))_{i \in N}$$

is the unique core surplus.

An example is any Böhm–Bawerk market where the competitive price is unique. For the class of Böhm–Bawerk markets (binary buyer and binary sellers), the converse implication also holds: the core surplus distribution is unique *only if* the vector of marginal contributions is feasible.[19] On the other hand, as soon as we allow sellers and/or buyers to sell or buy more than one unit, the fact that the core is unique utilitywise does not imply that the vector of marginal contributions is feasible. An example follows.

[18] And $v(i) = 0$, as is always the case in this chapter because $v(S)$ is a *net* surplus. The general property is: if $v(N) = v(N \setminus \{i\}) + v(i)$, then $\pi_i = v(i)$.

[19] The convex games discussed in Section 7.4 are another class of games where the same equivalence is true. See Exercise 7.7.

Assume four sellers, each owning one unit; the reservation price (cost) of this unit is zero for s_1, s_2, s_3, and 1 for s_4. Assume three buyers; buyer 1 wants at most two units and would pay \$2 for each $(\partial u_1(1) = \partial u_1(2) = 2)$; buyers 2 and 3 want at most one unit, and $\partial u_2(1) = 2$, $\partial u_3(1) = 1$. One checks that the six units of total surplus $(v(S, B) = 6)$ are uniquely divided in the core as

$$\sigma_1 = \sigma_2 = \sigma_3 = \beta_2 = 1, \qquad \beta_1 = 2, \qquad \sigma_4 = \beta_3 = 0.$$

On the other hand, $v(N) - v(N \setminus \{1\}) = 6 - 3 = 3$.

Remark 2.4

We could easily extend the model captured by Definitions 2.1 and 2.2 to the case where each agent can be either a seller or a buyer of the traded good. A typical utility would be $u_i(y_i) + t_i$, where $y_i = 0$ if agent i's net trade is zero, $y_i > 0$ if she buys y_i units, and $y_i < 0$ if she sells. Under the assumption that u_i is concave and increasing for all y_i, agent i's demand remains a downward sloping function $d_i(p)$. If p^* is a price such that $d_i(p^*) = 0$, this agent will be a seller for $p \geq p^*$ and a buyer for $p \leq p^*$.

However, with nonconvex preferences (nonconcave utilities), the strict segregation of buyers from sellers will be of crucial importance to the core stability, as shown in Section 2.7.

2.6. DECREASING MARGINAL COSTS: EFFICIENCY

Within the market structure defined in Section 2.5, this section explores the consequences of dropping the convexity assumptions on utilities (decreasing marginal utilities) and costs (increasing marginal costs). We work alternatively in the model with a divisible good or with an indivisible good. The first, and most fundamental, observation is this: when firms have decreasing marginal costs, efficiency (Pareto optimality) demands that a single firm operates, a situation often called a *natural monopoly*.

The formal statement requires one more definition. We say that a cost function c has *increasing returns to scale* (or decreasing average costs) if we have

$$\text{for all } q, q' \text{ both nonnegative:} \quad q < q' \Rightarrow \frac{c(q')}{q'} < \frac{c(q)}{q}.$$

The definition of nondecreasing returns to scale (or nonincreasing average costs) is similar. These definitions, as well as Lemma 2.4, may be equally applied to a divisible good or to one that comes in indivisible units.

Lemma 2.4
(*i*) *Let $c_1, \ldots, c_n$ be n cost functions with nondecreasing (resp. increasing) returns to scale. Then, if we can freely combine these technologies, an efficient organization uses the cheapest single technology at any given level of production:*

$$\text{for all } q \geq 0: \quad \inf_{i=1,\ldots,n} c_i(q)$$

$$= \inf\left\{ \sum_{i=1}^{n} c_i(q_i) \middle/ \sum_{i=1}^{n} q_i = q \text{ and } q_i \geq 0, \quad \text{all } i \right\}, \quad (12)$$

(*resp. the only efficient organization is to use a single technology, namely, one of the cheapest technologies at the given level of production*).

(*ii*) *A cost function c_i with nonincreasing (resp. decreasing) marginal costs has nondecreasing (resp. increasing) returns to scale. However, a cost function with increasing returns to scale may have sometimes increasing marginal costs.*

Proof of Statement (i). Pick a profile of output demands $q_1, \ldots, q_n$ all positive ($q_i > 0$) and compare the returns $c_i(q_i)/q_i$, for $i = 1, \ldots, n$. Up to relabeling our functions, we may assume

$$\frac{c_1(q_1)}{q_1} \leq \frac{c_2(q_2)}{q_2} \leq \cdots \leq \frac{c_n(q_n)}{q_n}. \quad (13)$$

Denote $q = \Sigma_i \, q_i$ and invoke the assumption of nondecreasing (resp. increasing) returns:

$$c_1(q) \leq \frac{q}{q_1} \cdot c_1(q_1) \quad \left(\text{resp. } c_1(q) < \frac{q}{q_1} c_1(q_1) \right).$$

On the other hand, (13) implies

$$\frac{q}{q_1} \cdot c_1(q_1) = c_1(q_1) + \sum_{i=2}^{n} \frac{q_i}{q_1} \cdot c_1(q_1) \leq c_1(q_1) + \sum_{i=2}^{n} c_i(q_i).$$

The two above inequalities are now combined to yield the desired conclusion.

If we only have a subset of (at least two) positive demands q_i, the argument is adapted at once by restricting attention to those demands.

Proof of Statement (ii). Consider the inequality expressing nondecreasing returns to scale between q and q', with $0 \le q' < q$:

$$\frac{c(q)}{q} \le \frac{c(q')}{q'} \Leftrightarrow c(q') \ge \frac{q'}{q} \cdot c(q) + \frac{(q - q')}{q} c(0)$$

(recall our convention $c(0) = 0$). Now c has nonincreasing marginal costs if and only if it is a concave function.[20]

Moreover, if c has decreasing marginal costs, the right-hand inequality is strict, implying increasing returns as desired. The following cost function has increasing returns to scale but not decreasing marginal costs:

q	1	2	3	∞
c	6	8	11	11
ac	6	4	$3\frac{2}{3}$	0
∂c	6	2	3	0

where we write $ac(q) = \dfrac{c(q)}{q}$.   Q.E.D.

With n technologies identical to c, property (12) is equivalent to *subadditivity* (resp. strict subadditivity) *of costs:*

$$c\left(\sum_{i=1}^{n} q_i \right) \le \sum_{i=1}^{n} c(q_i) \quad \text{for all } q_1, \ldots, q_n \tag{14}$$

(resp. with a strict inequality). However, note that subadditivity of each cost function is not enough to guarantee (12) in Lemma 2.4 when the technologies are different. Here is an example with discrete quantities[21]:

q	1	2	3	∞
c_1	7	14	20	21
c_2	10	12	20	21

$c_1(1) + c_2(2) = 19 < 20 = \inf\{c_1(3), c_2(3)\}.$

---

[20] If q is real-valued, a concave function has right and left derivatives at every q and they satisfy $c'_-(q_1) \ge c'_+(q_2)$ for all q_1, q_2 with $q_1 < q_2$.
[21] The example is converted into one with continuous quantities by simple linear interpolation.

Therefore, subadditivity is a weaker property than nondecreasing returns to scale.

Remark 2.5

If all c_i have nonincreasing marginal costs, property (i) in Lemma 2.4 can be somewhat strengthened. The only efficient organization of production will involve a *single* active technology (namely, any one of the cheapest technologies at the given level of production) in all cases except one: when all active technologies have constant and equal returns from 0 to q. Thus nonincreasing marginal costs is enough to guarantee a single active firm if no cost function is linear between 0 and q.

We illustrate now the collapse of the competitive equilibrium. A good produced under decreasing marginal costs yields a natural monopoly (Lemma 2.4): in any efficient allocation, we expect only one firm to be active. So let us consider a market with a single firm with decreasing marginal costs and a downward sloping aggregate demand curve δ (Figure 2.2).

Efficiency requires production of q^* units, but no price signal can induce the monopolist to behave competitively (i.e., to maximize profit while viewing this price as given) and produce exactly q^*. In fact, a competitive monopolist facing a price p will either produce at capacity (if p exceeds the average cost at Q—i.e., the lowest conceivable average cost) or not produce at all (if p is below $ac(Q)$). If our monopolist has $\cup$-shaped marginal costs and average costs (as in Figure 2.3), the situation is hardly more satisfactory. A competitive monopolist facing a price above its minimal average cost will answer by supplying a quantity above the quantity $\bar{q}$ where marginal cost is minimal (see Figure 2.3: facing the price p^*, the firm would supply q, not q^*). The discrepancy between the quantity competitively supplied and that required by efficiency is only amplified when several firms are present; in the example of Figure 2.3, if n firms own the technology depicted, the price p^* generates the supply $n \cdot q$ (even though efficiency still commands producing q^*)!

If there is no hope to decentralize the firm's behavior by means of a price, can we at least decentralize the buyer's behavior by the price p^*? Surely this price will generate the right amount of demand as $q^* = \delta(p^*)$. However, it will not bring enough revenue to cover the costs of production: $q^* \cdot p^* = q^* \cdot \partial c(q^*) < q^* \cdot ac(q^*) = c(q^*)$.[22]

---

[22] The inequality $\partial c(q) < ac(q)$ follows since the derivative of $ac(q)$ is negative (average costs are decreasing, by Lemma 2.4). See also the numerical example below.

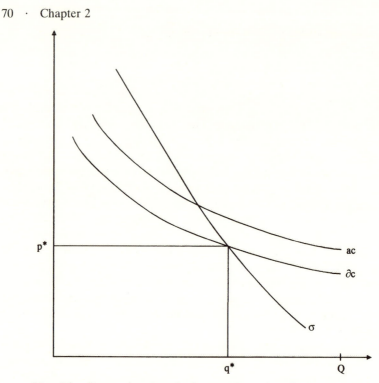

Fig. 2.2. Decreasing marginal cost: the price signal does not work.

One familiar answer to this second difficulty is to replace the single price by a two-part tariff consisting of a fixed fee f plus a (linear) price p: an agent pays $f + p^* \cdot q$ for q units of the good. An efficient allocation obtains if a "lump-sum fee" $f = [c(q^*) - q^* \cdot \partial c(q^*)]/m$ is charged to *all m* consumers present in the market, whether or not they buy any amount of the good (indeed, the unavoidable fee does not affect anyone's decision to buy or not, so every agent chooses the quantity he buys according to p^*). Such a "price" respects one essential feature of a competitive decentralization, namely, the anonymity of the price signal. On the other hand, it destroys what is perhaps the second most fundamental feature of a price signal, namely, *voluntariness of trade*. If showing up at the marketplace is costly whether or not I buy anything, then I will think twice before deciding to show up, and the lump-sum fee will effectively be levied only on those who buy something. Therefore, lump-sum fees are not consistent with a decentralization by a price signal (on the buyers' side) in a market with free entry and exit of agents. It is only meaningful when the public authority

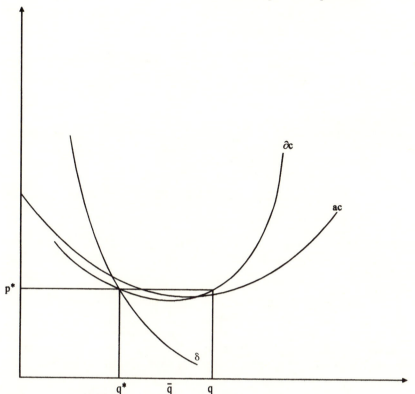

Fig. 2.3. ∪-shaped marginal cost: the price signal does not work.

regulating the market has the power to levy taxes. Then one must normatively justify taxing agents who derive no surplus from the good in question (as when a pacifist is forced to pay taxes some of which will go to national defense); this requires an ethical judgment on the type of goods being allocated, and goes beyond the logic of microeconomic equilibrium analysis. See further discussion of coercive taxes in Sections 5.4 and 5.6.

If we now insist on voluntary trade, can we achieve an efficient outcome by means of an anonymous price signal? Sometime we can, if we use a two-part tariff (or more generally, a nonlinear price) as illustrated in the example below. Sometimes we cannot, as in the case of a binary demand, because in that case any anonymous price signal is equivalent to an ordinary price (and the above discussion shows that efficiency and voluntariness of trade are generally incompatible).

In the example, consumers can buy up to two units of the good, and the marginal costs of production are

q	1	2	3	4	5	6	7	8	9
∂c	50	40	35	30	26	24	23	22	21

The seven agents and their (decreasing) marginal utilities for one unit and for two units are

	u_1	u_2	u_3	u_4	u_5	u_6	u_7
$\partial u(1)$	60	50	40	38	22	15	10
$\partial u(2)$	30	28	23	20	10	10	0

Efficiency (i.e., maximization of the total surplus $\sum_{i=1}^{7} u_i(q_i) - c(\sum_i q_i)$) requires production of six units at a total cost of 205, allocated as follows:

$$2 \text{ units for agents 1 and 2,} \quad 1 \text{ unit for agents 3 and 4}$$

(for a total surplus of $246 - 205 = 41$). A uniform price p yields the efficient demand if (and only if) p is between 23 and 28, bringing a revenue of at most 168, namely, 37 short from covering costs (note that this deficit is almost as large as the net surplus).

Next consider a two-part tariff with a fixed fee $f = 10$ and variable price 27.5. Thus the actual price is 37.5 for one unit and 65 for two units. This yields the efficient demand (check, e.g., $u_2(2) - 65 > u_2(1) - 37.5$, 0 and $u_3(1) - 37.5 > 0$, $u_3(2) - 65$, and so on), and exactly covers the costs.

Note that the example is not robust to a small change of the data such as the one reducing $u_4(1)$ from 38 to 35 (leaving everything else intact). In Exercise 2.6, we check that a two-part tariff (viz. the most general format of a nonlinear price in this example) generating the efficient demand in that case cannot bring enough revenue.

Under decreasing marginal costs of production, the competitive price logic does not work. As technical efficiency alone requires a natural monopoly (Lemma 2.4), and as economic efficiency is compatible neither with a price-taking behavior by the monopolist nor with voluntary trade of the consumers, we must resort to core analysis to explain how a profit-seeking (privately owned) firm will be forced by the competitive pressure of other potential entrants to implement an efficient outcome (and, as it turns out, to forgo any positive profit; this is the argument developed in the next section).

Alternatively, we can call upon the political authority to regulate the behavior of the natural monopoly (e.g., by nationalizing it). In that case, the potential users of the technology take control of the production process and pricing policy, with both an efficiency objective and a concern for fairness; indeed, economic efficiency alone is generally compatible with several pricing policies (see the above example) and the choice of one of them is a normative issue. High in the list of normative arguments of our regulating authority is the core stability interpreted as a no subsidization property (see below).[23] The normative discussion of a regulated monopoly is the subject of Chapter 5. See Sections 5.4 and 5.5 for the case of increasing returns.

Whatever our interpretation of the core (positive, as a contestability property, or normative, as nonsubsidization; see Chapter 5), voluntariness of trade will be saved, but anonymity of trade will be lost.

2.7. DECREASING MARGINAL COSTS: THE CORE

Consider a market with n firms, all exhibiting decreasing marginal costs, and m consumers whose utilities are virtually arbitrary: marginal utilities may decrease, increase, or do both. We simply assume that utilities are nondecreasing in the quantity of good consumed, eventually constant (consumers become satiated), and continuous if the good is infinitely divisible. A crucial assumption is that firms and consumers are distinct: the same agent cannot be both a seller of the good in some allocations and a buyer of the good in some other allocations. Thus consumers do not own initially any amount of the traded good, and firms do not derive any utility from buying one more unit of that good. We call this situation a segregated market.[24]

We can evaluate precisely the effect on the firm's profits of the competitive pressure conveyed by the core stability property (Definition 2.2). The result will then be compared to Lemma 2.1, describing the core of an oligopoly under increasing marginal costs.

Lemma 2.5. Consider a segregated market with n nonincreasing marginal costs firms, where (one of) the efficient (joint surplus-maximiz-

---

[23] Another normative argument is, precisely, the anonymity of trades, leading to the test of no envy that plays a central role in the discussions of common property (Chapters 4 and 5). The above examples show that anonymity of trades is frequently incompatible with efficiency; we show in the next section that it is also routinely incompatible with the core property (when it happens to be compatible with efficiency). See the numerical examples.

[24] In the next section, we show that the core analysis is very different in nonsegregated markets; in particular, the core may be empty.

ing) quantity traded is q^*, and where firm 1 is among the most efficient firms at this level of production:

$$c_1(q^*) = \min_{i=2,3,\ldots} (c_i(q^*)) - \varepsilon$$

where $\varepsilon \geq 0$ represents firm 1's technical advantage at level q^*

(*hence an efficient organization of production is to have firm 1 active and all other firms passive*).

The core of this economy is nonempty. In all core allocations, all firms $2, 3, \ldots$ make no profit, whereas firm 1's profit is at most ε. There is a core allocation where firm 1 makes zero profit, too.

Suppose, moreover, that firm 1 is ε_0 less costly than all other firms at every level of production $(c_1(q) + \varepsilon_0 \leq c_i(q)$ for all $q \geq 0$, all $i \geq 2)$. Then there is a core allocation where firm 1's profit is $\min\{\varepsilon_0, v(N)\}$ where $v(N)$ is the maximal joint surplus.

Corollary 1. (*Two is enough to wipe out the firms' profits*). *If at least two firms are efficient in at least one efficient allocation $(c_1(q^*) = c_2(q^*) = \min_i c_i(q^*))$, then none of the firm's profits is positive in any core allocation. An example is the case where all firms (two or more) own a copy of the same technology.*

The proof of Lemma 2.5 is given in the Appendix. Corollary 1 is related to the general fact that "free entry wipes out the firms' profits," discussed at the end of Section 2.5 (and requiring no assumption whatsoever on the technology common to all firms).

The case of a monopoly ($n = 1$) is also covered by Lemma 2.5, by making ε_0 arbitrarily larger than $v(N)$. In contrast with the increasing marginal costs case (Lemma 2.2), the core puts no lower bound on the monopolist's profit. Yet, it is still the case that collusion can only pay.

Corollary 2. If a single firm has a technological advantage (at every level) no smaller than the efficient joint surplus $v(N)$ in the economy (for instance, it is a monopoly supplier), then its profit in a core allocation can be anywhere from zero to $v(N)$.

Lemma 2.5 and Corollary 1 provide a powerful positive argument for the deregulation of the allocation of goods produced under decreasing marginal costs. The single supplier of the good (the natural monopoly) will not pocket any "rent" as long as it faces the threat of entry by at least one competitor with equal technological ability. Moreover, a cost advantage is the only way our monopolist can cash any profit; thus, the incentives to improve productivity are even greater than in the increasing marginal costs case.

However, core analysis provides also potent arguments for regulation. Compare Lemma 2.5 with Lemma 2.1: if all firms have increasing marginal costs, and if at least two of them are active in some efficient allocation, the core reduces to the set (often a singleton) of competitive allocations. The (active) firms still make a positive profit, however, even if they all have the same technology. By contrast, decreasing marginal costs firms can only make a positive profit if they have a technical advantage (and their advantage is an upper bound on their profit). Another important difference, on the consumer side, concerns the distribution of surplus allowed by the core: under increasing marginal costs, it is tightly determined by the (often unique) competitive price; under decreasing marginal costs, a wide range of surplus distributions is possible, as the examples discussed below demonstrate.[25]

A large core, naturally, is less useful than a deterministic solution. This is the point of the numerical example below; a more systematic discussion is in Section 5.4. Another limit of core analysis is the possibility of an empty core (the "destructive competition" discussed in the next section). Both limits provide strong arguments for regulation. They are both amplified in the case of a monopoly producing multiple goods.

Three Identical Consumers and Two Identical Firms

q	1	2	3	4	5	6	7	8	9
∂u_i	5	4	3	0	.	.	.	.	0
∂c	5	4	4	3	3	2	2	1	1

$i = 1, 2, 3.$

Efficiency commands producing 9 units, for a net surplus of 11. By Lemma 2.5, we have $\sigma_1 = \sigma_2 = 0$, and

$$v(s, b_i) = 0, \qquad v(s, b_i, b_j) = 3, \qquad v(s, b_1, b_2, b_3) = 11$$

(where s stands for a firm).

One core allocation respects anonymity of trade: every consumer pays 8.33, for a surplus share $\beta_i = 3.67$ for all i. On the other hand, the

---

[25] Of course, the equality of the core and the competitive equilibrium depends on the restrictive assumption of a binary demand in Lemma 2.1 (by contrast, Lemma 2.5 makes no restrictive assumption on consumers' preferences). If demand is not binary, the core under increasing marginal costs is larger than the competitive set; see section 2.5.

surplus distribution $\beta_1 = 0$, $\beta_2 = 3$, $\beta_3 = 8$ is in the core as well! Therefore, the core fails the "equal treatment of equals" property.

Two Different Firms and a Binary Demand

q	1	2	3	4	5	6	7	...
∂u	20	19	18	17	16	15	14	...
∂c_1	18	17.5	17	16.5	16	15.5	15	...
∂c_2	19	18	17.5	17	16.5	16	15.5	...

We have seven agents or more and two firms; the first firm has a technological advantage. Efficiency commands serving four (or five) consumers, for a total surplus $v(N) = 5$. Because $v(N) = v(s_1 : s_2, b_1 b_2 b_3 b_4)$, we know that $\beta_i = 0$ for all $i \geq 5$. At $q^* = 4$, we have $c_1(q^*) = 69$, $c_2(q^*) = 71.5$; therefore, the firm's profit is bounded as follows:

$$0 \leq \sigma_1 \leq 2.5, \qquad \sigma_2 = 0.$$

The highest surplus share for s_1 in the core is $\sigma_1 = 2.5$; corresponding surplus distributions among consumers include

$$\beta_1 = 2.5, \qquad \beta_2 = \beta_3 = \beta_4 = 0 \quad \text{(most advantageous to } b_1)$$

and

$$\beta_1 = 1, \qquad \beta_2 = 1.5, \qquad \beta_3 = \beta_4 = 0 \quad \text{(most advantageous to } b_2).$$

The smallest surplus share for s_1 in the core is $\sigma_1 = 0$; corresponding surplus distributions include

$$\beta_1 = 3.5, \quad \beta_2 = 1, \quad \beta_3 = 0.5, \quad \beta_4 = 0 \quad \text{(most advantageous to } b_1)$$

and

$$\beta_1 = 2, \quad \beta_2 = 1.5, \quad \beta_3 = 1.5, \quad \beta_4 = 0 \quad \text{(most advantageous to } b_3).$$

Clearly, all core allocations involve nonanonymous trade (consumer b_4 loses money if he pays one-fourth of total cost). The closest to anonymous trade in the core obtains when the first three consumers pay $17\frac{1}{3}$ and b_4 pays 17 (hence $\beta_1 = 2.67$, $\beta_2 = 1.67$, $\beta_3 = 0.67$, $\beta_4 = 0$). This core allocation is the one constructed in the proof of Lemma 2.5.[26] Note that in this economy, efficiency is compatible with voluntary trade.

[26] See Appendix 2.3. Exercise: prove this claim.

Remark 2.6. *Contestability*

The above discussion of the core requires discrimination among consumers (nonanonymous trade). This may be utterly infeasible, as when the consumers are very numerous and individually negligible (think of the traded good as electric power and of consumers as households). In this case, the anonymity of trade becomes a feasibility constraint and the core analysis must be adapted accordingly. This leads to the notion of a contestable allocation.

Say that we have n decreasing marginal costs firms $c_1, \ldots, c_n$ and a downward sloping aggregate demand curve δ. For simplicity, we restrict attention to uniform prices.[27] Consider an allocation where firm i^* sets a price p^* and serves the corresponding demand $\delta(p^*)$. We call this allocation *contestable* if

  (i) $ac_i * (\delta(p^*)) \leq p^*$ (firm i^* covers its cost),
  (ii) for all $i \neq i^*$, all $p < p^*$, and all $q \leq \delta(p)$: $p \leq ac_i(q)$.

Condition (ii) says that another firm cannot profitably enter the market by charging a price not higher than p^* and serving a portion of the demand thus generated. Given decreasing average costs (Lemma 2.4), this amounts to the inequality $p \leq ac_i(\delta(p))$. If firm i^* has no technological advantage (e.g., if all firms have the same cost function c), these two conditions say that firm i^* makes no profit $(ac_i * (\delta(p^*)) = p^*)$, and that p^* is the lowest price at which firm i^* covers its cost (the lowest intersection of the curves ac_i and d). This is, of course, the surplus-maximizing allocation given the uniform pricing constraints. If firm i^* does have a productivity edge, a contestable allocation is compatible with a positive profit; see Exercise 2.7.

2.8. NONCONVEX PREFERENCES AND EMPTY CORES

Firms with decreasing marginal costs are the simplest example of agents with nonconvex preferences. Other examples include firms with $\cup$-shaped marginal costs (a more realistic assumption; see the comments in Section 2.1) and buyers with increasing (or $\cap$-shaped) marginal utilities.

Naturally, the lack of a competitive equilibrium is a robust phenomenon: price-taking behavior simply yields discontinuous demands or supply functions even if a single agent has nonconvex preferences.

A superficial reading of Lemma 2.5 could suggest that nonemptiness of the core, too, is a very robust phenomenon; after all, the core is

---

[27] If the demand is binary, this entails no loss of generality.

nonempty whenever all firms have increasing marginal costs, and for arbitrary (quasi-linear) preferences of the buyers. The examples offered in this section (and the additional exercises) are meant to show that Lemma 2.5 is in fact a tight result. Seemingly small deviations from its assumptions allow for markets with empty cores. We give four simple examples of such "destructive" competition.

First, allow ∪-shaped marginal costs: this is compatible with subadditivity of costs ((14)), hence with a natural monopoly (see Lemma 2.4 as well as Remark 2.8 in Appendix 2.3); yet *Faulhaber's* ultrasimple example (which follows) shows that the core may easily be empty. Second, allow decreasing marginal costs firms and increasing marginal costs firms to coexist; then the *mixed convexity* example shows that the core, again, may be empty. Third, remove the assumption of segregated markets, that is to say, allow some agents to act as either buyers or sellers, depending on the allocation.[28] In this case, Scarf's celebrated example warns us, again, of the possibility of an empty core. We give it in full in Chapter 5 (see Exercise 5.12). The simpler example proposed here has nonsegregated markets and nonconvex preferences. Scarf's own example has convex preferences.

Faulhaber's Example (Faulhaber [1975])

We have three identical binary consumers who value one unit of an (indivisible) good at $a (and get no further utility from consuming two units). Assume $a \geq 6$. On the other side of the (segregated) market are two firms with identical cost functions:

q	1	2	3
∂c	6	3	5

Check the subadditivity of costs (despite ∪-shaped average costs, ac goes down from 6 to 4.5, then up to 4.67). By Lemma 2.5 (more precisely, by its variant with subadditive costs; see Remark 2.8 in Appendix 2.3), in a core allocation each firm gets zero profit. Total net surplus obtains by producing three units of the good (costs are subadditive):

$$v(N) = 3a - (6 + 3 + 5) = 3a - 14.$$

A coalition of two consumers and one firm generates $2a - (6 + 3) = 2a - 9$ dollars of surplus. Let $\beta_1, \beta_2, \beta_3$ be the surplus distribution

---

[28] Interpretation: an agent initially owns several units of the traded good; for a high enough price, he is willing to sell, but if the price is low enough, he will buy.

among consumers in a core allocation. The following system must hold:

$$\beta_1 + \beta_2 + \beta_3 = 3a - 14 \quad \text{(because firms make zero profit)},$$
$$\beta_i + \beta_j \geq 2a - 9 \qquad\qquad \text{for all pairs } i, j.$$

Summing up the three inequalities yields a contradiction, as $3(2a - 9) > 2(3a - 14)$.

See Exercise 2.10 for a variant of Faulhaber's example where the good is perfectly divisible, as well as the interesting replication of the indivisible example. Also, Exercise 2.9 has consumers with $\cap$-shaped marginal utilities (and firms with increasing marginal costs).

Two Examples with Mixed Convexities

We have two identical binary consumers b_1, b_2 with value $7 for one unit of an indivisible good. Consumers' preferences are thus convex. On the other side, the two firms s_1, s_2 can produce up to two units each; one of them has increasing marginal costs, one of them has decreasing marginal costs.[29]

q	1	2	3
∂c_1	4	7	∞
∂c_2	7	3	1

Compute the surplus accruing to, respectively, the grand coalition $N = \{s_1\, s_2, b_1, b_2\}$, and three smaller coalitions:

$$v(N) = v(s_2, b_1, b_2) = 7 + 7 - (7 + 3) = 4$$

(firm 2 produces two units),

$$v(s_1, b_1) = v(s_1, b_2) = 7 - 4 = 3.$$

Suppose that the surplus allocation $\sigma_1, \sigma_2, \beta_1, \beta_2$ corresponds to a core allocation. We must have

$$\sigma_1 + \beta_1 \geq 3, \qquad \sigma_1 + \beta_2 \geq 3, \qquad \sigma_2 + \beta_1 + \beta_2 \geq 4, \qquad \sigma_2 \geq 0$$

---

[29] By Lemma 2,5, the core *is* nonempty if both firms have decreasing marginal costs ; if both have increasing marginal costs, then even a competitive equilibrium exists. See Lemma 2.3.

and

$$\sigma_1 + \sigma_2 + \beta_1 + \beta_2 = 4,$$

which yields a contradiction, again by summing up the four inequalities.

We give for completeness a symmetrical example with two binary sellers owning one unit of indivisible good each, and two buyers, one of them with increasing marginal utility, the other with decreasing marginal utility:

	q	1	2
s_1	∂c	1	∞
s_2	∂c	3	∞
b_1	∂u_1	4	1
b_2	∂u_2	3	4

(15)

Pareto optimality requires that b_2 buys two units, with a corresponding total surplus $v(N) = 7 - (1 + 3) = 3$. We have similarly,

$$v(s_1, b_1) = 3, \qquad v(s_2, b_1) = 1, \qquad v(s_1, s_2) = 3.$$

Hence the core is empty as above.[30]

Exercises 2.12 and 2.13 give variants of the above examples with an indivisible and with a perfectly divisible good.

A Nonsegregated Market

This nonsegregated market has three identical agents who can either buy or sell one unit at most. Each agent has ∩-shaped marginal utility: $\partial u(1) = 2$, $\partial u(2) = 3$, $\partial u(q) = 0$ for $q \geq 3$, and owns initially one unit. An agent cannot buy and sell at the same time; he will buy one unit if $p < 2.5$ and sell one unit if $p > 2.5$. Clearly, total surplus from trade is \$1 (when one agent sells a unit to another agent), and any two-person coalition can achieve this surplus: the situation is the familiar game "divide the dollar among three under majority voting"; hence the core is empty. Recall from Remark 2.4 that with convex preferences, a nonsegregated market where each agent acts as buyer or seller has a nonempty core (and even a competitive equilibrium). Exercise 2.14 gives a variant of the above example.

---

[30] By summing up the inequalities $\sigma_1 + \beta_1 \geq 3$, $\sigma_2 + \beta_1 \geq 1$, $\sigma_1 + \sigma_2 + \beta_2 \geq 3$.

2.9. TRADING GAMES IN THE BÖHM–BAWERK MARKET

In this ultrasimple market (Section 2.3), it is easy to design a strategic game that approximately implements a competitive allocation at all profiles. However, this optimistic statement must be qualified if the market contains full-size sellers and/or buyers. In this section, we look only at the Böhm–Bawerk (BB) market. Trading games that implement the competitive equilibrium allocation in the more general context with full-fledged demands and supply have been proposed by Shapley and Shubik [1977], Dubey [1982], and many others. In those mechanisms, the agent's message consists of a (simplified) demand function. Their description is beyond the scope of this volume. See, however, the discussion of several impossibility results in Section 3.8.

Consider a BB market with m buyers and n sellers and utilities:

$$u_1 \geq u_2 \geq \cdots \geq u_m, \qquad v_1 \leq v_2 \leq \cdots \leq v_n. \qquad (16)$$

Given a utility profile (16), denote by $[p_-, p_+]$ the interval of competitive prices:

$$p_- = \sup(u_{q^*+1}, v_{q^*}), \qquad p_+ = \inf(u_{q^*}, v_{q^*+1}), \qquad (17)$$

where q^* is the largest index such that $u_{q^*} \geq v_{q^*}$. If $u_1 < v_1$, we set $p_- = +\infty$ and $p_+ = 0$. Recall that to every price p, such that $p_- \leq p \leq p_+$, we associate a competitive allocation (or, equivalently, a core allocation) by letting the first q^* buyers trade at price p with the first q^* sellers. In the case where $u_{q^*+1} = v_{q^*+1}$, it is a matter of indifference to b_{q^*+1} and s_{q^*+1} whether or not they do trade. By convention, we assume that these two agents do not trade.

Definition 2.3. In the BB market with m buyers and n sellers, pick a number λ, $0 \leq \lambda \leq 1$, and define the λ-competitive mechanism as follows: each buyer and each seller reports his or her own utility; the mechanism performs no trade whatsoever if each buyer reports a utility lower than that of each seller ($\bar{u}_i < \bar{s}_j$ for all i, j). Otherwise, the mechanism computes the interval of competitive prices according to the reported utilities (these reports are first rearranged in increasing order as in (16)) and implements the competitive allocation with price $p = \lambda p_+ + (1 - \lambda) p_-$.

For instance, if we have *one* seller and m buyers, the 1-competitive mechanism if a *first-price auction* where the seller on one hand, and all the buyers on the other hand, simultaneously submit their bids. If the highest buyer's bid is lower than the lowest seller's bid, no trade takes place; otherwise, the object is awarded to (one of) the highest bidder, who pays the value of his bid to the seller. Similarly, the 0-competitive mechanism is a *second-price auction* where, if trade occurs, (one of) the highest bidder gets the object and pays the *second-highest* bid to the seller.

Notice a slight ambiguity in Definition 2.3: in case several buyers (or sellers) have equal utility u_{q^*} (or v_{q^*}). How this ambiguity is resolved (e.g., by means of a fixed priority ordering of the buyers and of the sellers) is irrelevant to the subsequent strategic analysis.

Lemma 2.6.   For all λ, $0 \leq \lambda \leq 1$, and for all utility profiles, the set of Nash equilibrium outcomes of the λ-competitive mechanism consists of all competitive allocations and of the no trade outcome.

Proof.   The message profile where all buyers report $\tilde{u}_j = 0$ and all sellers report a large $\tilde{v}_i$ (larger than the largest utility u_j) is always a Nash equilibrium where no trade takes place. Consider next an equilibrium where some trade takes place, and denote by $\tilde{u}_{q^*}, \tilde{v}_{q^*}$ the corresponding critical (reported) utilities (as in (17)). Thus b_{q^*} and s_{q^*} do trade. We have $\tilde{u}_{q^*} \geq \tilde{v}_{q^*}$. In fact, $\tilde{u}_{q^*} = \tilde{v}_{q^*}$ (or s_{q^*} could raise her price a little), and this is the equilibrium price p. Note that a seller such that $v_i > p$ or a buyer such that $u_j < p$ does not trade in this equilibrium. Next, pick a buyer b_j such that $u_j > p$ and check that b_j must be trading in this equilibrium outcome. Otherwise, b_j can improve upon his equilibrium net profit (zero) by reporting $\tilde{u}_j = \frac{1}{2}(p + u_j)$ (which yields at $\frac{1}{2}(u_j - p)$ units of net profit). Similarly, a seller such that $v_i < p$ must be trading in this equilibrium. Therefore, the equilibrium allocation is competitive.

Conversely, it is an easy matter to check that every competitive allocation is a Nash equilibrium outcome of the λ-mechanism.   Q.E.D.

Lemma 2.6 reads: any λ-competitive mechanism implements, via Nash equilibrium behavior, the set of competitive allocations as well as the "no trade" outcome. Nash equilibrium behavior results from decentralized behavior via the myopic adjustment process known as Cournot tatonnement; in this sense, it is a meaningful equilibrium prediction when the players choose their strategies independently and, moreover, have no information about utilities other than their own. See the discussion of Cournot tatonnement in standard textbooks (e.g., Friedman [1986], Moulin [1986]); see also the more general definition of adaptive behavior by Milgrom and Roberts [1991].

If we can assume that the market participants know each other's preferences, we can use the two extreme competitive mechanisms for deterministic implementation.

Lemma 2.7. In the 1-competitive mechanism, every seller has a dominant strategy, namely, to report truthfully his or her utility. On the other hand, buyers may find it profitable to report a lower value than their true utility (they never find it profitable to overreport). Every Nash equilibrium of the mechanism where (i) sellers report truthfully their utility, (ii) buyers do not overreport (equivalently, buyers do not use a dominated strategy), yields a competitive allocation with price p_- (namely, the 0 competitive allocation).

In the 0-competitive mechanism, reporting truthfully is a dominant strategy for every buyer. Sellers never underreport (a report below v_j is a dominated strategy for seller j, if v_j is her true utility). Every Nash equilibrium where buyers report truthfully and sellers do not underreport yields a competitive allocation with price p_+.

The formal proof is omitted. It also shows that in every λ-competitive mechanism, with $0 < \lambda < 1$, both buyers and sellers may profit from misreporting.

Lemma 2.7 can be summarized as "the p_+-competitive mechanism *implements* the p_--competitive allocation and vice versa." "Implementation" refers to decentralized behavior with Nash equilibrium when no player is using a dominated strategy. This is a broadly accepted concept of strategic rationality, provided our players' actions are decentralized (Section 1.6). An example will illustrate some important features and limitations of this concept.

Consider a first-price auction (the 1-competitive mechanism) with one seller and three buyers and utilities:

$$v = 1000, \qquad u_1 = 1500, \qquad u_2 = 1300, \qquad u_3 = 800.$$

The truthful report $\bar{v} = 1000$ is a dominant strategy for the seller: no matter how the buyers bid, the seller cannot improve his profit by reporting more or less than a valuation of 1000 for the object. Buyers, on the other hand, do not want to report truthfully, for this prevents the possibility of making any profit at all (note that with two or more sellers in the market, a truthful report by a buyer may yield a positive profit in the 1-competitive mechanism). The strategic dilemma of each buyer is this: if I underreport too much, I have little chance to win the object, but a small underreport brings a small profit if I win. The model of a game with incomplete information is well suited to analyze this trade-off (see, e.g., Myerson [1991]). In our simpler model (a game with complete information), we can single out a unique Nash equilibrium only if buyer b_1 is fully informed of other buyers' utility. As overreporting is a dominated strategy for b_2, buyer b_1 knows that no bid from b_2 or b_3 can exceed \$1300, and therefore he can "safely" bid \$1301 to win the object and secure \$199 of profit (we assume that the smallest incremental bid is \$1 and identify the allocation where b_1 gets the object at price \$1301 with the p-competitive allocation). Of course, a bid $\tilde{u}_1$ below \$1300 cannot be part of a Nash equilibrium (if b_1 gets the object, then b_2 can make a positive profit by overbidding b_1 a little bit; if b_1 does not get the object, then b_1 is better off by overbidding whomever wins the object by a little bit).

Going back to the seller, we see that the assumption of a decentralized choice of bids by s and by b_1 (each of them selecting a bid independently and secretly) is crucial. For if the seller's bid $\tilde{v}$ can be chosen and shown to b_1 before he makes his choice, then s would bid $\tilde{v} = \$1499$ (assuming as usual that \$1 is enough to motivate b_1 into buying at \$1499). Thus the commitment power (the opportunity to choose one's strategy and make this irrevocable choice known to the other players) of the seller allows him to reap all the surplus (to get his best competitive allocation). Symmetrically, if b_1 has the power to first commit himself, he will secure the 0-competitive allocation by bidding \$1301. In between these two extreme allocations lie all the Nash equilibrium outcomes described in Lemma 2.6: $\tilde{v} = p = \tilde{u}_1 > \tilde{u}_2, \tilde{u}_3$, where $1300 \leq p \leq 1500$.

Note a corollary of Lemma 2.6: if the competitive price is unique ($p_- = p_+$), then any λ-competitive mechanism yields a unique Nash equilibrium outcome. Lemma 2.7 tells more: in the 0-mechanism (and in the 1-mechanism), truthtelling is then the dominant-strategy equilibrium (this statement holds true for any λ-mechanism as well). It turns out that we can design a very simple mechanism, called the pivotal mechanism,[31] with the property that (i) truthtelling is a dominant strategy for all players at all profiles, and (ii) the corresponding equilibrium outcome is "approximately" a competitive allocation. The catch is that the mechanism is not "feasible," that is to say, the sellers receive more than what the buyers pay: some outside banker must subsidize this market. This sounds like a cheap trick: if we can pour manna from heaven onto the market participants, why would they bother trading in

---

[31] This mechanism proves useful in many other contexts; see, e.g., Moulin [1988], Chapter 8. Its only drawback is to require the transferable utility context.

the first place? The answer is that the budget deficit of the pivotal mechanism is expected to be small if the number of market participants is large. Moreover, we will define (immediately after definition 2.4) a relative of the pivotal mechanism, called the "Dida" mechanism, where the deficit is replaced by a more realistic (and often smaller) budget surplus.

Definition 2.4   The Pivotal Mechanism for the Böhm–Bawerk Market.
Each buyer and each seller reports his or her utility. If the (at least one) competitive allocation (at the reported profile) involves no trade, then no trade takes place. If, on the other hand, there is an interval $[p_-, p_+]$ of competitive prices (at the reported profile) such that (at least) q^* sellers trade, the pivotal mechanism implements those trades but each buyer pays p_- and each seller receives p_+. The deficit $q^* \cdot (p_+ - p_-)$ is paid by an outside "banker."

Thus each buyer (resp. seller) receives precisely the same outcome as in the 0-competitive mechanism (resp. the 1-mechanism). By Lemma 2.7, truthtelling is a dominant strategy for all players; this is the property called *strategy-proofness* in Section 1.6.

Another equivalent definition of the pivotal mechanism is that (i) efficient trading takes place, and (ii) agent i receives his marginal contribution $mc_i = v(N) - v(N \setminus \{i\})$. (Exercise: check that this definition yields precisely the same mechanism, except for the unimportant fact that the $(q^* + 1)$th buyer and seller may engage in an irrelevant trade.) Under this more general definition, the strategy-proofness of the pivotal mechanism holds true as well.[32]

We turn to a related mechanism, dubbed the Dida (direct double auction) mechanism by its inventor, Preston McAfee. This mechanism, again, does not equalize the expenses of the buyers with the income of the sellers. This time the buyers pay a little more than the sellers, and the budget surplus is swallowed by an outside agent (e.g., as a salary for enforcing the rules of the mechanism).

Definition 2.5.   The Dida Mechanism. We have at least two sellers and at least two buyers ($m, n \geq 2$). Sellers and buyers report their utilities. If the (at least one) competitive allocation (at the reported profile) involves no trade, then no trade takes place. Otherwise, in every (reported) competitive allocation, at least q^* sellers and buyers are

---

[32] Exercise: prove this claim.

active.[33] Distinguish two cases:

(a) Suppose $q^* < \min\{m, n\}$ and $v_{q^*} \le p \le u_{q^*}$, where $p = \frac{1}{2}(u_{q^*+1} + v_{q^*+1})$. Then the first q^* buyers trade with the first q^* sellers at price p (note that p is a reported competitive price).

(b) Suppose $q^* = \min\{m, n\}$ or $p < v_{q^*}$ or $u_{q^*} < p$. Then the first $q^* - 1$ buyers buy a good at price u_{q^*} and the first $q^* - 1$ sellers sell a good at price v_{q^*}.

Lemma 2.8. *The Dida mechanism is strategy-proof (truthful report of one's utility is a dominant strategy for every player at every profile). In case (a), the mechanism selects a competitive allocation (and generates no monetary surplus). In case (b), the outcome is inefficient on two accounts. First, b_{q^*} and s_{q^*} are left inactive; second, there is a monetary surplus as ($q^* - 1$) buyers pay more than ($q^* - 1$) sellers. Total surplus loss in case (b) is $q^* \cdot (u_{q^*} - v_{q^*})$.*

Proof. See Exercise 2.18. The Dida mechanism is a more realistic decentralization device than the pivotal one (because it is easier to find a volunteer arbitrator if he is to get some money than if he is to give some money). Its budget surplus may be larger in case (b) than the budget deficit of the pivotal (as $u_{q^*} - v_{q^*} \ge p_+ - p_-$), but it is definitely smaller in case (a). The number $q^* \cdot (u_{q^*} - v_{q^*})$ is not very large anyway.[34] However, recall that Dida, unlike Pivotal, is not defined if we have only one seller and/or only one buyer.

There exists no strategy-proof mechanism implementing an efficient outcome of the BB market and generating no surplus and no deficit at all profiles. See Exercise 2.16.

APPENDIX TO CHAPTER 2

A2.1. Proof of Lemma 2.1

In the transferable utility context, an efficient allocation is described by the allocation of the traded good (see Section 2.2). Pick any efficient allocation where at least two firms are active. Let S^* denote the set of active firms; if s_i is in S^*, say that firm s_i sells q_i units of the good

---

[33] Recall that, after ordering the (reported) utilities as $\bar{u}_1 \ge \cdots \ge \bar{u}_m$, $\bar{v}_1 \le \bar{v}_2 \le \cdots \le \bar{v}_n$, the number q^* is the smallest index such that $\bar{u}_{q^*} \ge \bar{v}_{q^*}$ and $\bar{u}_{q^*+1} \le \bar{v}_{q^*+1}$, if such a number exists. If, on the other hand, $\bar{u}_j > \bar{v}_i$ for all i, j, we have $q^* = \min\{m, n\}$.

[34] It converges to zero as $\min(n, m)$ goes to infinity if all utilities u_i, v_j are independently and identically distributed; see McAfee [1994].

($q_i > 0$). Similarly, let B^* denote the set of active buyers at that allocation. By feasibility, we have

$$|B^*| = \sum_{s_i \in S^*} q_i. \qquad (18)$$

Now consider a core allocation where the traded good is allocated as just described, and where seller s_i receives $\$r_i$, while buyer j pays $\$p_j$. First, observe that for $s_i \notin S^*$, we have $r_i = 0$, and for $b_j \notin B^*$, we have $p_j = 0$. This results, as usual, from the equality $v(N) = v(N \setminus \{s_i\}) = v(N \setminus \{b_j\})$, and the observation that $v(N) - v(N \setminus \{s_i\})$ is an upper bound on s_i's profit (see the end of Section 2.5).

Now for any $s_i \in S^*$, we choose an arbitrary coalition M_i, $M_i \subseteq B^*$, with q_i buyers. The inequality

$$r_i \geq \sum_{j:\, b_j \in M_i} p_j \qquad (19)$$

follows from the core property; its violation implies an objection by coalition $\{s_i, M_i\}$. Assume a coalition M_i has been chosen for all i such that $s_i \in S^*$, so as to partition B^* ((18) implies that this is possible). Summing up the above inequalities over $s_i \in S^*$, we obtain an equality (denoting by S, B respectively the set of all sellers and of all buyers, by feasibility we have $\sum_S r_i = \sum_B p_j$; moreover, $r_i = 0$ in $S \setminus S^*$ and $p_j = 0$ in $B \setminus B^*$). Therefore (19) is an equality for all $s_i \in S^*$ and all coalition M_i in B^* with q_i elements. As $|M_i| < |B^*|$ (because S^* contains at least two sellers), it follows that p_j does not depend on j: each buyer pays the same price p. Moreover, equality (19) yields $r_i = q_i \cdot p$ and our allocation is competitive after all.

To complete the proof of Lemma 2.1, it remains to check that if a core allocation exists where firm s_1, say, is the only active firm, then this allocation is competitive as well. By assumption, at least one other firm can produce at least one unit at cost c. This implies that all active buyers pay c; hence, we have a competitive allocation. Q.E.D.

A2.2. Proof of Lemma 2.2

As in Remark 2.2, denote by q^* the largest solution of system (7). Then the lowest competitive price is

$$p_- = \sup\{\partial u(q^* + 1), \partial c(q^*)\}.$$

Consider an active buyer i. Buyer i's profit in the core is bounded above by his marginal surplus:

$$\beta_i \le v(N) - v(N \setminus \{b_i\}).$$

Upon permuting b_i and b_{q^*+1} in an efficient allocation,

$$v(N) + \partial u(q^* + 1) - \partial u(i) \le v(N \setminus \{b_i\}).$$

Upon making i an inactive agent in an efficient allocation, we get

$$v(N) + \partial c(q^*) - \partial u(i) \le v(N \setminus \{b_i\}).$$

Combining the above three inequalities:

$$\beta_i \le \partial u(i) - p_- \quad \text{for } i = 1, \ldots, q^*.$$

These inequalities imply the following lower bound on the monopolist's profit σ in the core:

$$\sigma \ge v(N) - \left(\sum_{i=1}^{q^*} \partial u(i) \right) + q^* \cdot p_- = q^* \cdot p_- - c(q^*).$$

Thus σ is bounded below by its worst competitive profit. Conversely, we know that every competitive allocation is a core allocation as well.

To complete the proof of Lemma 2.2, consider the surplus allocation $\sigma = v(N)$, $\beta_i = 0$, for all i (that is, buyer i pays precisely $\partial u(i)$ to the monopolist). This is a core allocation, for a coalition containing only buyers can generate no surplus on its own.   Q.E.D.

A2.3. Proof of Lemma 2.5

STEP 1

We assume first that all cost functions are equal to c and $n \ge 2$. Denote by $N \setminus S_i$ the coalition containing all consumers and all firms but firm s_i. By Lemma 2.4 and the assumption $n \ge 2$, we have

$$v(N) = v(N \setminus \{s_i\}) \quad \textit{for all } i = 1, \ldots, n.$$

As discussed at the end of Section 2.5, this implies (as in property (11)) that all firms make zero profit in any core allocation.

Now we prove that the core is nonempty by explicitly constructing an allocation in the core. Let B be the set of all consumers. For all $j \in B$, we define a function α_j as follows[35]:

$$\text{for all } \lambda \ge 0 \quad \alpha_j(\lambda) = \max_{q \ge 0} \{u_j(q) - \lambda \cdot q\}.$$

---

[35] This expression is well defined because u_i is eventually flat.

Clearly, α_j is continuous and nonincreasing in λ, and $\alpha_j(\lambda)$ goes to zero as λ goes to infinity (recall our convention $u_i(0) = 0$). Moreover, $\sum_{j \in B} \alpha_j(0) \geq v(N)$, since c is nonnegative. Therefore, there is a number λ^* such that

$$\sum_{j \in B} \alpha_j(\lambda^*) = v(N). \tag{20}$$

We claim that the division of total surplus where consumer j receives $\alpha_j(\lambda^*)$ (and firms get nothing) comes from a core allocation. Suppose not; there is a coalition T of consumers such that

$$\sum_{j \in T} \alpha_j(\lambda^*) < v(s_1, T), \tag{21}$$

where $v(s_1, T)$ is the maximal surplus when a firm produces for the consumers in T. A corresponding optimal allocation is q_j, $j \in T$ [36]:

$$v(s_1, T) = \sum_T u_j(q_j) - c\left(\sum_T q_j\right).$$

By definition of $\alpha_j(\lambda^*)$, we have

$$u_j(q_j) - \lambda^* \cdot q_j \leq \alpha_j(\lambda^*) \quad \text{for all } j \in T.$$

Therefore, (21) implies

$$c\left(\sum_T q_j\right) < \lambda^* \cdot \left(\sum_T q_j\right).$$

As c is concave and nonnegative, the above inequality implies

$$\text{for all } x \geq 0 \quad c\left(x + \sum_T q_j\right) - c\left(\sum_T q_j\right) < \lambda^* \cdot x. \tag{22}$$

Now we choose,[37] for all j in $B \setminus T$, a quantity q_j such that

$$\alpha_j(\lambda^*) = u_j(q_j) - \lambda^* q_j \quad \text{all } j \in B \setminus T.$$

[36] Its existence is guaranteed by our continuity assumptions on utility functions, and the fact that a concave function is continuous.
[37] Again, by our continuity assumptions.

Summing up these with inequality (21), we get

$$\sum_B \alpha_j(\lambda^*) < \sum_B u_j(q_j) - \left(c\left(\sum_T q_j\right) + \lambda^* \cdot \left(\sum_{B\setminus T} q_j\right) \right).$$

From (22), we deduce

$$c\left(\sum_T q_j\right) + \lambda^* \cdot \left(\sum_{B\setminus T} q_j\right) > c\left(\sum_B q_j\right);$$

hence,

$$\sum_B \alpha_j(\lambda^*) < \sum_B u_j(q_j) - c\left(\sum_B q_j\right),$$

in contradiction of (20). The proof of Step 1 is complete.

Remark 2.7

The core allocation just constructed has normative appeal in its own right when the technology is the common property of all consumers; it will be generalized in Section 5.5.

Remark 2.8

If the technology c is subadditive (property (14)) but marginal costs do not decrease everywhere, the argument showing that all firms get zero profit in the core is still valid, but the proof about the existence of the core is not. Faulhaber's example (Section 2.8) has an empty core despite subadditive costs.

STEP 2

Consider the general case where each firm s_i has a different cost function c_i. Assume, as in the statement of Lemma 2.5, that firm 1 is the most efficient at level q^*, with technical advantage ε. Then we have

$$v(N \setminus s_i) = v(N) \quad \text{for } i = 2, \ldots, n,$$

$$v(N \setminus s_1) \geq v(N) - \varepsilon.$$

(To show the latter, pick an efficient allocation where q^* is produced at cost $c_1(q^*)$; coalition $N \setminus s_1$ can produce q^* at cost $c_1(q^*) + \varepsilon$.) As in Step 1, this implies that, in the core, firms $2, \ldots, n$ make zero profit, whereas firm 1 makes at most ε.

Next, define the concave function $c = \min_{i=1,\ldots,n}\{c_i\}$ (concavity is preserved by the min operation). By Step 1, if c is the common cost function of the n firms, the core contains an allocation where all firms make zero profit. Such an allocation is also in the core with the true cost functions c_i, because the surplus available to any coalition can only decrease when we replace c by c_i.

STEP 3

It remains to show the last statement of Lemma 2.5. Assume $c_1(y) + \varepsilon_0 \leq c_i(y)$ for $y \geq 0$, all $i \geq 2$. Define a (concave) cost function c: $c(y) = c_1(y) + \varepsilon_0$. In the economy where each firm has cost function c, pick, as in Step 1, a core allocation where all firms get zero. Clearly, the allocation where firm 1 gets the profit ε_0 and everything else is unchanged is feasible in the initial economy (as $v(N; c) = v(N; c_1, \ldots, c_n) - \varepsilon_0$). It is a core allocation as well, because $v(M; c) = v(M; c_1, \ldots, c_n) - \varepsilon_0$ if M contains s_1, and $v(M; c) \geq v(M; c_1, \ldots, c_n)$ if M does not contain s_1.

Remark 2.9

The above proof is equally valid when the traded good comes in indivisible units or is perfectly divisible.

EXERCISES ON CHAPTER 2

Exercise 2.1. Simple Markets with Convex and Nonconvex Preferences

In each economy, compute the competitive prices (if any), the core allocations, and check whether the Edgeworth proposition holds or not. Notice that sellers are always permitted to buy (market is not segregated).

(a) Each seller owns one unit. Two buyers.

	s_1	s_2	b_1	b_2
$\partial u(1)$	0	3	2	2
$\partial u(2)$	2	2	6	1

(b) Each seller owns one unit. Two buyers.

	s_1	s_2	b_1	b_2
$\partial u(1)$	0	3	2	5
$\partial u(2)$	2	2	6	0

.

(c) Same as in question (b).

	s_1	s_2	b_1	b_2
$\partial u(1)$	1	3	4	3
$\partial u(2)$	0	0	1	4

.

(d) Seller s_1 owns one unit, seller s_2 owns two units. Two buyers.

	s_1	s_2	b_1	b_2
$\partial u(1)$	0	4	6	5
$\partial u(2)$	2	2	2	5

$\partial u(3) = 0.$

(e) Same as in question (d).

	s_1	s_2	b_1	b_2
$\partial u(1)$	0	4	6	6
$\partial u(2)$	2	2	2	5

$\partial u(3) = 0.$

(f) Same as in question (e).

	s_1	s_2	b_1	b_2
$\partial u(1)$	0	2	3	3
$\partial u(2)$	1	1	1	2
$\partial u(3)$	0	0	1	1

.

Exercise 2.2. The Core of the Böhm–Bawerk Market

In the market described in Section 2.3, prove that the set of competitive prices covers the interval

$$I = \left[\sup\{u_{q^*+1}, v_{q^*}\}, \inf\{u_{q^*}, v_{q^*+1}\}\right].$$

(*Hint:* You may adapt the proof of Lemma 2.3.) Then show that the set of core allocations and of competitive allocations coincide.

Exercise 2.3. A Variant of Lemma 2.2

The demand is binary and generates a downward sloping demand curve δ. We have n firms, $i = 1, \ldots, n$, with nondecreasing marginal costs. The good is divisible and the functions δ and ∂c_i are all continuous (the exercise can be easily adapted to the case of indivisible units of good).

Let p^* be the competitive price and q^* the efficient level of trade. We assume that only firm 1 is active in at least one competitive allocation. Show that this is equivalent to

$$\delta(p^*) = q^*, \qquad \partial c_1(q^*) = p^*, \qquad \partial c_i(0) \geq p^* \quad \text{for } i = 2, \ldots, n.$$

In Figure 2.4 is represented a typical configuration with an aggregate

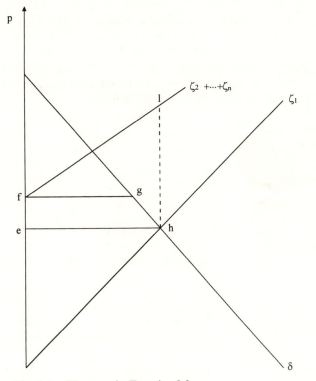

Fig. 2.4.   The core in Exercise 2.3.

supply ζ_{-1} by the firms $2, \ldots, n$. Show that in a core allocation, firm 1's profit is at most $\pi^* + \alpha$ and at least π^*, where π^* is the competitive profit $(0eh)$ and α is the area $(efgh)$.

Thus the largest extra profit α (beyond the competitive profit π^*) is smaller than (ehk) (corresponding to the highest profit of a real monopolist) and smaller than firm 1's technological advantage (area $(ehlf)$). Contrast this conclusion with Lemma 2.5.

In particular, if a single competitor of firm 1 (say, firm 2) is able to produce one single unit at price p^*, the core profit of firm 1 equals the competitive profit.

Exercise 2.4. The Numerical Example in Section 2.5 Continued

(a) The good is divisible. We have n identical buyers and m identical sellers with the following demand and supply curves:

$$\delta_i(p) = \frac{1}{n}(100 - p), \qquad \zeta_j(p) = \frac{4}{n} \cdot p.$$

Compute the competitive equilibrium allocation and check that the total profit of sellers and that of buyers remains constant.

(b) In the above economy, show that the core has the "equal treatment property": all buyers (resp. sellers) get the same share of surplus. Show that the identical profit β of each buyer and the identical profit σ of each seller vary within the following bounds:

$$\beta = 4000 \cdot \frac{\lambda}{n}, \qquad \sigma = 4000 \cdot \frac{\lambda}{n},$$

where λ is such that

$$\frac{4}{5 + \varepsilon} \leq \lambda \leq \frac{4}{5 - \varepsilon'}.$$

In the above inequality, the small numbers ε and ε' are defined as follows:

$$\varepsilon = \inf\left\{ \frac{k}{l} \cdot \frac{m}{n} - 1 \,\middle|\, k, l \text{ integers such that } \frac{k}{l} \cdot \frac{m}{n} > 1 \text{ and } k \leq n \right\},$$

$$\varepsilon' = \inf\left\{ 1 - \frac{k}{l} \cdot \frac{m}{n} \,\middle|\, k, l \text{ integers such that } \frac{k}{l} \cdot \frac{m}{n} < 1 \text{ and } l \leq m \right\}.$$

For instance, if $m = n$, we have $\varepsilon = 1/(n - 1)$, $\varepsilon' = 1/n$. If $n = m \cdot n'$ for some integer n', we have $\varepsilon = 1/(n - r)$, $\varepsilon' = 1/n$. Give the value of $\varepsilon, \varepsilon'$ if n and m are relatively prime.

(c) Consider an economy with n identical buyers and m identical sellers as follows:

$$\delta_i(p) = 100 - p, \qquad \zeta_j(p) = 4p. \tag{23}$$

Answer the same questions as in (a) and notice that the total surplus in the economy increases as we add either one more buyer or one more seller. Show that the profit of each seller (resp. of each buyer) increases as we add one more buyer (resp. one more seller), but decreases as we add one more seller (resp. one more buyer). What about total profit of the sellers (resp. of the buyers) when we add one more seller (resp. one more buyer)?

Exercise 2.5. An Example with Indivisible Goods

There is a single firm with increasing marginal costs and seven agents willing to consume at most two units of the indivisible good:

q	1	2	3	4	5	6	7	8	9
∂c	12	13	15	17	18	20	22	25	27

,

	u_1	u_2	u_3	u_4	u_5	u_6	u_7
$\partial u(1)$	60	50	40	38	22	15	10
$\partial u(2)$	30	28	23	20	10	10	0

.

Note that this utility profile appears at the end of Section 2.6 (with a decreasing marginal costs function). Compute the efficient trade, the competitive price(s) and competitive allocations, and the bounds of the profit for each player in the core.

Exercise 2.6. Where the Anonymity of Trade May or May Not Be Compatible with Core Stability

We consider the economy discussed at the end of Section 2.6. We have the same seven consumers as in Exercise 2.5, but the cost function now exhibits decreasing marginal costs:

q	1	2	3	4	5	6	7	8	9
∂c	50	40	35	30	26	24	23	22	21

.

(a) Assume two identical firms with this cost function and show that the two-part tariff described at the end of Section 2.6 yields a core allocation. Find all two-part tariffs yielding a core allocation, and describe all corresponding core allocations. Show that not all core allocations can be obtained in this way.

(b) Next, change the utility profile by lowering $u_4(1)$ to 35 (instead of 38), everything else being unchanged. Show that a two-part tariff inducing the efficient trade cannot raise enough revenue to cover the costs (the maximal revenue it can generate is 196). Hence it cannot pick a core allocation. Find a core allocation that is "as close as possible" to a two-part tariff allocation.

Exercise 2.7. On Contestability (as defined in Remark 2.6)

(a) Consider n firms with decreasing (nonincreasing) marginal costs and a downward sloping demand function. We assume that each (decreasing) average cost curve intersects once (from below) the demand curve; this intersection takes place at price p_i for firm i.

Suppose $p_1 \le p_2 \le p_i$ for all $i \ge 2$. Show that every allocation where firm 1 charges a price p, $p_1 \le p \le p_2$, and serves the corresponding demand is contestable. Check that there are not other contestable allocations.

(b) Suppose our n firms have $\cup$-shaped marginal costs, such that the lowest average cost is reached below the demand curve:

$$\text{if } q_i^* \text{ solves } \min_q ac_i(q), \quad \text{then } q_i^* < \delta(ac(q_i^*)).$$

Suppose that firms 1, 2, and 3 have the lowest possible average cost:

$$p^* = \min_q ac_1(q) = \min_q ac_2(q)$$

$$= \min_q ac_3(q) \le \min_q ac_i(q) \quad \text{for all } i \ge 4.$$

In the (exceptional) cases where $q_1^* + q_2^* = \delta(p^*)$ or $q_1^* + q_2^* + q_3^* = \delta(p^*)$, show that we can speak of a contestable allocation at price p^* where the demand is split between two or three firms (in the formal definition of Remark 2.6, condition (i) must be adapted). If, on the other hand, $\delta(p^*)$ cannot be written as the exact sum of a subset of numbers q_i^*, $i = 1, 2, 3$, show that no contestable allocation exists.

Exercise 2.8. The Two Examples with Mixed Convexities at the End of Section 2.9

In the first example, we have two binary buyers with utility 7 for (one unit of) the good and two sellers, one with convex, one with nonconvex preferences. Recall $\partial c_1(1) = 4$, $\partial c_1(2) = 7$, $\partial c_2(1) = 7$, $\partial c_2(2) = 3$. The core is empty.

(a) Add one buyer with utility u for (one unit of) the good. For what values of u is the core still empty? Describe the shape and size of the core as u varies.

(b) Add one seller endowed with one good and a utility v for the good. For what values of v is the core still empty? Describe its shape and size as v varies.

(c) Answer a similar pair of questions for the second example with mixed convexities (of the buyers), given by (15).

Exercise 2.9. Buyers with ∩-Shaped Marginal Utilities

This is a variant of Faulhaber's and Scarf's examples where sellers (firms) but not buyers have convex preferences. The good comes in indivisible units, and three identical buyers have the following marginal utilities:

q	1	2	3
∂u	1	6	3

(a) Suppose three sellers own one unit each and derive no utility from the good (they "produce" one unit at zero cost). Show that the core is empty.

(b) Is the conclusion of question (a) affected if we have two sellers instead of three? Four sellers instead of three? What if we have two sellers each owning two units?

(c) Suppose each buyer is also a seller, and can sell up to two units at zero cost (interpretation: the agent owns initially two units and their marginal utility is $0, 0, 1, 6, 3$). Show that the core is empty, too, as in Scarf's example. What if each agent can sell at most one unit at zero cost?

Exercise 2.10. Faulhaber's Example with Divisible Goods

(a) Assume three identical buyers with marginal utility equal to 30 up to $q = 3$ and zero afterward: $u_i(q_i) = \min\{30q_i, 90\}$, $i = 1, 2, 3$. Assume two identical sellers with $\cup$-shaped marginal costs

$$\partial c_j(q) = 30 - 9q + q^2.$$

Show that the core is empty.

(b) Assume three identical sellers with marginal cost 10 up to $q = 4$ and $+\infty$ afterward. Assume three identical buyers with $\cap$-shaped marginal utilities

$$\partial u_i(q) = \max\left\{10 + 4q - \frac{q^2}{3}, 0\right\}.$$

Show that the core is empty.

(c) We have three consumers with utility functions

$$u_1(q_1) = \min\{2q_1, 24\}, \qquad u_2(q_2) = \min\{\tfrac{3}{2}q_2, 12\}, \qquad u_3 = u_2,$$

and two identical firms with cost function

$$
\begin{aligned}
c(q) &= 10 \quad \text{if } 0 < q \le 10, \\
&= 20 \quad \text{if } 10 < q < 20, \\
&= 30 \quad \text{if } 20 \le q < 30,
\end{aligned}
$$

and so on. Show that the core is empty.

Exercise 2.11. Replicating Faulhaber's Example

(a) We have n identical buyers with utility 7 for (one unit of) the good and n identical firms with the marginal cost

$$\partial c(1) = 6, \qquad \partial c(2) = 3, \qquad \partial c(3) = 5, \qquad \partial c(4) = +\infty.$$

Show that if n is odd and $n \ge 3$, the core is empty. Show that if n is even and $n \ge 4$, the core contains a single allocation where all buyers share equally the total surplus.

(b) Discuss, according to a, the existence and size of the core when n identical buyers have utility, a, $a \ge 0$.

Exercise 2.12. Empty Cores when Marginal Costs are not Monotonic

(a) Two identical firms have the following cost function (indivisible units):

q	1	2	3	4	5	6
∂c	1.5	5	0	0	3	∞

The demand side is made of five binary buyers with the following demand graph:

q	1	2	3	4	5
∂u	3.5	3	2.5	2	1.5

Compute the efficient allocation, and check that four buyers are active. Show that the core is empty (check first that no firm can make a positive profit in a core allocation).

(b) This segregated market has three identical binary buyers who value one object at \$17.5. Three firms can produce the good at the following costs:

q	1	2	3	4
∂c_1	20	10	10	10
∂c_2	20	10	10	10
∂c_3	10	10	∞	∞

Show that the core is empty.

Exercise 2.13.   Market with Convex and Nonconvex Buyers

This is a divisible-good version of the example (15) above. Two identical firms have the following increasing marginal costs:

$$\partial c(q) = q \quad \text{if } 0 \leq q \leq 10,$$
$$= \infty \quad \text{if } 10 < q.$$

One buyer has convex preferences:

$$\partial u_1(q_1) = 20 - q_1 \quad \text{if } 0 \leq q_1 \leq 10,$$
$$= 0 \quad \text{if } 10 < q_1.$$

The second buyer has increasing marginal utilities:

$$\partial u_2(q_2) = 5 + \tfrac{2}{5}q_2 \quad \text{if } 0 \leq q_2 \leq +\infty.$$

Compute the efficient allocation (where both buyers are active). Show that the core is empty.

Exercise 2.14.   A Nonsegregated Market

Agents 1 and 2 can both be buyers or sellers. Agents 3 and 4 can only buy. Their marginal utilities are as follows (where negative quantities are sales and positive quantities are purchases):

q	-2	-1	1	2	3
∂u_1	1	2	8	0	0
∂u_2	∞	1	2	0	0
∂u_3	∞	∞	5	2	0
∂u_4	∞	∞	4	3	0

Show that the core is empty. (*Hint:* Compute $v(N)$.) Then compute $v(14)$ and $v(23)$ to deduce that agent 1's surplus share is at most 4. Next, compute $v(12)$ and $v(134)$ to deduce that agent 1's surplus share is at least 5.

Exercise 2.15.   An Example of the Pivotal Mechanism

The economy is as in question (c) of Exercise 2.4. There are n identical buyers and n identical sellers with demand and supply given by (23).

Consider the direct mechanism where each buyer reports a (nonincreasing) demand curve, each seller reports a (nondecreasing) supply curve, and an efficient trade (at those reported preferences) is implemented. Moreover, the monetary transfers are adjusted in such a way that agent i's net profit is exactly $mc_i = v(N) - v(N \setminus \{i\})$ (for all i, seller or buyer).

(a) Show that the mechanism is strategy-proof (this property is independent from the particular choice of the economy).

(b) Compute the deficit generated by this mechanism in this economy. Check that it is the fraction $4(m + n - 1)/[(4m + n - 1) \cdot (4m + n - 4)]$ of total surplus. Therefore, for large m and/or large n, it is a vanishingly small fraction of the surplus.

Exercise 2.16.   An Impossibility Result

(a) In a Böhm–Bawerk market with one seller and one buyer, show that the following mechanism is strategy-proof and "budget-balanced" (generates neither surplus nor deficit). Fix an arbitrary price p_0. Each agent reports his/her utility for the object. If u (reported by buyer) and v (reported by seller) are such that $v \leq p_0 \leq u$, then trade takes place at price p_0. Otherwise, no trade (and no transfer of money) takes place.

(b) In the same market, show that there is no strategy-proof and budget-balanced mechanism that implements an efficient allocation at all utility profiles. (*Hint:* Suppose such a mechanism exists. It takes the following form: if $u < v$, no trade, no transfer of money; if $v < u$, trade at price $p(u, v)$.) Show that $v \leq p(u, v) \leq u$. Moreover, if we fix u, v such that $v < u$, check the following inequalities:

$$v' < u \Rightarrow p(u, v) - v \geq p(u, v') - v,$$

$$v < u' \Rightarrow u - p(u, v) \geq u - p(u', v).$$

Conclude that $p(u, v)$ is constant for $v < u$ and deduce a contradiction.

Exercise 2.17. Proof of Lemma 2.8

Consider a seller with reservation price v_i, and a certain profile of reports $u_1, \ldots, u_m, v_1, \ldots, v_{i-1}, v_{i+1}, \ldots, v_n$ by the other participants. We must show that agent i cannot benefit from misreporting. The notation p and q^* follow Definition 2.5.

When i reports truthfully v_i, distinguish three cases. If $v_i < v_{q^*}$, then she will sell for sure at price p or v_{q^*}, but never less than v_{q^*}. A misreport to $v_i' < v_{q^*}$ will change nothing to the outcome; a misreport to $v_i' \geq v_{q^*}$ may keep her active or not, but if she sells, the price will not be higher (check the various configurations). The second case is $v_i > v_{q^*}$, where seller i will not sell if she reports truthfully. Then the only misreport allowing her to sell would be $v_i' \leq v_{q^*}$, implying a revenue of at most v_{q^*} (as $p < v_{q^*+1}$), hence no more than v_i.

The third case is $v_i = v_{q^*}$. If $i = i^*$, then she sells at a profit if we are in case (a) or does not sell if we are in case (b). In the latter case, she can only sell by lowering her report and then selling at a loss. In the former case, raising her report can only keep her active at the same price or make her inactive. It remains to consider the cases where $v_i = v_{q^*}$, but because several sellers report the same value v_{q^*}, i could be ranked before q^* or after q^* depending upon the (unspecified) tiebreaking rule. If i is ranked before q^*, the argument is the same as in the case $v_i < v_{q^*}$. If i is ranked after q^*, we must have $p < v_{q^*+1} = v_{q^*}$; hence, trade occurs at price v_{q^*} and seller i will not benefit by misreporting in such a way that she becomes active.

Exercise 2.18. *Aggregate Supply Curve with Increasing Marginal Costs*

The good is produced in indivisible units (q_i is an integer) and firm i has a nondecreasing marginal cost c_i. Recall that its supply curve ζ_i is defined by

$$\text{for all } q \geq 0, \qquad \zeta_i(p) = q \Leftrightarrow \partial c_i(q) \leq p < \partial c_i(q + 1).$$

Denote by c the joint cost function of the firms $i = 1, 2, \ldots, n$:

$$\text{for all } q \geq 0, \qquad c(q) = \inf\left\{ \sum_{i=1}^{n} c_i(q_i) \,\Big|\, \sum_{i=1}^{n} q_i = q \right\}.$$

Show that the supply curve of c is the sum of individual supply curves: $\zeta = \sum_{i=1}^{n} \zeta_i$.

Core and Competitive Equilibrium: Multiple Goods

3.1. INTRODUCTION

Both concepts of the core and of the competitive equilibrium are defined in a much broader class of markets than the reading of Chapter 2 would suggest. Those markets may involve an arbitrary finite (or infinite)[1] number of different (heterogeneous) goods; each one of these goods may be perfectly divisible, or it may come in indivisible units.

On the agents side, individual preferences are not restricted by the severe assumption of quasi-linearity: we usually assume that preferences are monotonic, although that assumption is not essential; see Section 3.6. In fact, even the transitivity of preferences and their completeness can be relaxed without much ado.[2] Finally we can also accommodate an infinite number of agents; see the discussion of the Edgeworth proposition in Section 3.7.

Despite the formidable generalization of the context, the three main properties of the core and the equilibrium in the markets with one good and money are preserved: all competitive allocations are stable in the core sense (Lemma 3.6); when individual preferences are convex and continuous, a competitive equilibrium exists (Theorem 3.4); when the endowment of each individual agent is negligibly small relative to the overall endowment of the economy, the core and the set of competitive allocations coincide.

The last statement is known as the *Edgeworth proposition*, and is the hardest of the three. Recall from Chapter 2 that the coincidence of the core and of the competitive set occurs in a market with one good and money when goods are indivisible and all agents initially own at most one unit of the good (as in the Böhm–Bawerk market; Section 2.3), or when buyers each want at most one unit and at least two sellers with convex preferences are active (Section 2.4). If the traded good is divisible, strict equality of the core and of the competitive set does not hold, but it does hold in the limit when a large number of identical sellers face a large number of identical buyers (see an example in

---

[1] The only example of an infinite number of goods discussed in this volume is the land economy; see Section 4.5.

[2] Although we shall not discuss it in this volume. See Shafer [1976].

Section 2.5). In markets with multiple divisible goods, a similar (and much more general) statement holds; it is the subject of Section 3.7.

The first half of Chapter 3 deals with four important special classes of markets with multiple heterogeneous goods, where the structure of the core and the competitive set is both simple and interesting. The *house barter* model is presented in Section 3.2. Its unique feature is that the core and competitive set contain a single allocation (also easy to compute), and the corresponding direct revelation mechanism is strategy-proof; therefore the equilibrium allocation is particularly easy to implement. By way of contrast, Section 3.3 presents the *marriage* market, where the core stability is always feasible but no core allocation can be interpreted as a competitive equilibrium. The marriage market is interpreted as a market for heterogeneous indivisible goods resembling the house barter model; the lesson is that the competitive equilibrium is not a robust concept in the presence of indivisibilities.

In Sections 3.4 and 3.5, we discuss two direct generalizations of the Böhm–Bawerk market (Section 2.3). In the first one, called the *bilateral assignment* model, each seller offers one indivisible good (a different house for each seller) and each buyer wants to consume at most one good (each buyer wants at most one house, she is willing to pay a different price for each different house on sale). As sellers and buyers are segregated (a seller would not or cannot trade houses with another seller), the core has a simple lattice structure and we can easily compute the best core allocation from the buyers' point of view or the best allocation from the sellers' point of view. In the general *assignment market* of Section 3.5, sellers and buyers are no longer segregated. If each sellers offers a single house for sale, the equality of the core and of the competitive set if still guaranteed, but the Edgeworth proposition disappears as soon as there are some "capitalists," namely, sellers who initially own more than one indivisible good.

The general exchange economy with (finitely many) divisible goods (often called an Arrow–Debreu economy) is defined in Section 3.6, where the main existence result of the competitive equilibrium is stated. After discussing the Edgeworth proposition in Section 3.7, we review in Section 3.8, several disappointing impossibility results on the implementation of the competitive set by a decentralized mechanism.

3.2. HOUSE BARTER

This model was originally introduced by Shapley and Scarf [1974]. We have n agents, and n houses (all different). An agent can consume only one house (she can live in one place at a time) and cannot be homeless. Each agent ranks the houses from his first choice to his last choice, without ever being indifferent between two houses.[3] Initially, each agent owns a house (so there is a one-to-one assignment of houses to agents). Agents can exchange houses (within a given coalition, all rearrangements of their initial houses are feasible), but there is no medium of exchange such as money.

Remarkably, the house barter model *always* yields a unique core allocation (a rare occurrence when the domain of individual preferences is so large!).

Here is an example with four agents. Initially, agent i owns house h_i. The four individual preferences are as follows:

	u_1	u_2	u_3	u_4
top	h_3	h_4	h_1	h_3
	h_2	h_1	h_4	h_2
	h_4	h_2	h_3	h_1
bottom	h_1	h_3	h_2	h_4

$$(1)$$

Agents 1 and 3 can swap houses and end up with their top choice. Therefore (a) the initial allocation is not Pareto optimal, and (b) any allocation in the core must give h_3 to agent 1 and h_1 to agent 3. For instance, the Pareto optimal allocation

$$h_2 \text{ to } 1, \qquad h_4 \text{ to } 2, \qquad h_1 \text{ to } 3, \qquad h_3 \text{ to } 4$$

is "blocked" by an objection of coalition $\{1, 3\}$ that is able to guarantee, of its own resources, a better house to agent 1 and the same house to agent 3 (remember that an objection need not improve the welfare of every coalition member, but only of some members without deteriorating that of any member).

Now there are only two allocations where 1 and 3 swap houses, namely, the one where 2 and 4 keep their initial house, and the one where they swap. As the swap is Pareto improving, we are left with a unique candidate for a core allocation, namely,

$$h_3 \text{ to } 1, \qquad h_4 \text{ to } 2, \qquad h_1 \text{ to } 3, \qquad h_2 \text{ to } 4$$

It is easy to check that this allocation is not threatened by any other coalition.

---

[3] This assumption of strict preferences is important. When we allow indifferences in an agent's preferences, only some of the results stated in this section survive; see Exercise 3.4.

Lemma 3.1. The Top Trading Cycle Algorithm. The following algorithm defines the unique core outcome of the house trading model among the agents of $N = \{1, 2, \ldots, n\}$.

For each agent $i \in N$, let $\sigma(i) \in N$ be the owner of agent i's top house. Find a cycle of the mapping σ that is, a sequence $i_1, \ldots, i_K$ such that $\sigma(i_k) = i_{k+1}$, $k = 1, \ldots, K-1$ and $\sigma(i_K) = i_1$ and exchange houses along a cycle (so agent i_k gets the house owned by agent i_{k+1}, with the convention $i_{K+1} = i_1$). Note that the cycle may be of length 1 (when an agent's top choice is the house she initially owns), in which case this agent will simply not trade. For any mapping σ, there exists at least one such cycle, called a "top trading cycle."

Call N_1 the union of all top trading cycles of σ and let all agents of N_1 get their top choice. Repeat the operation within $N \setminus N_1$ ($\sigma_1(i)$ is the owner of agent i's top house within those owned by the agents in $N \setminus N_1$, and exchange houses along all top trading cycles of σ_1). Call N_2 the union of all top trading cycles of σ_1 and repeat the operation within $N \setminus (N_1 \cup N_2)$, and so on.

Proof. Consider a top trading cycle of σ. The agents in this cycle can all achieve their top choice without the help of anyone outside the cycle; thus, a core outcome must serve N_1 precisely as in the algorithm. Given that N_1 is served in this way, a coalition in $N \setminus N_1$ cannot be better than when everyone gets her first choice in $N \setminus N_1$; on the other hand, a cycle of σ_1 in $N \setminus N_1$ can, as a coalition, guarantee his top choice in $N \setminus N_1$ to each of its members. Hence, in a core outcome, the agents of N_2 are served as in the algorithm (they get their top choice in $N \setminus N_1$), and so on. Thus the algorithm defines the only conceivable core outcome.

It remains to check that this outcome, call it z^*, is indeed in the core. Consider a coalition S and an allocation z_S feasible by exchanges within S, and guaranteeing to all in S at least the same level of welfare as that awarded by z^*. If the intersection of S with N_1 is not empty, it must consist of the union of some cycles of σ (otherwise the top choice of someone in $S \cap N_1$ belongs to an agent outside S); and z_S coincides with z^* on $S \cap N_1$. Repeating this argument we find that $S \cap N_2$ consists of the (possibly empty) union of cycles of σ_2, and z_S coincides with z^* on $S \cap N_2$, and so on. Q.E.D.

Notice that a looser algorithm following an arbitrary sequence of Pareto improving trading cycles (that may allocate second choice or worse to some agents) would not always reach the core. This can be seen in the simple example (1). Say that agents 1 and 2 exchange houses first (a Pareto improving move). The resulting allocation is

$$h_1 \to 2, \qquad h_2 \to 1, \qquad h_3 \to 3, \qquad h_4 \to 4,$$

from which any sequence of Pareto improving trades (e.g., 2 and 4 trade houses, then 3 and 4 trade) results in the allocation

$$h_1 \to 3, \qquad h_2 \to 1, \qquad h_3 \to 4, \qquad h_4 \to 2,$$

not the core allocation! Thus agent 1 made a mistake (from which agent 4 benefited) by accepting to trade for a less than top house with agent 2.

Remark 3.1

An additional feature of the core allocation achieved by the top trading cycle algorithm is this: any other allocation is blocked by an objection of a coalition using the very allocation that this coalition gets in the core. See Exercise 3.1.

The top trading cycle algorithm may take up to n rounds (where the first round consists of finding N_1 and exchanging houses along the cycles of N_1, the second round does the same for N_2, and so on). Yet the number of rounds does not reflect the intensity of trading in the core. Quite the contrary, we may have a single round where everyone trades (when σ has a single cycle comprising all agents in N), whereas we have n rounds only if the initial allocation is Pareto optimal and no trade takes place.[4] The successive rounds correspond, in fact, to the relative value of houses in the competitive equilibrium.

Lemma 3.2. *Let T be the number of rounds in the algorithm of Lemma 3.1, so that $N_1 \cup N_2 \cup \cdots \cup N_T$ is a partition of N. Then a competitive equilibrium price obtains as follows. Pick a decreasing sequence $q_1 > q_2 > \cdots > q_T$ and let the price of all houses in N_t be q_t, for all $t = 1, \ldots, T$. Moreover, the core outcome is also the unique competitive equilibrium outcome.*

Proof. There is certain oddity in talking about competitive price in a context where there is actually no medium of exchange. Yet we may create a fictitious unit of account (fiat money), call prices in this unit, and derive a (very real) budget constraint for each market participant. Consider the core allocation z^* and a price vector as described in the

---

[4] See Exercise 3.2.

Lemma. An agent N_t owns a house that is worth q_t, and thus she can afford any house in N_t, $N_{t+1}, \ldots, N_T$. Allocation z^* gives her top choice in $N_t \cup N_{t+1} \cup \cdots \cup N_T$, as required by the competitive equilibrium (maximization of her utility within the limits of her budget constraint).

Check now that a competitive equilibrium allocation must be precisely the core allocation z^*. Let p be a competitive price and z a corresponding competitive allocation. In z, houses are exchanged along certain trading cycles $M_1, M_2, \ldots, M_S$: the agents in M_s exchange houses in circular fashion (i_k gets the house owned by i_{k+1}). Ultimately, no money changes hands; hence, the price of all houses within a given cycle must be the same: all houses traded in M_s have the same price q_s. Without loss of generality, assume $q_1 \geq q_2 \geq \cdots \geq q_S$. Then the agents in M_1 get their top choice (so M_1 is a top trading cycle), those in M_2 get their top choice among the houses owned by the agents in $N \setminus M_1$ (clearly, if $q_1 = q_2$, M_2 is a top trading cycle in N as well; and even if $q_1 > q_2$, M_2 may be a top trading cycle in N), those in M_3 get their top choice among houses owned in $N \setminus (M_1 \cup M_2)$ and so on. The conscientious reader will check that this algorithm picks precisely the core outcome z^*, because N_1 is, for instance, the union $M_1 \cup M_2 \cup M_3$, N_2 is the union $M_4 \cup M_5$, and so on. The minor variations in the competitive price do not change the final allocation.

Q.E.D.

The miraculous coincidence of the core and/or competitive equilibrium set into a single outcome is not the end of the story. Look at the direct mechanism that mechanically computes the outcome z^* from the reports of individual preferences. This mechanism is strategy-proof in the strong sense that even a joint misreport by a coalition cannot be profitable.

Lemma 3.3.   The direct revelation mechanism implementing the unique core allocation is nonmanipulable: no coalition can gain (that is, make at least one coalition member better off and none worse off) by jointly misreporting preferences.

Proof.   For the true preference profile, denote, as in Lemma 3.2, by $N_1, \ldots, N_T$ the partition of N resulting from the top trading cycle algorithm. Let S be a manipulating coalition and let t^*, $1 \leq t^* \leq t$, be the smallest index such that $S \cap N_{t^*} \neq \varnothing$. Clearly, when the agents in N^*, $N^* = \bigcup_{t=1}^{t^*-1} N_t$, report truthfully, they get their core house no matter what the agents outside N^* report (this is clear from the definition of the algorithm). Therefore, when S misreport, but $N \setminus S$ is truthful, the best house that an agent in S can hope for is her top choice among those houses not owned initially by an agent in N^*; but this is precisely what all agents in $S \cap N_{t^*}$ get by reporting truthfully! Thus, if the misreport by S is profitable to S, the agents in $S \cap N_{t^*}$ actually get no more and no less than their core house. We can now repeat the argument to show that the agents in $S \cap N_{t^*+1}$ get their core house as well, and so on. In the end, the manipulation does not strictly improve the welfare of anyone in S.

Q.E.D.

Remark 3.2

Another source of potential manipulation by coalitions consists of exchanging houses *prior* to the implementation of the core mechanism. Here is an example with three agents and three houses. Initially, agent i owns house h_i, and their preferences are as follows:

u_1	u_2	u_3
h_3	h_1	h_1
	h_2	h_2
	h_3	h_3

With truthful report of preferences and of initial property rights, the first top trading cycle involves 1 and 3 and the resulting core allocation is

$$h_1 \to 3, \qquad h_2 \to 2, \qquad h_3 \to 1.$$

However, if agents 1 and 2 trade houses (not always a Pareto improving move) before showing up to play the core mechanism, the resulting allocation is

$$h_1 \to 2, \qquad h_2 \to 3, \qquad h_3 \to 1;$$

hence, agent 2 strictly benefits and agent 1 does not lose. As it turns out, such a manipulation can never be strictly profitable for all members of the coalition (trading houses prior to the mechanism). See Exercise 3.3. Notice also that the coalition (12) could achieve the same outcome by (i) misreporting their preferences, and (ii) exchanging houses ex post. (Exercise: prove this claim.)

To conclude this section, we discuss the robustness of the house barter model in two directions. First, allow the individual preferences to exhibit indifferences. Then a fair amount of the above results are preserved, in particular:

(a) *There is at least one competitive allocation and the competitive set is obtained by running the top trading cycle algorithms (exploring all the options created by indifferences).*

(b) *The core is a subset, possibly strict and possibly empty, of the set of competitive allocations. All core allocations yield the same utility distributions.*

Property (a) is straightforward (its proof follows the argument in the proof of Lemma 3.2). Property (b) (due to Wako [1991]; its proof is the subject of Exercise 3.5) is more surprising: we normally expect the core to be a superset of the competitive set! Here is a three-agent example showing that the core may be empty:

u_1	u_2	u_3
h_2	h_1, h_3	h_2
h_1, h_3	h_2	h_1
		h_3

Initially, agent i owns h_i. Using top trading cycles, we find two competitive allocations, namely,

$$z: \quad h_1 \rightarrow 2, \quad h_2 \rightarrow 1, \quad h_3 \rightarrow 3,$$

and

$$z': \quad h_1 \rightarrow 1, \quad h_2 \rightarrow 3, \quad h_3 \rightarrow 2.$$

Now z is blocked by an objection of $\{2, 3\}$ and z' is blocked by an objection of $\{1, 2\}$. Examples where the core contains a unique allocation and the competitive set contains at least one more allocation are given in Exercise 3.4.

The second way to generalize the house barter model is more radical. Suppose that we allow the agents to trade two kinds of goods, say, houses and cars. Assume that each agent initially owns one car and one house and consumes exactly one car and one house. Suppose, moreover, that individual preferences are separable between houses and cars (which car I own does not affect my ranking of the various houses and vice versa). Then it is not clear whether the core of the economy will be empty or not (no systematic result is available yet). Of course, if we allow nonseparable preferences, the core may be empty. (Exercise: give an example.)

3.3. THE MARRIAGE MARKET

This important model[5] is another instance where the core concept is more successful than the competitive equilibrium. Recall from Section 2.6 that the competitive concept collapses when a good is produced under increasing returns to scale; in that case, the core is a reasonable substitute, which, however, may cut a large subset of the Pareto optimal frontier (and may easily be empty, too; see Section 2.8). By contrast, in the marriage market, the core is always nonempty and often small; moreover, its end points (in a sense to be made precise soon) are easy to compute.

We have a set M of men and a set W of women. Each man (and each woman) has strict preferences over his (her) potential spouses. Marriages are exclusively monogamous and heterosexual.[6] We assume for simplicity that there is the same number of men and of women. This is of no consequence from a technical standpoint: all key results are maintained when M and W are of different size, and we view "remaining single" as an option that each agent ranks among the set of his potential mates (so that some mates are less desirable than the celibacy).[7]

A *matching* is a one-to-one pairing of each man to a woman. The core stability property follows naturally from the fact that each individual owns himself or herself. Say that a certain matching is proposed; if a certain pair of one man and one woman prefers one another over their proposed mate, then the matching in question is not in the core. Call a matching *stable* if no pair of one man and one woman can object as above; this is enough for core stability because of the assumption of strict preferences. In particular, pairwise stability implies Pareto optimality.[8]

We can interpret the marriage market as an exchange economy in which (a) each man and each woman owns a personalized indivisible good, (b) each man (resp. woman) wants to consume at most one of the goods initially held by women (resp. by men) and derives no utility from

---

[5] For a survey of its practical applications to several job-matching problems, see Roth and Sotomayor [1990].

[6] Other examples are a set of firms and a set of computer specialists (when each firm needs exactly one computer specialist), or a set of pianists and a set of violinists when there is a market for piano–violin duos (but not for trios).

[7] See Exercise 3.10 for more discussion. Of course, in realistic models of marriage, the celibacy is always an important option; e.g., there is a market for solo pianists or solo violinists.

[8] If the matching σ is Pareto inferior to the matching σ', there is a man m preferring $w' = \sigma'(m)$ over $w = \sigma(m)$. Then w' prefers $m = \sigma'(w')$ over $m' = \sigma(w')$ (remember that w' is never indifferent between any two men); hence the pair (m, w') blocks the matching σ.

the goods held by men (resp. women). There is no medium of exchange such as money.[9]

To discover a core stable matching, a simple heuristic can be helpful. Start with an arbitrary matching. If we find a blocking pair, match them and match their former mates together. Continue until, hopefully, we reach a stable outcome. Here is an example with four men, four women, and the following preferences:

m_1	m_2	m_3	m_4
w_4	w_4	w_3	w_4
*	w_3	w_1	w_1
*	*	*	*
*	*	*	*

w_1	w_2	w_3	w_4
m_2	*	m_1	m_1
m_1	*	m_2	*
m_4	*	*	*
m_3	*	*	*

$$(2)$$

Start from the matching (m_i, w_i), $i = 1, 2, 3, 4$. Here (m_1, w_4) object; so we marry them and (forcibly) marry w_1 to m_4. Now (m_2, w_3) have an objection; we marry them, as well as m_3 to w_2. Hence

$$(m_1, w_4) \quad (m_2, w_3) \quad (m_3, w_2) \quad (m_4, w_1).$$

Check that this is a stable matching: for instance, although w_2 may prefer any other man to m_3, she cannot convince any of these to switch. In fact, this matching is the only matching in the core (out of 24 possible matchings; one can check directly that there is no other core matching, but it is much easier to apply Theorem 3.1 by checking that the M-optimal and the W-optimal matchings coincide).

Observe that this core matching can be interpreted as a competitive equilibrium. Set the same price for a man and a woman if they are married and choose these prices as follows:

$$p_{m_3} = p_{w_2} < p_{m_4} = p_{w_1} < p_{m_2} = p_{w_3} < p_{m_1} = p_{w_4}.$$

Then check that the price of any woman (resp. man) preferred by man m (resp. woman w) to his (her) core mate is higher than his (her) own price, so that no one can afford to buy a preferred mate (and everyone can afford to buy his/her current mate).

However, this interpretation of the core matching as a competitive allocation is often impossible. A simple example with three men and

---

[9] When we add money to the picture, the bilateral assignment model of Section 3.4 obtains.

three women is as follows:

m_1	m_2	m_3
w_1	w_1	w_2
*	w_2	w_3
*	w_3	w_1

w_1	w_2	w_3
m_3	m_2	m_3
m_1	*	*
m_2	*	*

(3)

Here the unique stable matching is

$$(m_1, w_1) \quad (m_2, w_2) \quad (m_3, w_3).$$

(Indeed, any matching containing (m_2, w_1) is blocked by (m_1, w_1) and any matching containing (m_3, w_1) is blocked by (m_3, w_3); thus, we must have (m_1, w_1) in the core. In the market reduced to m_2, m_3, w_2, w_3, the matching (m_2, w_2) (m_3, w_3) gives everyone his/her top choice.) Yet if there exists a competitive price system, it must satisfy

$$p_{m_i} = p_{w_i} \quad \text{for } i = 1, 2, 3$$

(so that married agents can afford each other),

$p_{w_1} < p_{m_3}$ so w_1 cannot afford m_3,

$p_{m_3} < p_{w_2}$ so m_3 cannot afford w_2,

$p_{m_2} < p_{w_1}$ so m_2 cannot afford w_1.

These equalities and inequalities are inconsistent.

Our last example, before stating some general results about the marriage market, is meant to illustrate (i) that the core may contain several different matchings, and (ii) that the naive algorithm used above may not converge to a stable matching:

m_1	m_2	m_3
w_2	w_1	w_1
w_1	w_3	w_2
w_3	w_2	w_3

w_1	w_2	w_3
m_1	m_3	m_1
m_3	m_1	m_3
m_2	m_2	m_2

(4)

Start with the initial matching

$$(m_1, w_1) \quad (m_2, w_2) \quad (m_3, w_3).$$

There are objecting pairs, namely, (m_1, w_2) and (m_3, w_2). If we choose to satisfy (m_1, w_2), we reach

$$(m_1, w_2) \qquad (m_2, w_1) \qquad (m_3, w_3),$$

where, again, two objecting pairs emerge: (m_3, w_2) and (m_3, w_1). Say that we choose (m_3, w_2) so as to reach

$$(m_1, w_3) \qquad (m_2, w_1) \qquad (m_3, w_2).$$

Here, again, we have a choice of objecting pairs; say that we choose (m_3, w_1) (the other objecting pair is (m_1, w_1)). We are now matching

$$(m_1, w_3) \qquad (m_2, w_2) \qquad (m_3, w_1),$$

from which the objection by (m_1, w_1) brings us back (after four steps) to the original matching! On the other hand, by making different choices of objecting pairs, we reach quickly the two stable matchings of this market, namely,

$$(m_1, w_2) \quad (m_2, w_3) \quad (m_3, w_1) \quad \text{(via the } (m_1, w_2) \text{ objection,} \atop \text{then the } (w_1, m_3) \text{ objection),}} \tag{5}$$

$$(m_1, w_1) \quad (m_2, w_3) \quad (m_3, w_2) \quad \text{(via the } (m_3, w_2) \text{ objection).} \tag{6}$$

Notice that every man and every woman has a different mate in these two matchings, and that all men prefer (5) to (6), whereas all women prefer (6) to (5). This feature is quite general.

Theorem 3.1. (Gale and Shapley [1962]). *In any marriage market with strict preferences:*

(a) There is at least one stable matching; the core is never empty.

(b) There is a stable matching, called the M-optimal matching, where every man gets the best of all his core mates (there is no stable matching where he is matched to a preferred woman) and every woman gets her worst core mate. There is a stable matching, called the W-optimal matching, where every woman gets the best of all her core mates and every man gets his worst core mate. The core contains a single matching if and only if the M-optimal matching and the W-optimal matching coincide.

(c) The M-optimal matching is computed by means of the Gale–Shapley algorithm where men propose (defined below).

(d) A statement symmetrical to (c) holds for the W-optimal matching.

Definition 3.1.   The Gale–Shapley Algorithm where Men Propose

STEP 1

Each man proposes to his first-choice woman. If a woman receives exactly one proposal, this man is called her engagee. If a woman receives more than one proposal, she keeps the proposer she likes best as her engagee and rejects the others. Men are now partitioned into engaged or rejected. The algorithm stops if all men are engaged otherwise, we go to the next step.

STEP 2

All rejected men propose now to their second-best choice. Each woman receiving new proposals keeps as her engagee the man she likes best among current proposer(s) and possibly former engagee, and rejects the others. (Thus a man previously engaged may now be rejected.) The algorithm stops if all men are engaged; otherwise, we go to the next step.

STEP 3

All rejected men propose now to their next choice, and women update their engagements according to new proposals (if any).

This continues until all men are finally engaged, at which point the engagement pattern turns into the final matching. Since each man proposes to any woman only once, the algorithm must stop after finitely many steps. The proof of Theorem 3.1 is in Appendix 3.1.

An example is the three-men, three-women market (4). If men propose, w_2 receives an offer from m_1 while w_1 receives two offers and keeps m_3. Next round, m_2 offers to w_3, and the algorithm stops on matching (5). If women propose, m_1 receives two offers, keeps w_1 and rejects w_3, while m_3 receives an offer from w_2. Next, w_3 offers to m_3, who still keeps w_2. After this second rebuttal, w_3 finally offers to m_2, and the algorithm stops on matching (6). A more complicated example of the Gate–Shapley algorithm is given in Exercise 3.6.

Remark 3.3

Many more interesting properties of stable matching are covered in Roth and Sotomayor [1990]. For instance, the *M*-optimal matching is shown to be weakly Pareto optimal from the men's point of view; see Exercise 3.7. More importantly, the core possesses a lattice structure by means of the following "supremum" and "infimum" operations. If μ and μ' are two stable matchings, construct the matching $\mu \vee \mu'$ as follows; match every man m to whomever he prefers from his mate in μ and his mate in μ'. If μ and μ' are two arbitrary matchings, this construction may not yield a one-to-one matching from M onto W, but if μ and μ' are both stable, it does. Moreover, $\mu \vee \mu'$ is a stable matching as well, and it matches every woman with whomever she likes least among her mate in μ and her mate in μ'; see Exercise 3.8.

The core of a marriage market is easy to estimate, because the M-optimal and W-optimal matchings are its two bounds (they give utility bounds for each agent). Moreover, the M-optimal matching can be used as a direct mechanism to implement a stable matching. Although this mechanism is not strategy-proof, its strategic properties are still very strong.

Lemma 3.4. In the direct mechanism where agents report their preferences after which the M-optimal matching is implemented, truthtelling is a dominant strategy for the men, although not for the women. Assuming that men report truthfully, the optimal manipulation of the women yields a stable matching.

The proof is omitted (see Roth and Sotomayor [1990]).

Consider, for example, the M-optimal mechanism in economy (4). If all report truthfully, the matching (5) results. A (small) manipulation by w_1 will be enough to bring about the matching (6), a strict improvement for both w_1 and w_2. Woman w_1 reports $m_1 > m_2 > m_3$, and the Gale–Shapley algorithm works as follows:

STEP 1: w_1 receives offers from m_2 and m_3 and (untruthfully) keeps m_2; w_2 has an offer from m_1.

STEP 2: m_3 offers to w_2, who therefore keeps that offer and rejects m_1.

STEP 3: m_1 offers to w_1, who keeps that offer and rejects m_2.

STEP 4: m_2 offers to w_3, and the matching (6) is reached.

Thus the gap between the M-optimal and the W-optimal matchings serves as an upper bound on the possible extent of strategic manipulation in these two direct mechanisms (implementing respectively the M-optimal and the W-optimal matchings).

To conclude this section, we stress that all these remarkable properties of marriage markets (foremost among them, the nonemptiness of the core) hinge upon the bilateral nature of the market, with an agent on one side of the market trading exclusively with agents on the other side of the market, one at a time. The markets discussed in the next section share this bilateral character, and indeed, share many stability properties with marriage markets. A core existence result (Example 7.3 in Section 7.3) will provide a completely general explanation of this fact.

However, as soon as we allow for (a) a genderless matching (the roommate problem, where a group of women must be partitioned in pairs), or (b) trilateral matchings (say when a pianist, a violonist, and a cello form a trio), or (c) many-to-one matchings (e.g., polygamous marriages or, more realistically, a firm choosing a set of employees to team up with), even the existence of a core allocation does not always hold, as demonstrated by means of examples in Exercise 3.11.

3.4. BILATERAL ASSIGNMENT

The set N of agents is divided into a set A of type-A agents a_i, $i = 1, \ldots, n$, and a set B of type-B agents b_j, $j = 1, \ldots, n'$. The sizes n and n' are arbitrary. A pair (a_i, b_i) containing one agent of each type can generate the cooperative surplus $v(i, j) \geq 0$. No other coalition can bring any more surplus; for instance, a group of type-A agents can bring no surplus, and a coalition containing several agents of each type generates surplus by forming pairs (a_i, b_j) and summing up the $v(i, j)$.

This bilateral structure is analogous to that of the marriage market (this will be confirmed by the parallel set of properties in both models), with the important difference that money is available to transfer utility (and utilities are quasi-linear). Two salient features of bilateral assignment markets will be of particular interest to us: the core is never empty and coincides with the set of competitive allocations; and the core contains an A-optimal allocation and a B-optimal allocation, both of them easy to compute. See Theorem 3.2.

Before the formal discussion, we describe a simple exchange economy generating a bilateral assignment problem. In a market à la Böhm–Bawerk, each seller brings a different indivisible good (a different house). Let A be the set of sellers. Each buyer wants at most one house, and the utility of buyer b_j for seller a_i's house (his willingness to pay) is u_{ji}. Denote by z_i the reservation price of seller i for her house. Then the cooperative surplus of coalition $\{a_i, b_j\}$ is

$$v(i, j) = \max\{u_{ji} - z_i, 0\}.$$

To fit the bilateral pattern, we must also assume that the market is segregated (sellers do not trade houses among themselves), as will be the case, for instance, if each seller does not care for any house but her own. Of course, the usual Böhm–Bawerk market is a particular case of this model where all houses happen to be identical.

Another useful interpretation of the bilateral assignment model involves joint production by complementary inputs. For instance, A is a set of pianists and B a set of violonist, and $v(i, j)$ represents the market value (expected profit) of the duo $\{a_i, b_j\}$.

We start our discussion by two simple examples. In the first one, $n = 2$, $n' = 3$, and the surplus $v(i, j)$ is given by the following matrix:

$$\begin{matrix} a_1 \\ a_2 \end{matrix} \begin{bmatrix} 5 & 3 & 3 \\ 4 & 3 & 2 \end{bmatrix}$$

$$b_1 \quad b_2 \quad b_3 \tag{7}$$

The optimal assignment[10] pairs (a_1, b_1) and (a_2, b_2) for a total surplus of 8. By adjusting the monetary transfers, these eight units of surplus can be split in any way we like among the five agents. We shall denote by α_i, β_j, respectively, the profit shares of agents a_i and b_j, respectively. The system of inequalities for core stability reflects two concerns: an agent can always refuse to cooperate and get zero profit; any pair (a_i, b_j) could secede and share its "own" surplus $v(i, j)$ as it pleases. Hence,

$$\alpha_i \geq 0, \qquad \beta_j \geq 0, \qquad i = 1, 2, \qquad j = 1, 2, 3,$$

$$\alpha_1 + \beta_1 \geq 5, \qquad \alpha_2 + \beta_2 \geq 3,$$

$$\alpha_1 + \beta_2 \geq 3, \qquad \alpha_2 + \beta_1 \geq 4, \tag{8}$$

$$\alpha_1 + \beta_3 \geq 3, \qquad \alpha_2 + \beta_3 \geq 2.$$

The two inequalities (8) plus inequality $\beta_3 \geq 0$ sum up to the feasibility constraint $\sum_i \alpha_i + \sum_j \beta_j = 8$; hence, they are all equalities:

$$\beta_3 = 0, \qquad \alpha_1 + \beta_1 = 5, \qquad \alpha_2 + \beta_2 = 3. \tag{9}$$

We now write the core system in terms of the pair α_1, α_2 only, which boils down to

$$\alpha_2 \leq 3 \leq \alpha_1 \leq \alpha_2 + 1.$$

This corresponds to a small triangle in the (α_1, α_2) plane (see Figure 3.1). The two interesting vertices of this triangle are:

- $(\alpha_1, \alpha_2) = (3, 2)$ (whence $\beta_1 = 2$, $\beta_2 = 1$, $\beta_3 = 0$), where both A agents get their worst core profit share, whereas all B agents get their best core profit share;

[10] The term "assignment" has the same meaning as "matching" in the marriage market (provided our description of matching includes the option to remain single; see note 7 above). We use a different word as a reminder that the current model involves monetary transfers, too.

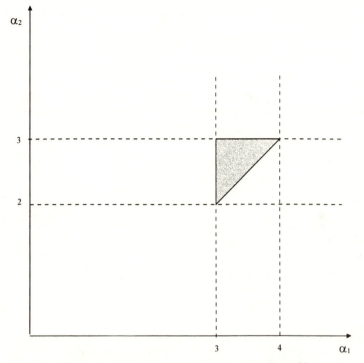

Fig. 3.1. The core of the bilateral assignment problem (7).

- $(\alpha_1, \alpha_2) = (4, 3)$ ($\beta_1 = 1$, $\beta_2 = \beta_3 = 0$), where both A agents get their best core profit, whereas all B agents get their worst.

In our next example, the set of core profit distributions becomes a single line and the analogy with the interval of competitive prices is even clearer. We have $n = n' = 3$, and the surplus matrix is as follows:

$$\begin{matrix} a_1 \\ a_2 \\ a_3 \end{matrix} \begin{bmatrix} 2 & 1 & 2 \\ 0 & 2 & 3 \\ 2 & 0 & 2 \end{bmatrix}$$

$$b_1 \quad b_2 \quad b_3 \qquad\qquad (10)$$

Here three surplus-maximizing assignments coexist:

	$(a_1, b_1),$	$(a_2, b_2),$	$(a_3, b_3),$
or	$(a_3, b_1),$	$(a_2, b_2),$	$(a_1, b_3),$
or	$(a_1, b_2).$	$(a_2, b_3),$	$(a_3, b_1).$

All three generate $v(N) = 6$ units of surplus. In a core allocation, we must have

$$\alpha_1 + \beta_1 \geq 2, \quad \alpha_2 + \beta_2 \geq 2, \quad \alpha_3 + \beta_3 \geq 2, \quad \text{and} \quad \sum_i \alpha_i + \sum_i \beta_i = 6;$$

hence, all three inequalities are equalities. Repeating the same argument for the other two optimal assignments, we get a system of seven equations (of which only five are independent):

$$\alpha_1 + \beta_1 = \alpha_1 + \beta_3 = \alpha_2 + \beta_2 = \alpha_3 + \beta_1 = \alpha_3 + \beta_3 = 2,$$

$$\alpha_1 + \beta_2 = 1, \qquad \alpha_2 + \beta_3 = 3.$$

Combined with the requirements α_i, $\beta_j \geq 0$, this determines finally a one-dimensional interval of core profit distributions:

$$(\alpha_1, \alpha_2, \alpha_3) = (\lambda, \lambda + 1, \lambda),$$

$$(\beta_1, \beta_2, \beta_3) = (2 - \lambda, 1 - \lambda, 2 - \lambda), \quad \text{where } 0 \leq \lambda \leq 1.$$

Observe once again, that the choice $\lambda = 0$ yields the best (worst) core allocation for type-B (type-A) agents, whereas the choice $\lambda = 1$ yields the best (worst) for type-A (type-B). To interpret this allocation as a competitive outcome, think of the "price" of a_1 (and of a_3) as λ, that of a_2 as $\lambda + 1$, the price of b_1 as $2 - \lambda$, etc. A type-B agent, say b_2, can buy the services of a type-A agent at the announced price (of the type-A agent). If he buys agent a_1, say, agent b_2 must pay him his price λ and she (agent b_2) can cash the residual profit $v(1,2) - \lambda$. In other words, the price of agent a_1 is the salary of his effort in cooperating in the creation of the surplus $v(i, j)$. The competitive equilibrium property says that b_2 is indifferent between buying a_1 [for net profit $v(a_1, b_2) - \lambda = 2 - \lambda$] or buying a_2 (for a net profit $v(a_2, b_2) - (\lambda + 1) = 2 - \lambda$], but b_2 positively refuses to buy a_3 [for a profit of $v(a_3, b_2) - \lambda = -\lambda!$]. This is as it should be, because the pair (a_3, b_2) appears in no optimal assignment, contrary to (a_1, b_2) and to (a_2, b_2).

Our next result shows that the two numerical examples are quite representative of the general configuration of the core. We need a couple of definitions first. For a pair of coalitions $A' \subseteq A$, $B' \subseteq B$, we denote by $v(A', B')$ the maximal cooperative surplus within $A' \cup B'$. This is the maximal sum $\Sigma v(i, j)$ when we assign the agents of A' to those of B'. We call an assignment of A to B *optimal* is it generates the maximal surplus $v(A, B) = v(N)$. Finally, we say that a *pair* (a_i, b_j) is optimal if it appears in some optimal assignment. For instance, there are two optimal pairs in example (7) and seven in example (10).

Lemma 3.5. Given is a bilateral assignment market (A, B, v). A nonnegative surplus distribution $(\alpha_i, \beta_j)_{i \in A, j \in B}$ is a core allocation if and only if, for all pairs (a_i, b_j) in $A \times B$, we have

$$\alpha_i + \beta_j \geq v(i, j), \quad \text{with equality when } (a_i, b_j) \text{ is an optimal pair.} \quad (11)$$

An agent who is not paired in at least one optimal assignment (for example, an agent who belongs to no optimal pair) receives zero. If α_i (resp. β_j) is the price of agent a_i (resp. b_j), then any core allocation is competitive: if (a_i, b_j) is an optimal pair, then a_i maximizes his profit by buying agent b_j; if, on the other hand, a_i (resp. b_j) is not part of any optimal pair, then he finds it optimal to remain alone.

(Here "a_i buys b_j" means that b_j sells his rights over the surplus generated by cooperating with a_i.)

Proof. Take a core allocation (α_i, β_j) and an optimal assignment: $\sum v(i, j) = v(N)$. By core stability, $\alpha_i + \beta_j > v(i, j)$ for all pairs (i, j) appearing in this assignment, and $\alpha_i, \beta_j \geq 0$ for all i, j not paired in this assignment. Summing up yields an equality because a core allocation is efficient. Hence, $\alpha_i + \beta_j = v(i, j)$ for all optimal pairs. Note that $\alpha_i = \beta_j = 0$ for any agent who is left out in at least one optimal assignment. The converse property (property (11) implies core stability) is just as easy to prove, once we realize that the surplus $v(A', B')$ of any coalition rests on the assignments within $A' \cup B'$.

Check the competitive interpretation of the core. Say that agent a_i considers buying agent b_j with corresponding profit $v(i, j) - \beta_j$. Distinguish two cases. If a_i is part of an optimal pair (i, j), then (11) implies

$$v(i, j) - \beta_j = \alpha_i \geq v(i, j') - \beta_{j'} \quad \text{for all } j'.$$

Thus a_i's "demand" consists of all agents b_j with whom she forms an optimal pair. Next, if a_i is not part of any optimal pair, we know that $\alpha_i = 0$; hence, (11) reads $v(i, j) - \beta_j \leq 0$ for all j. That is, a_i finds it optimal to remain single.   Q.E.D.

A consequence of Lemma 3.5 (quite useful in applications) is the equal-treatment property: if two agents (of the same type) generate the same surplus no matter whom they are paired with (e.g., $v(a_i, b_j) = v(a_{i'}, b_j)$ for all j), then these two agents receive the same surplus share in the core. (Exercise: prove this claim.)

In the two numerical examples above, the core has two end-points, one most favorable for each type of agent. This feature is quite general.

Theorem 3.2

(a) *The core of a bilateral assignment market* (A, B, v) *is never empty. Moreover, there is a (unique) core allocation, called the A-optimal allocation, where every type-A agent gets his largest core profit share (and where every type-B agent gets her smallest core profit share). A symmetrical statement holds upon exchanging A and B.*

(b) *In the A-optimal allocation, every type-A agent receives his marginal contribution to the overall surplus,*

$$\alpha_i = v(A, B) - v(A \setminus \{\alpha_i\}, B)$$

(and the profit shares of type-B agent are given by (11)). Similarly, in the B-optimal allocation, every type-B agent receives her marginal contribution to overall surplus.

Proof. To prove existence of a core allocation requires a duality argument; see Shapley and Shubik [1972]. We check that the core of the bilateral assignment market, like the core of the marriage market, has a lattice structure. Pick (α_i, β_j) and (α_i', β_i') both in the core and define

$$\underline{\alpha}_i = \min\{\alpha_i, \alpha_i'\}, \qquad \overline{\alpha}_i = \max\{\alpha_i, \alpha_i'\},$$

$$\underline{\beta}_j = \min\{\beta_j, \beta_j'\}, \qquad \overline{\beta}_j = \max\{\beta_j, \beta_j'\}.$$

Then both $(\underline{\alpha}_i, \overline{\beta}_j)$ and $(\overline{\alpha}_i, \underline{\beta}_j)$ are core allocation. To see this, fix a pair a_i, b_j and observe that

$$\left. \begin{array}{l} \alpha_i + \beta_j \geq v(i, j) \Rightarrow \alpha_i + \overline{\beta}_j \geq v(i, j) \\ \alpha_i' + \beta_j' \geq v(i, j) \Rightarrow \alpha_i' + \overline{\beta}_j \geq v(i, j) \end{array} \right\} \Rightarrow \underline{\alpha}_i + \overline{\beta}_j \geq v(i, j).$$

Next, pick an optimal pair (a_i, b_j). We have

$$\alpha_i + \beta_j = \alpha_i' + \beta_j' \Rightarrow \{\alpha_i \leq \alpha_i' \Leftrightarrow \beta_j' \leq \beta_j\}.$$

Therefore, either $\alpha_i = \underline{\alpha}_i$ and $\beta_j = \overline{\beta}_j$, or $\alpha_i' = \underline{\alpha}_i$ and $\beta_j' = \overline{\beta}_j$. In both cases, $\underline{\alpha}_i + \overline{\beta}_j = v(i, j)$ holds. Finally, if a_i is not paired in at least one optimal assignment, then $\alpha_i = \alpha_i' = 0$ (by Lemma 3.5), so that $\underline{\alpha}_i = 0$ as well.

The lattice property, and the fact that the core is a closed subset of $\mathbb{R}^{A \cup B}$, imply at once the existence of an A-optimal and a B-optimal assignment. The difficult proof of statement (b) is also omitted; see Leonard [1983] or Demange and Gale [1985].   Q.E.D.

Theorems 3.1 and 3.2 underscore the analogies of the two bilateral markets (one with and one without money). When the bilateral assignment has sellers for one type of agents and buyers for the other type (as described at the beginning of this section), we can use the buyer-optimal allocation as a direct mechanism (where agents report their utilities for the goods). Statement (b) in Theorem 3.2 says that this mechanism coincides with the privotal mechanism for the buyers (see Section 2.9; this mechanism generalizes the O-competitive mechanism for the Böhm–Bawerk market; see Definition 2.3). In particular, the mechanism is strategy-proof for the buyers (compare with Lemma 3.4). Of course, a symmetrical statement holds if we use the seller-optimal mechanism.

A further similarity of the two bilateral markets is this: in the buyer-optimal mechanism, the optimal manipulation by the sellers yields the seller-optimal core allocation (see Demange and Gale [1985]).

A last remark about the two bilateral markets in this section and the previous one: the core concept is particularly convincing when it involves only small coalitions where transaction costs associated with direct agreements are minimal. In bilateral exchange models, only the smallest of all coalitions (namely, pairs of agents) are needed to achieve a core outcome (in particular, Pareto optimally must hold when no coalition of two agents can object). A very general theorem about coalition structures (formally stated in Section 7.3) implies that a bilateral exchange game must have a nonempty core no matter what the profile of preferences. Therefore, the two models are robust to a very broad generalization of the preference domains (e.g., allowing indifferences in the marriage market, or removing the transferable utility assumption in bilateral assignment).

3.5. ASSIGNMENT ECONOMIES

In a general exchange economy, many goods are exchanged and trade is multilateral: every agent can act as a seller on some goods and as a buyer on other goods. The assignment economies are fully general models of exchanges except for two restrictions: (i) the nature of the goods: there are indivisible goods (in arbitrary number; these goods can be all different) and money, and (ii) individual preferences: each agent wants to consume *at most* one unit of *at most* one indivisible good (so we think of these goods as houses, or cars), and his preferences are quasi-linear (utility is transferable via monetary transfers; of course, every agent has unbounded reserves of cash). Apart from these (admittedly severe) restrictions on individual preferences, the ownership structure is arbitrary: an agent may own no indivisible good initially (then call him a buyer) or he may own several goods (call him a seller for convenience, but keep in mind that he may also buy a new house).[11] Thus the model of an assignment economy generalizes all markets of Chapter 2 where the (homogeneous) good is indivisible and demand is binary (Sections 2.3 and 2.4), by introducing several heterogeneous goods and losing the bilateral segregation between sellers and buyers (that was so essential to the discussion of Sections 3.3 and 3.4).

In this section, we pay special attention to the ownership structure. It turns out that if, in the initial distribution of houses, no single agent owns more that one house (we speak of an economy *without capitalists*), then the core and the set of competitive allocations *coincide* (for any specification of individual utilities). Thus the Edgeworth proposition holds, even though each market participant holds a significant fraction of the global resources (this should be contrasted with the case of divisible goods, where the Edgeworth proposition requires vanishingly small market participants: see Section 3.7). On the other hand, if the assignment economy has some capitalists (who own initially more houses than they care to consume), then the competitive set is, as usual, a (generally) strict subset of the core.[12]

An assignment economy is a triple (N, H, μ, u), where N is a set of agents; H is a set of houses, to which we add the "null" house, denoted by $h_\varnothing$, with the convention that an agent who consumes no house (lives in the street!) is actually consuming $h_\varnothing$; μ is a mapping from H into N describing the initial property rights ($\mu(h) = i$ means that agent i owns house h; of course, i may own several or no houses initially); and u is a utility profile giving the utility (willingness to pay) $u_i(h)$ of every agent i for every house h (with the normalization convention $u_i(h_\varnothing) = 0$). We always assume $u_i(h) \geq 0$ for all i, h. As usual, the final utility $u_i(h) + t_i$ means that agent i consumes house h and receives a monetary transfer t_i.

We start by describing the set of Pareto optimal allocations (it is, of course, independent from the ownership structure). Because each agent

---

[11] In that case, he has no more use for the houses he owned initially, so he is likely to sell them if there is any demand.

[12] Moreover, this discrepancy cannot be removed by a replication argument, unlike in the divisible goods context; see Exercise 3.18.

wants to consume only one house,[13] maximizing the joint utility amounts to choosing the best assignment of agents to houses, where *assignment* is a mapping σ from the set N of agents into the set $H \cup h_\varnothing$ of houses, such that at most one agent is attached to every real house (or formally: $\sigma(i) = \sigma(j)$ only if $\sigma(i) = \sigma(j) = h_\varnothing$).

Here is an example with two houses, four agents, and the following utilities:

$$
\begin{array}{c|c|c|c|c}
 & u_1 & u_2 & u_3 & u_4 \\
\hline
h_1 & 3 & 6 & 8 & 4 \\
\hline
h_2 & 5 & 4 & 6 & 7 \\
\end{array}
\tag{12}
$$

The assignment $1 \to h_2$, $2 \to h_1$, $3, 4 \to h_\varnothing$ yields the surplus $u_1(h_1) + u_2(h_2) = 5 + 6 = 11$. In fact, the unique efficient (optimal) assignment is $1, 2 \to h_\varnothing$, $3 \to h_1$, $4 \to h_2$, yielding a surplus $u_3(h_1) + u_4(h_2) = 15$. The efficient allocations consist of an optimal assignment and an arbitrary vector of transfers t, such that $\sum_{i \in N} t_i = 0$. Before computing the competitive allocations, a formal definition will be useful. Pick an optimal assignment σ. Let $p = (p_h)_{h \in H \cup h_\varnothing}$ be a vector of prices (one for each house).

Definition 3.2. If σ is an optimal assignment, we say that a pair (σ, p) is a competitive assignment if

(a) $p_h \geq 0$ for all h, $p_{h_\varnothing} = 0$,
(b) for all $i \in N$, we have

$$\text{for all } h \in H \cup h_\varnothing: \quad u_i(h) - p_h \leq u_i(\sigma(i)) - p_{\sigma(i)}. \tag{13}$$

The corresponding competitive allocation assigns $\sigma(i)$ to agent i and the (positive or negative) monetary transfer

$$t_i = \left\{ \sum_{h:\, \mu(h) = i} p_h \right\} - p_{\sigma(i)}. \tag{14}$$

Condition (12) says that agent i maximizes her utility by demanding the house $\sigma(i)$ given the price vector p. Notice that the initial ownership structure (given by μ) does not enter inequalities (13). The explanation is the linearity of utilities with respect to money; whether agent i buys the house $\sigma(i)$ or any other house h, she will cash in the revenue

---

[13] Or equivalently, the consumption of h_1, h_2 is worth $\max\{u_i(h_1), u_i(h_2)\}$ to agent i.

$\sum_{h:\ \mu(h)=i} p_h$ from selling her initial endowment. So the two terms cancel on both sides of (13), whence comes an important simplification in assignment economies: *the set of competitive prices does not depend on the initial property rights over the houses.* In fact, as long as there are no more (*nonnull*) houses than agents ($|H| \le |N|$), the set of competitive prices does not depend on the choice of the optimal assignment (if there is a choice).[14]

Back to the numerical example (12), where the optimal assignment is unique, we write the system (13):

$$p_1 \le 8 \quad \text{and} \quad 6 - p_2 \le 8 - p_1: \qquad \text{agent 3 demands } h_1,$$
$$p_2 \le 7 \quad \text{and} \quad 4 - p_1 \le 7 - p_2: \qquad \text{agent 4 demands } h_2,$$
$$3 - p_1 \le 0 \quad \text{and} \quad 5 - p_2 \le 0: \qquad \text{agent 1 demands the empty house},$$
$$4 - p_2 \le 0 \quad \text{and} \quad 6 - p_1 \le 0: \qquad \text{agent 2 demands the empty house}$$

This yields a typical pentagon of competitive prices:

$$6 \le p_1 \le 8, \qquad 5 \le p_2 \le 7, \qquad p_1 \le p_2 + 2. \tag{15}$$

Fix now the initial property rights as: agent 4 owns h_1 and agent 2 owns h_2. The resulting competitive allocation is

$$t_1 = 0: \qquad \text{agent 1 stays put},$$

$$t_2 = p_2: \qquad \text{agent 2 sells his house},$$

$$u_3(h_1) + t_3 = 8 - p_1,$$

$$u_4(h_2) + t_4 = 7 + p_1 - p_2.$$

Here agent 4 wants the price differential between h_1 and h_2 to be as small as possible, which is not to the liking of agents 2 and 3.

Next we compute the core allocations. Consider an allocation with the optimal assignment σ and the net transfer t_i for agent i (by feasibility, we have $\sum_i t_i = 0$). Suppose this allocation is in the core.

---

[14] To see this, pick σ, σ' both optimal and write (by (13)) $u_i(\sigma(i)) - p_{\sigma(i)} \ge u_i(\sigma'(i)) - p_{\sigma'(i)}$. Summing up, we obtain an equality if the sets of houses allocated under σ and under σ' coincide, which is surely true if H contains at most $|N|$ houses, and the houses are all valuable to everyone ($u_i(h) > 0$ for all i, all nonempty H). Thus, $p_{\sigma(i)} - p_{\sigma'(i)}$ is the same for all competitive prices p. The claim follows (we omit the details). On the other hand, if H contains more houses than $|N|$, the competitive prices may vary with the optimal assignment. An example with $|N| = 1$ is easily constructed.

First, coalition {234} can generate the whole surplus without the help of agent 1:

$$v(N) = v(234) = 7 \Rightarrow t_1 = 0$$

(recall that the marginal contribution $v(N) - v(N \setminus \{i\})$ is always an upper bound of agent i's core profit; Section 2.5). Next, we invoke core stability for the following coalitions:

$$\{3\}: \qquad v(3) = 0 \leq 8 + t_3 \Rightarrow -t_3 \leq 8,$$

$$\{12\}: \qquad v(12) = 1 \leq t_2 - 4 \Rightarrow 5 \leq t_2,$$

$$\{23\}: \qquad v(23) = 2 \leq t_2 + (8 + t_3) - 4 \Rightarrow -t_3 - t_2 \leq 2,$$

$$\{24\}: \qquad v(24) = 5 \leq t_2 + (7 + t_4) - (4 + 4) \Rightarrow 6 \leq -t_3,$$

$$\{34\}: \qquad v(34) = 4 \leq (8 + t_3) + (7 + t_4) - 4 \Rightarrow t_2 \leq 7.$$

Upon setting $t_2 = p_2$ and $t_3 = -p_1$, our core allocation equals the p-competitive allocation (system (13) holds). We have shown that all core allocations are competitive allocations as well. The converse inclusion (competitive set is a subset of the core) holds true with great generality, and it can, of course, be checked directly on our example.

Theorem 3.3

(a) *In an assignment exchange economy where no agent owns more than one house (there is no "capitalist"), the competitive set and the core are nonempty and coincide.*

(b) *In an assignment economy with arbitrary ownership of the houses, a competitive allocation exists, and every competitive allocation is in the core as well. The competitive set may be a strict subset of the core.*

Statement (a), due to Quinzii [1984], is proven in Appendix 3.1 (Quinzii's general proof does not assume quasi-linear preferences). The existence result in statement (b) will be derived from Theorem 4.1, and the inclusion of the competitive set in the core is proven just like Lemma 3.6 (see also Exercise 3.17). Finally, we give some examples showing that the competitive set may be a strict subset of the core.

The simplest example is a monopoly configuration. Say that agent 1 (the monopolist) owns all the houses. There is a core allocation where the monopolist keeps the entire surplus (this is always true in a game where all coalitions not containing agent 1 can generate no surplus).

Clearly, this core allocation involves price discrimination (hence cannot be competitive) if the goods are homogeneous (as already discussed in Section 2.4: Lemma 2.2).

The strict inclusion of the competitive set in the core may also happen without a monopoly configuration. Our next example demonstrates this fact. It involves heterogeneous goods.[15] To avoid the trivial case of two disconnected monopoly markets (seller 1 sells cars and seller 2 sells boats; one-half of the buyers cares only for cars, the other half cares only for boats), we construct an example where two sellers compete for the business of the same buyers. We have two sellers and three buyers. Seller s_1 owns two identical white houses. Seller s_2 owns one blue house. None of the sellers derives any utility from a house; buyer b_1 prefers a white house over a blue one, whereas buyers b_2, b_3 prefer a blue one. Hence a utility profile as follows:

	s_1	s_2	b_1	b_2	b_3
White	0	0	4	2	2
Blue	0	0	2	4	4

(16)

An efficient allocation assigns the white house to b_1, the other white house to b_3 (for instance), and the blue house to b_2, generating a net surplus of 10. The following vector of payments:

$$b_1 \text{ pays 3 and } b_3 \text{ pays 1 to } s_1, \qquad b_2 \text{ pays 3 to } s_2$$

yields a core allocation with the surplus distribution

$$4 \text{ for } s_1, \qquad 3 \text{ for } s_2, \qquad 1 \text{ for } b_1, \qquad 1 \text{ for } b_2, \qquad 1 \text{ for } b_3. \quad (17)$$

This core allocation, however, cannot be a competitive allocation: b_1 and b_3 pay a different price for the same object (a white house). Seller s_1 can price discriminate against b_1 because, unlike b_3, he cannot threaten to make a deal with s_2 on his blue house (s_2 gets 3 for his house, which is more than what b_1 is willing to pay for it). Notice that b_3 is just indifferent between her current allocation and paying the same price as b_2 (via 3) for the blue house, thus preventing s_1 from charging her (b_3) more (lest she go to s_2 and form an objection). By contrast, in any competitive allocation, b_2 and b_3 still get the same net surplus, but b_1 always gets two more units that b_2 and b_3; the singularity of his preference turns to his advantage (a general feature of competitive allocations; see Section 4.6 for more examples of this).

---

[15] By necessity. Recall Lemma 2.1: with homogeneous goods, "two is enough for competition."

Exercise 3.16 gives the details of the competitive set and the core in Example (16).

3.6. ARROW–DEBREU ECONOMIES: DIVISIBLE GOODS AND CONVEX PREFERENCES

We present now the canonical version of exchange economies where preferences are no longer restricted by the transferable utility context and where agents consume arbitrary amounts of several divisible or nondivisible goods. The considerable gain in the generality of the model has a notational cost.

Let a finite-dimensional vector space $\mathbb{R}^K$ represent the space of consumption vectors. We have K different goods (or commodities), and a vector $z = (z^1, \ldots, z^k, \ldots, z^K)$ in $\mathbb{R}^K_+$ (the positive orthant of $\mathbb{R}^K$) represents the consumption of an agent: z^k units of good k, for $k = 1, \ldots, K$. Agent i's initial endowment is a vector ω_i in $\mathbb{R}^K_+$, and a feasible allocation (corresponding to a feasible trade among the n agents) is a list of consumption vectors $z_1, \ldots, z_i, \ldots, z_n$ (one for each agent), such that

$$z_i \in \mathbb{R}^K_+ \quad \text{for all } i \quad \text{and} \quad \sum_{i \in N} z_i = \sum_{i \in N} \omega_i. \tag{18}$$

Definition (18) corresponds to perfectly divisible commodities (such as wine or labor), but it is easily adapted to take indivisibilities into account. If good k comes in indivisible units (such as cars or appliances), simply restrict z_i^k and ω_i^k to be integers for all i.

Agent i's preferences are a preordering (i.e., a complete transitive and reflexive binary relation) R_i on the consumption set $\mathbb{R}^K_+$. We assume throughout this and the following sections that preferences are *monotonic*, namely, that an agent is always strictly better off when her consumption of *every* commodity increases:

$$\text{for all } z, z_* \quad \{z^k \geq z^k_* \quad \text{for all } k\} \Rightarrow \{z R_i z_*\}$$

$$\text{and} \quad \{z^k > z^k_* \quad \text{for all } k\} \Rightarrow \{z P_i z_*\},$$

where P_i is the strict preference associated with the relation R_i ($z P_i z_*$ iff $z R_i z_*$ holds and $z_* R_i z$ fails). Under monotonicity of preferences, the "goods" are indeed desirable and agents are never satiated.[16]

---

[16] This is a realistic assumption only if the goods allocated can be stored (like money); but it makes little sense to assume that agents are never satiated of perishable food. Taking the possibility of satiation into account is not difficult; it merely adds unimportant technical complications.

We define the two concepts of core stability and of competitive equilibrium. An exchange economy with K goods among n agents is described by a list $(R_1, \ldots, R_n; \omega_1, \ldots, \omega_n)$, where $R_1, \ldots, R_n$ is the preference profile and $\omega_1, \ldots, \omega_n$ is the endowment profile (the private ownership structure). An allocation $(z_1, \ldots, z_n)$ is a *core allocation* if it is feasible ((18)) and if there is no coalition S (a subset of $N = \{1, \ldots, n\}$, possibly N itself) that could improve upon its members' welfare by allocating its own resources as it sees fit:

$$\text{There exists } z_i' \text{ for all } i \in S \text{ such that } \left\{ \sum_{i \in S} z_i' = \sum_{i \in S} \omega_i \right\}$$

and $\{z_i' \, R_i \, z_i$ for all $i \in S$, with at least one strict preference$\}$.   (19)

Notice that the above property for $S = N$ corresponds to the Pareto optimality property.

A pair $(p; (z_1, \ldots, z_n))$ made of a price vector p (p is a nonnull vector of $\mathbb{R}_+^K$)) and an allocation is a *competitive equilibrium* if the allocation is feasible ((18)) and, moreover, for all i, we have

$$p \cdot z_i \leq p \cdot \omega_i \quad \text{and for all } z' \in \mathbb{R}_+^K: \quad \{p \cdot z' \leq p \cdot \omega_i \Rightarrow z_i \, R_i \, z'\} \quad (20)$$

(where $p \cdot z$ is the scalar product of $\mathbb{R}^K$).

We speak of competitive price, competitive allocation, and competitive set (i.e., the set of competitive allocations). Of course, all definitions are unaffected by the presence of constraints limiting the set of feasible allocations (such as indivisibilities of some of the goods).

The following simple result (often called the first fundamental theorem of welfare economics) points to one very interesting feature of the competitive equilibrium: its definition suggest cooperations by decentralized behavior, where each agent faces the competitive price and independently selects his optimal consumption, given his budget constraint[17]; on the other hand, it is always a stable outcome of cooperation by direct agreements.

Lemma 3.6. Under the assumption of monotonic preferences, a competitive allocation is a core allocation. In particular, it is a Pareto optimal allocation.

---

[17] Recall that, strictly speaking, the competitive concept is not given as a decentralized game where all the decision power is fully distributed among individual agents; some kind of coordination is needed to pick the right price, and the presence of a disinterested auctioneer does not prevent price manipulations. See Sections 2.9 and 3.8.

Proof. Take an arbitrary consumption vector z and agent i. If $p \cdot z < p \cdot \omega_i$, we can find z' above z ($z'^k > z^k$ for all k) and such that $p \cdot z' < p \cdot \omega_i$. By monotonicity of preferences, $z' P_i z$, and by definition (20), $z_i R_i z'$. Thus z_i is strictly preferred by agent i to every vector z, such that $p \cdot z < p \cdot \omega_i$. In particular, this implies $p \cdot z_i = p \cdot \omega_i$.

Now suppose a coalition S forms an objection against the competitive allocation $(z_1, \ldots, z_n)$. Without loss of generality, take $S = \{1, 2, \ldots, s\}$ and assume agent 1 strictly benefits in the objection $(z'_1, \ldots, z'_s)$:

$$\sum_1^s z'_i = \sum_1^s \omega_i, \qquad z'_1 P_1 z_1, \qquad z'_i R_i z_i \quad \text{for } i = 2, \ldots, s.$$

By the argument above, we have $p \cdot z'_i \geq p \cdot \omega_i$, and by definition (20), $p \cdot z'_1 > p \cdot \omega_1$. Therefore,

$$p \cdot \left(\sum_1^s z'_i \right) = \sum_1^s p \cdot z'_i > \sum_1^s p \cdot \omega_i = p \cdot \left(\sum_1^s \omega_i \right),$$

contradicting the feasibility of the objection. Q.E.D.

Notice that with quasi-linear preferences, the conclusion of Lemma 3.6 holds true even is utilities are only monotonic in money; see Exercise 3.17.

The central proposition of the competitive equilibrium theory says that an equilibrium always exists when preferences are convex and continuous. A preference ordering R_i over $\mathbb{R}_+^K$ is called *convex* if the upper contour set $U_{z_0} = \{z \in \mathbb{R}_+^K / z R_i z_0\}$ is a convex subset of $\mathbb{R}_+^K$ for all z_0. It is called *continuous* if the upper contour set U_{z_0} as well as the lower contour set $L_{z_0} = \{z \in \mathbb{R}_+^K / z_0 R_i z\}$ are closed subsets of $\mathbb{R}_+^K$ for all z_0.

Theorem 3.4. *In an exchange economy with perfectly divisible goods, a competitive equilibrium exists if each individual preference is convex and continuous, and, moreover, at least one of the two following assumptions holds: each agent is endowed with a positive amount of every good; all preferences are strictly monotonic.*

We give the intuition of the familiar proof in the "Edgeworth box" representation of a two-agent, two-good economy. For a general proof and a detailed discussion of our assumptions, see Mas-Colell [1985] or Border [1985]. In the Edgeworth box, two systems of coordinates are overlapping. In Figure 3.2, the point A represents an allocation for *both* agents: agent 1 who lives in the (usual) coordinate system Oab is

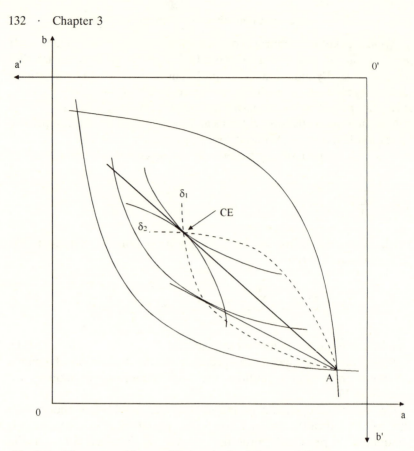

Fig. 3.2. The competitive equilibrium in the Edgeworth box.

consuming $\overline{OA}$, whereas agent 2 who lives in the coordinate system $O'a'b'$ consumes $\overline{O'A}$. In particular, the point O' is the allocation where agent 1 consumes all the resources and agent 2 none (and O represents the symmetrical allocation where agent 2 eats everything).

The curve $\delta_i(i = 1, 2)$ represents the demand of agent i: for a given price (p_X, p_Y), the corresponding budget line $p_X \cdot x_i + p_Y \cdot y_i = p_X \cdot \omega_i^X + p_Y \cdot \omega_i^Y$ goes through A and cuts δ_i at the allocation $\delta_i(p)$. The competitive equilibrium is at the intersection of both curves.

In Figures 3.3 and 3.4 are represented the competitive equilibrium (and the demand curves) in two simple cases: first, when both agents have linear preferences (represented by $u_i(z_i) = u_i \cdot z_i$); second, when both have Leontief preferences (represented by the utility function $u_i(z_i) = \min\{x_i/\lambda_i^1, y_i/\lambda_i^2\}$) In each case, the competitive allocation is unique, or they all yield the same utility vector (case of Figure 3.4b).

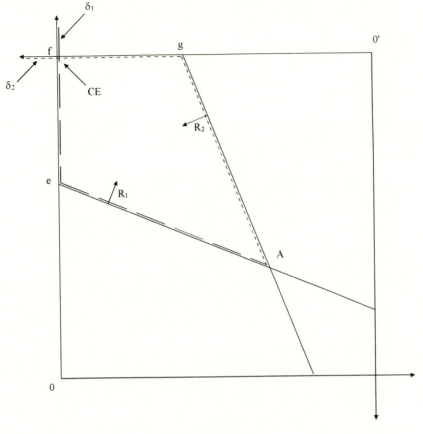

(a)

Fig. 3.3.   Linear preferences in the Edgeworth box.

In Theorem 3.4, the crucial assumption are the continuity and convexity of preferences: the discussion in Sections 2.6 and 2.8 illustrated
how quickly the competitive equilibrium may disappear when we allow
for nonconvex preferences (e.g., when one of the agents is a firm with
decreasing marginal costs). Our next example shows why we need an
additional "boundary" assumption: each ω_i is strictly positive, and/or
each preference ordering is strictly monotonic.

The example is a two-agent, two-good economy with $\omega_1 = (2, 1)$,
$\omega_2 = (0, 1)$, $u_1(z_1) = x_1$, $u_2(z_2) = x_2 + y_2$. Consider a price p such that
p_X and p_Y are both positive. The agent 1 sells his endowment of good Y
and demands more good X than the economy contains. If $p_X = 0$,

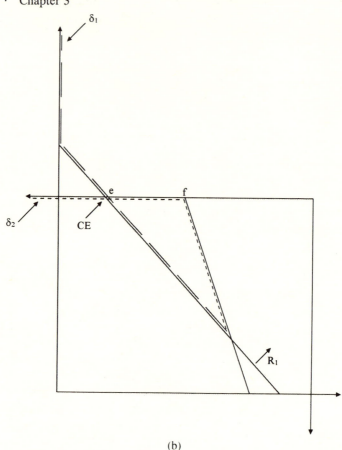

(b)

Fig. 3.3. Continued.

agent 1 demands an infinite amount of good X, and if $p_Y = 0$, agent 2 demands an infinite amount of good Y.

Remark 3.4 Constrained Competitive Equilibrium

A seemingly mild modification of the definition of a competitive equilibrium allows us to dispense with the boundary assumptions in Theorem 3.4. Call the pair $(p; z_1, \ldots, z_n)$ a *constrained competitive equilibrium* if this allocation is feasible and, moreover,

$$p \cdot z_i \leq p \cdot \omega_i \text{ and for all } z' \quad \{z' \leq \omega \text{ and } p \cdot z' \leq p \cdot \omega_i\} \Rightarrow z_i R_i z'.$$

Thus an agent's demand at any given price must stay within the set of

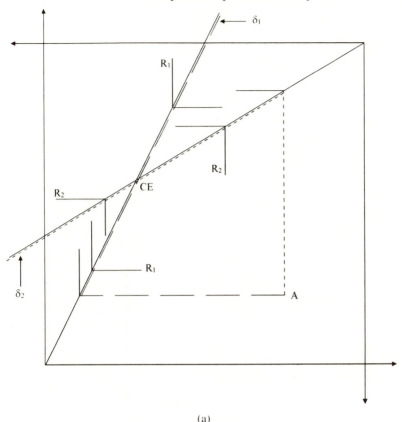

(a)

Fig. 3.4. Leontief preferences in the Edgeworth box.

feasible consumption vectors of this economy: nobody can demand resources that are not compatible with the total endowment of the economy. Clearly, Lemma 3.6 generalizes: the core of the economy contains all constrained competitive allocations (if preferences are monotonic).

One shows that in any economy where the agents have convex and continuous preferences, a constrained competitive equilibrium exists. In the above example, the initial allocation is the unique constrained competitive allocation, for all prices p such that $p_X \geq p_Y > 0$.

One serious drawback of the constrained competitive notion, however, is the frequent multiplicity of the equilibrium allocations. In the economy with linear preferences represented in Figures 3.3, the constrained competitive set is the whole interval of the efficient allocations

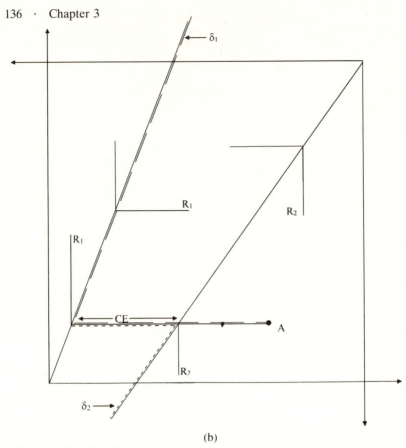

(b)

Fig. 3.4. Continued.

Pareto improving upon the initial allocation (the line *efg* in Figure 3.3a, and the interval *ef* in Figure 3.3b).

Let us return to the regular notion of competitive equilibrium. Our examples so far involve a unique competitive utility profile. This is by no means the general situation; economies with several competitive allocations, each with different utility distributions, are easily constructed. For instance, assume two goods, two agents with identical (Leontief) utilities $u_i(z) = \min\{x, y\}$ and initial endowments $\omega_i = (1, 2)$, $\omega_2 = (2, 1)$. Check that every price vector $p = (p_X, p_Y)$ sustains a competitive allocation, namely,

$$x_1 = y_1 = \frac{p_X + 2p_Y}{p_X + p_Y}, \qquad x_2 = y_2 = \frac{2p_X + p_Y}{p_X + p_Y}.$$

The largest the ratio p_X/p_Y, the better off is agent 2 and worse off is agent 1. Another example is depicted in Figure 3.4b.

In general, uniqueness of the competitive equilibrium price and allocation is guaranteed if each individual preference ordering satisfies the *gross substitutability* assumption, namely,

the demand $\delta_i(p_1, \ldots, p_K)$ is unique for all p and varies continuously with p; the quantity δ_{ik} of good k demanded by agent i is decreasing in p_k and increasing in $p_{k'}$, for all $k' \neq k$.

The competitive theory reaches much beyond the simple exchange economies described here. It now encompasses (a) economies with (increasing marginal costs) firms and consumers who own fractions (shares) of the firm; see an example in Section 5.2, (b) economies with nonmonotonic, noncomplete, and even nontransitive preferences (see Mas-Colell [1988]); see an example in Section 4.5, (c) economies with infinitely many commodities and/or infinitely many agents; see an example in Section 4.5.

3.7. THE EDGEWORTH PROPOSITION

The comparison of the core and the competitive set is the central theme of Chapters 2 and 3. Several important special cases where these two sets coincide have been discussed: in the transferable utility context, these are the oligopoly with binary demands (Section 2.4) and the assignment economy without capitalists (Section 3.5); the house barter model (Section 3.2) offers an example where preferences are not quasi-linear. A common feature of all these examples is the indivisibility of the traded goods.

With divisible goods, sharp equality of the core and the competitive set cannot be expected, except in a limiting sense. The Edgeworth proposition says that the core and the competitive set are approximately equal when goods are divisible and the initial endowment of each market participant is (negligibly) small, relative to the total endowment of the economy. For a rigorous statement of the proposition, the difficulty is to model the negligibility assumption. The most intuitive model, if not the most elegant from a mathematical standpoint, is the replication technique that we illustrate now.

Consider the two-agent, two-good economy depicted in the Edgeworth box of Figure 3.5. The contract curve joining O to O' is the set of Pareto optimal allocations, and the two indifference curves through A cut the core $[BC]$ on the contract curve. Now consider an economy with two copies of agent 1 and two of agent 2. Suppose a core allocation of

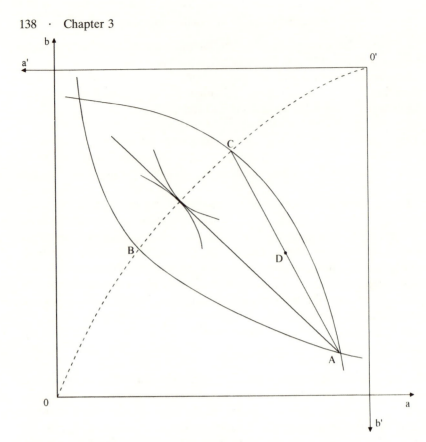

Fig. 3.5. The core shrinks under replication.

that economy gives the utility level of A to each type-2 agent. Then each type-2 agent receives the same vector of goods, and that vector must be C (because agent 2's preferences are strictly convex and by Pareto optimality). Now a coalition with both type-2 agents and one type-1 agent can give $\overrightarrow{OC}$ to agent 1 and $\overrightarrow{O'D}$ to each type-2 agent (where D is the midpoint of AC). This is a strict improvement for the type-2 agents; therefore, we conclude that in the economy replicated once, the core cuts a smaller segment on the contract curve, it "shrinks."

The replication method (used by Debreu and Scarf [1963] in their initial formalization of the Edgeworth proposition) starts with an arbitrary finite economy with convex preferences and replicates it k times (where k is an arbitrary integer); as k goes to infinity, the share of any one agent within total endowment becomes vanishingly small and one sees the core "shrink" toward the competitive set. See Appendix 3.2 for a heuristic argument of the proof.

In the presence of indivisible goods, the replication argument does not yield, in general, the convergence of the core toward the competitive set. Exercise 3.18 gives an example in an assignment economy with capitalists.

The most suitable model of the Edgeworth proposition uses a model of the exchange economy where the set of agents is at once a continuum (say, the interval $[0,1]$) and where the negligibility of each single agent corresponds to the technical property that the distribution of endowments has no atoms with respect to the Lebesgue measure on $[0,1]$; see Hildenbrand [1974].

3.8. TRADING GAMES

We extend the discussion of Section 2.9 to the context of divisible goods and arbitrary convex preferences. Just as we did in the (much simpler) Böhm–Bawerk market, we look at the direct competitive mechanism, where agents report (convex, continuous) preferences and a competitive allocation (at the reported preferences) is enforced. Of course, we cannot allow the agents to report nonconvex preferences because no competitive equilibrium would exist for some profiles of reports. In case the competitive set contains more than one allocation, the mechanism selects one of them by an arbitrary method (this will not affect the strategic equilibria of the direct revelation mechanism).

Lemma 3.7. (Notations as in Section 3.6.)

(a) In any direct competitive mechanism where preferences are convex and continuous (and players are not allowed to report any other kind of preferences), for each player we can find a (true) preference profile where he finds it profitable to misreport.

(b) The set of Nash equilibrium outcomes of a direct competitive mechanism reaches all feasible utility distributions where each agent gets at least her initial utility level. That is, for every feasible allocation $(z_1, \ldots, z_n)$ such that $u_i(z_i) \geq u_i(\omega_i)$ for all i, there is a Nash equilibrium of the direct competitive mechanism where each agent receives precisely the utility $u_i(z_i)$.

Proof. We use a two-person, two-good economy and show the two statements by means of figures. For statement (a), consider Figure 3.6. The point A represents the initial endowment, and z_1 represents the competitive allocation for the true preferences R_1, R_2. Given that agent 2 reports truthfully, agent 1 benefits by reporting a preference of which the indifference curve through ω_1 is a straight line like D_1. Indeed, if D_1 is close enough to the budget line (connecting ω_1 to z_1), then agent 2's best allocation on D_1 (namely, z_1') is preferred by agent 1 to z_1.

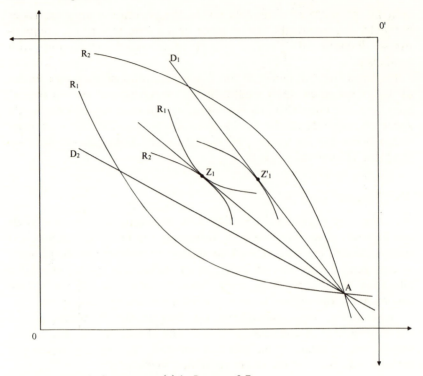

Fig. 3.6. Proof of statement (a) in Lemma 3.7.

Similarly, if agent 1 reports sincerely, agent 2 benefits by reporting an indifference curve through ω_1 such as the line D_2 (providing it is close enough to the budget line).

For statement (b), see Figure 3.7. We pick an arbitrary allocation z_1 in the area ABC. The figure shows the two indifference contours through z_1 and two curves Γ, Γ', with the property that Γ (resp. Γ') is outward tangent to the indifference contour of R_1 (resp. R_2) through z_1. It also shows the construction of a report R_2^* by agent 2 that will leave agent 1 with the choice of an allocation on the curve Γ; that is, by choosing carefully his report, agent 1 can achieve any allocation on Γ. The fact that Γ is an outward tangent to agent 1's indifference curve through z_1 means that agent 1 can do no better than z_1 if agent 2 reports R_2^*. Call R_1^* a similarly constructed report by agent 1 leaving agent 2 with the choice of an allocation on the curve Γ'. Then observe that z_1 is the competitive allocation at (R_1^*, R_2^*), by the very construction of R_i^*.   Q.E.D.

Lemma 3.7 is comparable to Lemmas 2.6 and 2.7: the direct competitive mechanism does not prevent manipulation, and the range of its

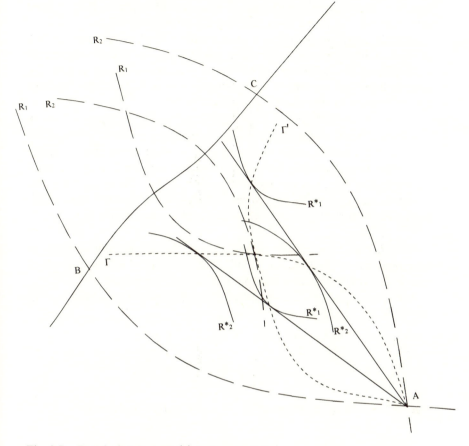

Fig. 3.7. Proof of statement (b) in Lemma 3.7.

Nash equilibrium outcomes is very large. Notice that a much less sophisticated mechanism, such as the agreement mechanism below, would do just as well (just as poorly):

> *agreement mechanism*: Each agent demands an allocation z_i. If demands are compatible ($\sum_{i=1}^{n} z_i \leq \sum_{i=1}^{n} \omega_i$), they are satisfied; otherwise, everyone keeps his initial endowment.

Alternatively, one can construct mechanisms where the set of Nash equilibrium outcomes boils down to the competitive outcome (with or without the "no trade" outcome). In those mechanisms, agents typically announce a price and a trade. All such mechanisms, like the above two (direct competitive and agreement mechanisms), hardly suggest a decentralized behavior in equilibrium: the outcome is a discontinuous function of messages, which makes a tatonnement procedure at least unrealistic and at worst divergent. See Dutta et al. [1993].

Strategy-proof mechanisms, on the contrary, capture a precise notion of decentralized behavior; therefore, any strategy-proof exchange mechanism, however inefficient, is worthy of our attention. Of course, we cannot hope for too much from strategy-proof mechanisms, as suggested already by the results of Section 2.9.

Lemma 3.8. (Hurwicz [1972]). *In an exchange economy with divisible goods and given initial endowments, consider a direct mechanism eliciting a convex preference ordering from all participants, and implementing a feasible allocation that (at the reported preferences) is Pareto optimal and makes no agent worst off than at his initial endowment. Then this mechanism cannot be strategy-proof (when true preferences are convex).*

Proof. The idea of the general proof, again, is captured by an example, due to Postlewaite [1979], with two agents and two goods X and Y. It generalizes to any number of agents and goods.

Agent 1 has initial endowment $\omega_1 = (0, 1)$ and utility function

$$u_1(x_1, y_1) = 3x_1 + y_1 \qquad \text{if } y_1 \geq \tfrac{1}{2},$$

$$= 3x_1 + 6y_1 - \tfrac{5}{2} \quad \text{if } y_1 \leq \tfrac{1}{2}.$$

Agent 2 has initial endowment $\omega_2 = (1, 0)$ and utility function

$$u_2(x_2, y_2) = x_2 + 3y_2 \qquad \text{if } x_2 \geq \tfrac{1}{2},$$

$$= 6x_2 + 3y_2 - \tfrac{5}{2} \quad \text{if } x_2 \leq \tfrac{1}{2}.$$

The indifference curves of either agent are piecewise linear, with a kink at $y_1 = \tfrac{1}{2}$ and $x_2 = \tfrac{1}{2}$, respectively. Figure 3.8 depicts the corresponding Edgeworth box. The imputations (i.e., Pareto optimal allocations to which no agent prefers his initial allocation ω_i) correspond to the line ABC, where $A = (\tfrac{1}{2}, \tfrac{1}{3})$ and $C = (\tfrac{2}{3}, \tfrac{1}{2})$.

Suppose agent 1 reports the linear preferences

$$v_1(x_1, y_1) = x_1 + 2y_1 \quad \text{all } x_1, y_1,$$

whereas agent 2 reports truthfully. The imputations at profile (v_1, u_2)

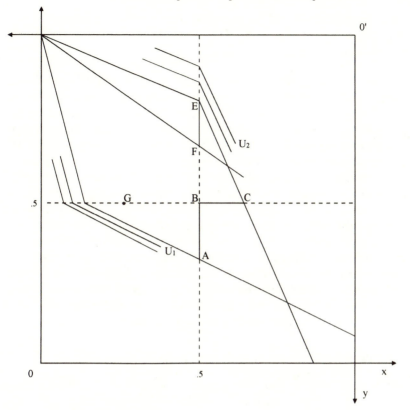

Fig. 3.8. Proof of Lemma 3.8.

are on the interval $[F, E]$, where $F = (\frac{1}{2}, \frac{3}{4})$, $E = (\frac{1}{2}, \frac{5}{6})$. They give to agent 1 at least the utility level $u_1(F) = \frac{9}{4}$.

Now consider a strategy-proof mechanism selecting an imputation at profile (u_1, u_2). Since reporting v_1 is not profitable for agent 1, our mechanism must pick an imputation that yields at least the utility $\frac{9}{4}$ to agent 1, namely, between B and C (B excluded, since $u_1(B) = 2$).

Symmetrically, if agent 2 reports the utility function

$$v_2(x_2, y_2) = 2x_2 + y_2,$$

whereas agent 1 reports truthfully, one checks that agent 2 enjoys at least the utility level

$$u_2(\tfrac{3}{4}, \tfrac{1}{2}) = \tfrac{9}{4} \quad \text{(corresponding to } G = (\tfrac{1}{4}, \tfrac{1}{2}) \text{ on the figure).}$$

Thus at profile (u_1, u_2), a strategy-proof mechanism must select an imputation between B and A (B excluded, since $u_2(B) = 2$). This contradiction concludes the proof.

Lemma 3.8 is hardly surprising: already in the Böhm–Bawerk market, there exists no efficient and strategy-proof mechanism (see Exercise 2.17 for a proof in the case of one seller and one buyer). In fact, in a two-person exchange economy with divisible goods, the only strategy-proof direct mechanism implementing a Pareto optimal allocation is a *dictatorial* mechanism where one agent eats the whole resources no matter what! (This rather discouraging result is proven by Zhou [1991].) Such a mechanism is incompatible with any individual property right.

Now we drop the efficiency requirement and ask: what direct mechanism is both strategy-proof and individually rational? As utility is no longer transferable via money, we cannot use the pivotal mechanism or any mechanism running a deficit that a benevolent arbitrator will pick up (or even running a surplus as the Dida mechanism does; see Section 2.9). A strategy-proof mechanism for the general domain of convex, continuous preferences must enforce a feasible trade for each set of reports. Here is an example in a two-good economy.

Fixed Price Exchange Mechanism (with Uniform Rationing)

We have n agents and two goods A, B. Agent i's endowment is $\omega_i = (\bar{a}_i, \bar{b}_i)$. We fix a price p for the exchange of one unit of good A to one unit of good B. Suppose first $n = 2$. Each agent demands an allocation $d_i = (a_i, b_i)$ satisfying the budget constraint

$$a_i + pb_i = \bar{a}_i + p\bar{b}_i \tag{21}$$

If $a_i \geq \bar{a}_i$ for $i = 1, 2$ or if $a_i \leq \bar{a}_i$ for $i = 1, 2$, no trade occurs. If $a_1 > \bar{a}_1, \bar{a}_2 > a_2$, set $\tau = \min\{a_1 - \bar{a}_1, \bar{a}_2 - a_2\}$; agent 1 buys τ units of good A from agent 2 in exchange for τ/p units of good B.

General case: $n \geq 2$. Agent i demands d_i satisfying (21). If $\Sigma_N a_i > \Sigma_N \bar{a}_i$ (excess demand of good A), we ration good A with the cap λ solving the equation

$$\sum_{i=1}^{n} \min\{\lambda, a_i\} = \sum_{i=1}^{n} \bar{a}_i.$$

Agent i receives $z_i = (\min\{\lambda; a_i\}, \sup\{b_i, \bar{b}_i + (\bar{a}_i - \lambda)/p\})$.

Lemma 3.9. When individual preferences are convex, the fixed price exchange mechanism is strategy-proof and individually rational. An agent's dominant strategy consists of demanding (one of) his preferred trade at the proposed fixed price.

The proof is identical to that of the strategy-proofness of the related uniform rationing mechanism for fair division; see Section 4.5 and Exercise 4.11.

The fixed price exchange mechanism is severely inefficient. As the trading price is fixed once and for all, chances are that we will fall very short of achieving an efficient outcome: efficiency occurs only when the fixed price p happens to be precisely a competitive price for the preference profile in question! Unfortunately, the class of strategy-proof mechanisms respecting individual property rights contains little more than the fixed price ones (as demonstrated by Barbera and Jackson [1995]; see Exercise 3.21 for a strategy-proof mechanism using two fixed prices).

So far we have discussed manipulation by misreporting one's preferences, but the participants of the mechanism may also consider the possibility of withholding part of their resources from consideration by the mechanism. If the commodities can be effectively concealed, an agent may hope to benefit from (i) withholding a certain fraction of his endowment from the mechanism, (ii) exchanging the rest via the mechanism, and (iii) consuming whatever the mechanism allocates him *plus* the commodities initially withheld.

A simple and devastating argument due to Postlewaite shows that manipulation by withholding is indeed profitable in *every* mechanism implementing an imputation of the reported profile (that is, for every such mechanism, there exist preference profiles at which some agents are better off if they manipulate).

Consider the economy used in the proof of Lemma 3.8. Fix a mechanism implementing an imputation of the reported profile. If our agents report truthfully, the mechanism selects an allocation of the line ABC in Figure 3.8. By the symmetry of utilities and endowments, we may assume that an allocation in the interval AB is selected. The agent 1 withholds 0.5 unit of good Y, whence the mechanism gives him at least (0.5) of good X and 0.25 of good Y; now his total consumption is at least $(0.5, 0.75)$ and is strictly preferred to B.

Note that the withholding manipulation is related to, but different from, the "throw away" manipulation. The latter occurs if, at the competitive equilibrium, it pays to destroy part of one's endowment.[18] See Exercise 3.20.

---

[18] This tactic is similar to that of a monopolist withholding its supply to raise the price.

In spite of the negative results reported here, the literature on strategic market games offers several plausible mechanisms of which the noncooperative equilibrium allocations coincides (or approximately coincides) with the competitive equilibrium allocations. See Shapely and Shubik [1977]. Mas-Collel [1980] and Dubey [1982].

APPENDIX TO CHAPTER 3

A3.1 Proof of Theorem 3.1 (Adapted from Roth and Sotomayor [1990] pp 28 and 33)

Let μ be the matching resulting from the Gale–Shapley algorithm where men propose. Suppose μ is not stable: some man m_i and woman w_j block. As m_i prefers w_j to whomever he is marrying in μ, he must have proposed to w_j earlier in the algorithm and been rejected by her in favor of some man m' she prefers to m_i. As the algorithm unfolds, woman w_j can only change from one engagee to one she prefers, so by transitivity of preferences, w_j ends up with a mate she strictly prefers to m_i. This contradicts our blocking assumption and μ is therefore stable.

We prove now that μ gives to each man his best achievable woman (where we mean that a certain woman is achievable for a certain man when there exists a stable matching where they are matched) and to each woman her worst achievable man. We prove the first statement by induction.

Suppose that up to a given step in the algorithm, no man has yet been rejected by a woman who is achievable for him. At this step, suppose woman w rejects man m in favor of man m', whom she keeps engaged (in particular, w prefers m' to m). We must show that w is not achievable for m.

We know that m' prefers w to any woman except for those who have previously rejected him, and hence (by the inductive assumption) are unachievable for him. Consider a hypothetical stable matching μ' that matches m to w and everyone else to an achievable mate. Then m' prefers w to his mate at μ' (by the induction assumption). So the matching μ' is unstable, since it is blocked by m' and w, who each prefer the other to their mate at μ'. Therefore, there is no stable matching that matches m and w, so they are unachievable for each other, which completes the proof.

We prove next that each woman gets her worst achievable mate. Suppose that w is matched with man m in μ, but there is a man m', achievable for w and to whom she prefers m. Then consider a stable matching μ' where w is matched with m'. In μ', m is matched with another woman w' to whom he prefers w (by the previous proof). Thus a contradiction: μ' is blocked by (m, w). The proof of the theorem is now complete.

A 3.2. Proof of Theorem 3.3

Existence of a Competitive Allocation. This follows from Theorem 4.1 about the existence of an envy-free allocation in the fair-assignment problem. Indeed, system (13) characterizing the competitive prices (independently of the private ownership structure) can be read as saying that the allocation to agent i of house $\sigma(i)$ and transfer $-p_{\sigma(i)}$ is envy-free. See Theorem 4.1 for details.

The Core and the Competitive Set Coincide. We prove the result for the case $q = n$. If $q < n$, we can simply endow agents $q + 1, \ldots, n$ with a worthless house $(u_i(h) = 0$ for all $i)$; upon checking that all worthless houses have zero price in any competitive equilibrium, we can then apply the result in the $q = n$ case.

That a competitive allocation is also a core allocation follows from a standard argument, reproduced here for completeness. Say that (σ, p) is competitive (so it satisfies system (13)) and let S be an objecting coalition. The objection consists of an assignment $i \to \tau(i)$ of the houses initially owned by S, and of a balanced vector of transfers t_i such that

$$u_i(\sigma(i)) - p_{\sigma(i)} + p_i \leq u_i(\tau(i)) + t_i \quad \text{for all } i \text{ in } S,$$

with at least one strict inequality.

From (13), we get

$$u_i(\sigma(i)) - p_{\sigma(i)} + p_i \geq u_i(\tau(i)) + p_{\tau(i)} + p_i;$$

hence, $t_i \geq p_i - p_{\tau(i)}$ for all i in S with at least one strict inequality. Summing up over i in S:

$$\sum_S t_i = 0 > \sum_S (t_i - p_{\tau(i)}) = 0,$$

as τ is a permutation of the houses owned by S. This is the desired contradiction.

We prove the converse property: a core allocation is a competitive allocation as well. We use the argument in Quinzii [1984]. Let (σ, τ) be a core allocation and denote

$$m_{ij} = u_i(\sigma(i)) - u_i(j) + t_i$$

and

$$p_1 = 0; \quad \text{for } i \geq 2: \quad p_i = \min\{m_{ii_1} + m_{i_1i_2} + \cdots + m_{i_t1}\},$$

where the minimum bears on all the sequences $(i, i_1, \ldots, i_t, 1)$ (with possible repetitions). Note that along a cycle $(i_0, i_1, \ldots, i_T = i_0)$, we have $\sum_0^{T-1} m_{i_l i_{l+1}} \geq 0$; otherwise, the coalition $\{i_0, i_1, \ldots, i_{T-1}\}$ can object. Therefore, p_i is well defined. Moreover, the inequality

$$m_{ij} + p_j \geq p_i \quad \text{for all } i, j \tag{22}$$

follows at once the definition of p.

To conclude the proof, consider the canonical partition of N into disjoint cycles of its bijection σ, where a cycle is a sequence without repetition $(i, \sigma(i), \ldots, \sigma^{K-1}(i))$ such that $\sigma^K(i) = i$. By the core property, on such a cycle we have

$$\sum_0^{K-1} t_{\sigma^k(i)} = 0 \tag{23}$$

(indeed, the sum must be nonnegative, or the cycle can object; by the overall budget balance of t, if the sum of a cycle is positive, that of some other cycle is negative). Invoke now (22) for the pairs $(\sigma^k(i), \sigma^{k+1}(i))$, $k = 0, \ldots, K - 1$, noticing $m_{j\sigma(j)} = t_j$. We get

$$t_{\sigma^k(i)} + p_{\sigma^{k+1}(i)} \geq p_{\sigma^k(i)} \quad \text{for } k = 0, \ldots, K - 1.$$

Summing up these inequalities gives an equation (by (23)); hence, all are equalities and we conclude

$$t_i + p_{\sigma(i)} = p_i \quad \text{for all } i.$$

It remains to combine this with (22) as follows:

$$t_i + p_{\sigma(i)} \leq m_{ij} + p_j = u_i(\sigma(i)) - u_i(j) + t_i + p_j \quad \text{for all } i, j,$$

or, equivalently,

$$u_i(\sigma(i)) - p_{\sigma(i)} + p_i \geq u_i(j) - p_j + p_i \quad \text{for all } i, j,$$

namely, p is a competitive price and (σ, t) its associated competitive allocation. Of course, a translation of p by a constant will ensure $p \geq 0$, without affecting the above inequalities.

A3.3. A Heuristic Argument for the Edgeworth Proposition (Mas-Collel [1985])

We start with an arbitrary exchange economy $(R_1, \ldots, R_n, \omega_1, \ldots, \omega_n)$, where each preference ordering R_i is continuous and *strictly* convex:

$$\{y_i \, R_i \, x_i, z_i \, R_i \, x_i \quad \text{and} \quad y_i \neq z_i\} \quad \text{imply} \quad [\lambda y_i + (1 - \lambda)z_i] \, P_i \, x_i$$

$$\text{for all } \lambda, 0 < \lambda < 1.$$

The p-replica economy E_p has $n \cdot p$ agents of n different types. There are p identical agents of type i, each endowed with the resources ω_i and the preference R_i. Thus, when p is large, each agent owns a small part (at most $1/p$) of total resources.

Check first the *equal-treatment property*. Consider a core allocation of E_p, say, $(x_{i,k})$, $i = 1, \ldots, n$, $k = 1, \ldots, p$, where (i, k) is the kth agent of type i. We claim $x_{i,k} = x_{i,k'}$, for all k, k'.

Pick in each sequence $(x_{i,k})_{k=1,\ldots,p}$ an element $x_{i,k(i)}$ where R_i is minimal: $x_{i,k} \, R_i \, x_{i,k(i)}$ for all k. By convexity of R_i, this implies

$$\left(\frac{1}{p} \sum_{k=1}^{p} x_{i,k} \right) R_i \, x_{i,k(i)} \quad \text{for all } i = 1, \ldots, n.$$

If for some i, all $x_{i,k}$ are not equal, this preference relation is strict, by strict convexity of R_i. In this case, the n-person coalition $(i, k(i))_{i=1,\ldots,n}$ could Pareto improve upon its utility by the allocation $y_{i,k(i)} = 1/p(\sum_{k=1}^{p} x_{jk})$. Indeed, $(y_{i,k(i)})_{i=1,\ldots,n}$ is a feasible reallocation of its initial endowments:

$$\sum_{i=1}^{n} y_{i,k(i)} = \frac{1}{p} \sum_{\substack{i=1,\ldots,n \\ k=1,\ldots,p}} x_{i,k} = \frac{1}{p} \sum_{\substack{i=1,\ldots,n \\ k=1,\ldots,p}} \omega_i = \sum_{i=1}^{n} \omega_i = \sum_{i=1}^{n} \omega_{i,k(i)}.$$

The proof of the equal-treatment property is now complete. Because of the equal-treatment property, a core allocation of E_p is the p-replica of a core allocation $(x_1, \ldots, x_n)$ of E. (Conversely, not all core allocations of the initial economy yield a core allocation of the replica economy!) We show now that this allocation must be a competitive equilibrium allocation if p is "very large." For simplicity, we assume $n = 2$ and two goods, so we can use a geometrical argument in the Edgeworth box.

Let (x_1, x_2) be a core allocation of E of which the p-replica is in the core of E_p for all p large enough. Allocation (x_1, x_2) is competitive if and only if the line from x_1 to ω_1 supports both upper contours of R_1 and R_2 at (x_1, x_2). If (x_1, x_2) is not competitive, this line cuts through the interior of these contour sets, and contains two points z_1, z_1' on both sides of x_1, such that $z_1 P_1 x_1$ and $(\omega - z_1') P_2 x_2$; see Figure 3.9.

By construction, $(z_1' - \omega_1) = \lambda(z_1 - \omega_1)$ for some positive number λ. By continuity of R_i, we can choose z_1, z_1' so that λ is a rational number, so there are two positive integers m_1, m_2 such that

$$m_1 \cdot (z_1 - \omega_1) = m_2 \cdot (z_1' - \omega_1). \tag{24}$$

Fix an integer $p \geq m_1 + m_2$ and consider in E_p a coalition with m_1 agents of type 1 and m_2 agents of type 2. The allocation z_1 for all type-1 agents, $\omega - z_1'$ for all type-2 agents is feasible for this coalition because (24) implies

$$m_1 \cdot z_1 + m_2 \cdot (\omega - z_1') = m_1 \cdot \omega_1 + m_2 \cdot \omega_2.$$

Moreover, every agent of the coalition is better off than at (x_1, x_2), which is the desired contradiction.

Note that smooth preferences are required in the above argument. (Exercise: why?) See Mas-Collel [1985] for details.

EXERCISES ON CHAPTER 3

Exercise 3.1. *External Stability of the Core in the House Barter Model (Wako [1991])*

Assume all preferences are strict. Show that if an allocation z is not the core allocation, there exists an objection by at least one coalition S where S uses its core allocation. In other words, all members of S prefer the core allocation to z, and S can achieve its part of the core allocation by means of its own endowments.

Exercise 3.2. *On the Number of Rounds in the Top Trading Cycle Algorithm (House Barter Model)*

Let T be the number of rounds.

(a) Show that $T = 1$ if and only if the economy has a unique Pareto optimal allocation.

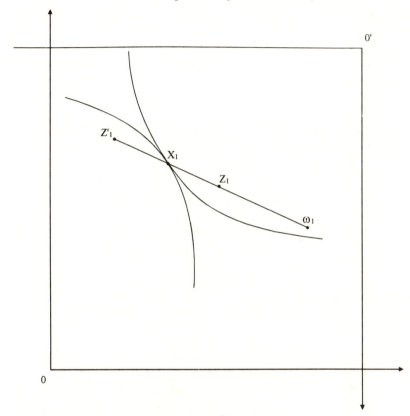

Fig. 3.9. Proof of the Edgeworth proposition.

(b) Show that $T = n$ (where n is the number of agents) only if the initial allocation is Pareto optimal (hence, it is the core allocation as well).

Exercise 3.3. Manipulation by Ex Ante Trading (House Barter)

All preferences are strict.

(a) Suppose a coalition manipulates the core mechanism by trading its initial houses *prior* to implementing the mechanism itself (henceforth reporting truthfully its preferences in the mechanism). Show that this coalition cannot make all its members strictly better off. Give an example where a three-person coalition can make two of its members strictly better off and the last equally well off in this fashion.

(b)* (Matt Kropf). Fix a preference profile and a coalition S. Fix once and for all the (truthful) report of coalition $N \setminus S$. Show that whatever allocation coalition S can achieve by swapping houses ex ante and reporting truthfully, it can also achieve by misreporting preferences and swapping houses ex post. Deduce that coalition S cannot achieve any more allocations by (i) swapping houses ex ante, (ii) misreporting preferences, and (iii) swapping houses ex post.

Exercise 3.4. House Barter when Indifferences are Possible

Initially, agent i owns h_i.
  (a) (Wako [1991]).

u_1	u_2	u_3
h_2, h_3	h_1, h_3	h_1, h_2
h_1	h_2	h_3

.

Show that we have five competitive allocations, two of which are also core allocations. The two core allocations coincide utilitywise.
  (b)

u_1	u_2	u_3	u_4
h_2, h_3	h_1	h_4	h_2
h_1	h_3	h_1	h_1
h_4	h_2	h_2	h_4
	h_4	h_3	h_3

.

Show that the core contains a single allocation and that the competitive set contains one more allocation (one that is Pareto inferior to the core allocation).
  (c) (Wako [1991])

u_1	u_2	u_3	u_4	u_5
h_2	h_3, h_4	h_1, h_3	h_5	h_1
h_1	h_2	h_4	h_4	h_4
h_3	h_1	h_2	h_3	h_2
h_4	h_5	h_5	h_2	h_3
h_5			h_1	h_5

.

Show that the allocation

$$h_1 \to 5, \qquad h_2 \to 1, \qquad h_3 \to 3, \qquad h_4 \to 2, \qquad h_5 \to 4$$

is the only core allocation. Show that there is one more competitive allocation, namely,

$$h_1 \to 3, \qquad h_2 \to 1, \qquad h_3 \to 2, \qquad h_4 \to 5, \qquad h_5 \to 4.$$

If we weaken the condition for core stability by requiring one objecting coalition to provide a strictly higher utility to all its members, we define a *weak core* larger than the core. Are both competitive allocations in the weak core? Show that the following noncompetitive allocation is in the weak core:

$$h_1 \to 1, \qquad h_2 \to 2, \qquad h_3 \to 3, \qquad h_5 \to 4, \qquad h_4 \to 5.$$

Find at least one more noncompetitive allocation in the weak core. (*Hint*: There is one "very close" to the core allocation.)

Exercise 3.5.   Uniqueness of the Core Utility in House Barter (Wako [1991])

Indifferences are possible. Show that two core allocations must yield the same utility to all agents. (*Hint*: Use the fact that a core allocation obtains by a top trading cycle algorithm.)

Exercise 3.6.   Computing (the End-Points of) the Core in a Marriage Market

(a) Suppose that preferences of the women are unanimous: they all rank the men with precisely the same ordering. Show that there is a unique stable matching and describe it.
(b) Consider the following market with four men and four women.

m_1	m_2	m_3	m_4		w_1	w_2	w_3	w_4
w_1	w_2	w_3	w_4		m_4	m_3	m_2	m_1
w_2	w_1	w_4	w_3	,	m_3	m_4	m_1	m_2
w_3	w_4	w_1	w_2		m_2	m_1	m_4	m_3
w_4	w_3	w_2	w_1		m_1	m_2	m_3	m_4

.

Find the M-optimal and W-optimal matchings. Is there any other stable matching? How many of them?

(c) We have five men, five women:

m_1	m_2	m_3	m_4	m_5
w_1	w_4	w_4	w_1	w_1
w_2	w_2	w_3	w_4	w_2
w_3	w_3	w_1	w_3	w_4
w_4	w_1	w_2	w_2	w_5
w_5	w_5	w_5	w_5	w_3

w_1	w_2	w_3	w_4	w_5
m_2	m_3	m_5	m_1	m_1
m_3	m_1	m_4	m_4	m_2
m_1	m_2	m_1	m_5	m_3
m_4	m_4	m_2	m_2	m_4
m_5	m_5	m_3	m_3	m_5

Apply the Gale–Shapley algorithm to find the two end-points of the core.

(d) In the markets of questions (b) and (c), find an optimal (joint) manipulation by the women in the direct mechanism implementing the M-optimal matching.

Exercise 3.7.   Pareto Optimality of the W-Optimal Matching

(a) Consider the Gale–Shapley algorithm where women propose. Show that any man who gets a proposal in the last step of the algorithm has not rejected any woman in any of the previous steps. Deduce that there is no matching that the women unanimously prefer to the outcome of the algorithm.

(b) In the following example, show that the W-optimal matching is not strongly Pareto optimal:

m_1	m_2	m_3
w_1	w_3	w_1
w_2	w_1	w_2
w_3	w_2	w_3

w_1	w_2	w_3
m_2	m_1	m_1
m_1	m_2	m_2
m_3	m_3	m_3

Exercise 3.8.   Lattice Property of the Core in a Marriage market

Consider a market with n men and n women and two stable matchings,

μ and μ'. For each man i, denote by $\mu(i)$ and $\mu'(i)$ his mate in μ and μ', respectively. Denote also by $\overline{\mu}(i)$ man i's preferred mate between $\mu(i)$ and $\mu'(i)$.

(a) Show that $\overline{\mu}$ is a matching (namely, is one-to-one). Show also that for any woman j, $\overline{\mu}^{-1}(j)$ is the worst of $\mu^{-1}(j)$ and $(\mu')^{-1}(j)$ (if these two differ).

(b) Show that $\overline{\mu}$ is a stable matching as well. Thus we have direct proof of the fact that there is a best stable matching for the men that is also the worst stable matching for the women.

Exercise 3.9. On Competitive Matchings (Marriage Market)

(a) In a given marriage market with n men and n women, consider a matching where man m_i and woman w_i are matched, for $i = 1, \ldots, n$. We call this matching "competitive" if there is a vector $p_1, \ldots, p_n$ of prices such that

$$\text{for all } i, j = 1, \ldots, n \quad \{m_i \text{ prefers } w_j \text{ to } w_i \Rightarrow p_j > p_i\}$$

$$\text{and} \quad \{w_i \text{ prefers } m_j \text{ to } m_i \Rightarrow p_j > p_i\}.$$

Show that this matching is stable. Show also that there is a pair (m_i, w_i) such that m_i is w_i's first choice and w_i is m_i's first choice.

(b)* Construct a marriage market with two or more competitive matchings. Construct a marriage market with (at least) one competitive matching and (at least one) stable but noncompetitive matching.

Exercise 3.10. Marriage Markets with the Option of Remaining Single*

We have n men and n' women, where n and n' are not necessarily equal, and each participant has the option of remaining single. He or she ranks the agents on the other side of the gender line as well as the "single" option. Thus the following preference:

$$m_1 > m_3 > s > m_4 > m_2$$

means that this woman's top choice is m_1, followed by m_3, but if she cannot marry m_1 or m_3, she prefers to stay single.

A matching is a mapping μ from $M \cup W$ into $M \cup W \cup \{s\}$, with the property

for all $m \in M$: $\mu(m) \neq s \Rightarrow \mu(m) \in W$ and $\mu(\mu(m)) = m$,

for all $w \in W$: $\mu(w) \neq s \Rightarrow \mu(w) \in M$ and $\mu(\mu(m)) = w$.

A matching is stable if (i) every nonsingle agent prefers his or her mate to remaining single, and (ii) there is no objection by one man and one woman (where objection are defined as usual).

Adapt the Gale–Shapley algorithm to take into account the "single" option and show that Theorem 3.1 is word-for-word preserved.

Exercise 3.11.   The Roommate Problem, and Other Matching Games with Empty Core

(a) The Roommate Problem (Gale and Shapley [1962]).   We have $2n$ agents. A matching consists of n pairs of agents. Each agent has a preference ordering over all $(2n - 1)$ potential roommates. A matching is (core) stable if no two agents prefer to share a room rather than stay with their current roommate. Note that the marriage model is a particular case of the roommate model where a roommate is interpreted as a spouse and each woman (resp. man) prefers any man to any woman (resp. any woman to any man).

In the following roommate problem with four agents a_i:

u_1	u_2	u_3	u_4
a_2	a_3	a_1	*
a_3	a_1	a_2	*
a_4	a_4	a_4	*

,

show that the core is empty.

(b) A Man–Woman–Child Market (Alkan [1988]).   There are n men, n women, and n children. A matching is a partition into n families, where a family consists of one man, one woman, and one child. Each agent ranks the various families that he might belong to. The following example shows that the core may be empty even with $n = 3$ and strict

preferences:

m_1	m_2	m_3
(w_1, c_3)	(w_2, c_3)	(w_3, c_3)
(w_2, c_3)	(w_2, c_2)	*
(w_1, c_1)	(w_3, c_3)	*
*	*	*

w_1	w_2	w_3
(m_1, c_1)	(m_2, c_3)	(m_2, c_3)
*	(m_1, c_3)	(m_3, c_3)
*	(m_2, c_2)	*
*	*	*

,

c_1	c_2	c_3
(m_1, w_1)	(m_2, w_2)	(m_1, w_3)
*	*	(m_2, w_3)
		(m_1, w_2)
		(m_3, w_3)
		*

.

To show that there is no stable matching, check successively (i) that all matchings giving m_1 (resp. m_2 and w_2) a better family than (m_1, w_1, c_1) (resp. (m_2, w_2, c_2)) are unstable, and (ii) that any matching not containing (m_1, w_1, c_1) (resp. (m_2, w_2, c_2)) is either blocked by (m_1, w_1, c_1) (resp. (m_2, w_2, c_2)) or is unstable. Then find a family that blocks any matching containing (m_1, w_1, c_1) and (m_2, w_2, c_2).

(c) Many-To-One Matching (Roth and Sotomayor [1990]). There are n firms and m workers. Each firm can hire as many workers as it wishes and has preferences over all subsets of workers; each worker can work in only one firm and has preferences over firms. Once again, a very simple example with two firms and three workers has an empty core. The preferences are as follows:

F_1	F_2
w_1, w_3	w_1, w_3
w_1, w_2	w_2, w_3
w_2, w_3	w_1, w_2
w_1	w_3
w_2	w_1
$\varnothing$	w_2

,

w_1	w_2	w_3
F_2	F_2	F_1
F_1	F_1	F_2
$\varnothing$	$\varnothing$	$\varnothing$

.

Check first that all matchings without unemployment are dominated. Then check the cases where one worker at least would not be employed.

Exercise 3.12. Computing the End-points of the Core in Bilateral Assignment

In the following bilateral assignment economies with three agents of each type, find the simplest form of the system of equalities and inequalities defining the core, and compute its A-optimal and its B-optimal allocations:

$$
\text{(a)} \quad \begin{bmatrix} 5 & 6 & 4 \\ 4 & 1 & 0 \\ 6 & 4 & 1 \end{bmatrix}
\qquad
\text{(b)} \quad \begin{bmatrix} 7 & 5 & 1 \\ 9 & 6 & 4 \\ 3 & 2 & 8 \end{bmatrix}
\qquad
\text{(c)} \quad \begin{bmatrix} 9 & 10 & 0 \\ 0 & 10 & 9 \\ 10 & 0 & 11 \end{bmatrix}.
$$

Exercise 3.13. An example of bilateral assignment

We have $A = B = \{1, 2, \ldots, n\}$, and the surplus function v is such that

$$\text{for all } i = 1, \ldots, n: \qquad v(i, i) = \max_j v(i, j) = \max_{i'} v(i', j).$$

Show that the end-points of the core leave no surplus whatsoever to one type of agents.

Exercise 3.14. Bilateral Assignment and Joint Production (Becker [1981])

Consider a production function $f(x, y)$ with two inputs X and Y, say labor and land. We have n landholders (agents b_j owns y_j units of land) and n workers (agent a_i owns x_i units of labor). Assume each plot of land can be worked by only one worker; thus the bilateral surplus:

$$v(i, j) = f(x_i, y_j) \quad \text{all } i, j = 1, \ldots, n.$$

Assume $x_1 \leq x_2 \leq \cdots \leq x_n$ and $y_1 \leq y_2 \leq \cdots \leq y_n$. The function f is nondecreasing in x and y and the two inputs are substitutes, in the following sense:

$$\text{for all } x, x', y, y': \qquad \{x \leq x', y \leq y'\} \Rightarrow f(x, y) + f(x', y')$$

$$\geq f(x, y') + f(x', y).$$

(a) Show that the assignment of landholder a_i to worker b_i is optimal.
(b) Show that in the landholder-optimal assignment, agent a_i receives

α_i and agent b_i receives β_i, given by the formulas

$$\alpha_i = f(x_i, y_i) - \beta_i,$$

$$\beta_i = \sum_{k=1}^{i-1} f(x_k, y_{k+1}) - f(x_y, y_k), \qquad \beta_1 = 0.$$

Exercise 3.15. *Generalization of the Gloves Market (Section 2.3)*

We have n identical type-A agents and $(m_1 + m_2)$ type-B agents. There are m_1 type-B agents of high quality and m_2 of low quality:

$$v(i, j) = 3 \quad \text{if } a_i \in A, \quad b_i \text{ is of high quality in } B,$$

$$v(i, j) = 2 \quad \text{if } a_i \in A, \quad b_j \text{ is of low quality in } B.$$

Compute the core of this game, discussing according to the following cases: $m_1 > n$; $m_1 = n$; $m_1 < n < m_1 + m_2$; $m_1 < n = m_1 + m_2$; $m_1 + m_2 < n$. (*Hint*: Use the equal treatment property.)

Exercise 3.16. *Computations for the Assignment Economy of Example (16)*

(a) Show that a competitive price takes the form $p_W = \lambda$, $p_B = \lambda + 2$, where λ is such that $0 \le \lambda \le 2$. The corresponding vector of surplus is

$$\text{Sellers:} \quad \sigma_1 = 2p, \quad \sigma_2 = p + 2:$$

$$\text{Buyers:} \quad \beta_1 = 4 - p, \quad \beta_2 = \beta_3 = 2 - p.$$

(b) To compute the core allocations of this economy, call q_i the payment of buyers i, $i = 1, 2, 3$, and show that seller 1 must receive $q_1 + q_3$, whereas seller 2 must receive q_2. Then check that such an allocation is in the core if and only if

$$q_2 = q_3 + 2, \quad 0 \le q_3 \le 2, \quad q_3 \le q_1 \le 4.$$

Deduce that, in the core, the surplus of buyer 1 may be larger or smaller than the surplus of buyers 2 and 3 (whereas it is always larger in the competitive set).

Exercise 3.17. *A Variant of Lemma 3.6*

Consider an economy with $K + 1$ divisible goods, with a distinguished good called money. All agents have quasi-linear utilities of the form

$u_i(y_i) + t_i$, where $y_i \in \mathbb{R}_+^K$ is agent i's nonmonetary consumption, and t_i is agent i's net cash transfer. All agents have unbounded reserves of cash. Agent i's initial endowment in the nonmonetary goods is denoted ω_i. We assume that ω_i is nonnegative, but the sign of t_i is arbitrary. A price vector is written as an element p of $\mathbb{R}_+^K$ (p_k is nonnegative for all k).

(a) Write the system of inequalities defining a competitive price and a competitive allocation.

(b) Show that every competitive allocation is a core allocation as well. Notice that we did *not* assume that the utility functions u_i are monotonic (but we did assume that all preferences are strictly monotonic in money).

Exercise 3.18. A Case where Replication does not Lead to the Edgeworth Proposition

This is an assignment economy (with capitalists and heterogeneous goods) where the core and the competitive set remain significantly different after any number of replications. We have five agents and three types of cars: red, blue, and white:

Agent	1	2	3	4	5
utility					
Red	0	0	10	10	0
Blue	10	0	0	0	10
White	0	10	0	0	0
owns	2 Red	2 Blue	1 White	∅	∅

(a) Compute the set of competitive prices and of competitive allocations.

(b) Show that the following allocation is in the core but is not a competitive allocation: everyone gets to consume his favorite type of car and receives the following net cash transfer:

$$x_1 = 5, \quad x_2 = 5, \quad x_3 = 0, \quad x_4 = -3, \quad x_5 = -7.$$

(c) Show that in the k-replica of our economy, the replica of this allocation is still in the core, and still not a competitive one.

(d)* Show that in the k-replica economy, the core has the equal-treatment property (formally defined at the beginning of Appendix 3.3).

Exercise 3.19. A Simple Two-Commodity Economy (Maschler [1976])

We have five traders, two divisible goods denoted X and Y, and money. Agents 1 and 2 own one unit of good X each, and agents 3, 4, 5 own q units of good Y each. Here q is positive but arbitrary. All agents have the same quasi-linear utility $u_i(x_i, y_i, t_i) = \min\{x_i, y_i\} + t_i$.

(a) Fix a coalition of agents S and write

$$S_X = S \cap \{1,2\}, \qquad S_Y = S \cap \{3,4,5\}.$$

Show that the net surplus that coalition S can generate on its own is

$$v(S) = \min\{|S_X|, q \cdot |S_Y|\}$$

(where, as usual, $|T|$ denotes the cardinality of T).

(b) Show that for q small enough and for q large enough, the core of this economy contains a single allocation. In particular, it treats equal agents equally.

(c) For $q = 0.6$, show that the core contains allocations treating equal agents unequally. Specifically, compute (i) all core allocations treating equals equally, (ii) a core allocation where the difference between player 1's and and 2's shares is maximal.

(d) For what values of q does the core contain allocations treating equals unequally?

(e) Compute the competitive price vectors and competitive allocation for all values of q.

Exercise 3.20. Manipulation by Destroying Endowments

(a) The Throw Away Paradox: Consider the two-person, two-good economy

$$\omega_1 = (0, 2), \qquad \omega_2 = (1, 1),$$

$$u_1(x_1, y_1) = \min\{x_1, y_1\}, \qquad u_2(x_2, y_2) = \tfrac{1}{10} \log(x_1) + \log(y_2).$$

Compute the (unique) competitive equilibrium. Next, drop agent 2's endowment to $\omega'_2 = (\tfrac{1}{2}, 1)$ and compute the new competitive equilibrium. What happens to agent 2's utility?

(b)* Fix the set of goods, the set of agents, and a profile of (monotonic, continuous) preferences. Pick a cardinal utility u_i, representing R_i for all i. Then consider the following mechanism. Each agent reports his/her endowment ω_i. Then the mechanism implements a (often unique) Pareto optimal allocation $z_1, \ldots, z_n$ such that

$$u_i(z_i) - u_i(\omega_i) = u_j(z_j) - u_j(\omega_j) \quad \text{for all } i, j = 1, \ldots, n.$$

Show that this mechanism is not manipulable by throwing away part of one's endowment. Show by an example that it is manipulable by withholding part of one's endowment.

Exercise 3.21.   Two-Fixed-Prices Mechanism (Barbera and Jackson [1995])

We have two agents and two goods X, Y. The initial endowments are $\omega_1 = (10, 10)$, $\omega_2 = (5, 5)$. The roles of the two agents differ. Agent 1 chooses to buy some good X or to sell some good X. These trades have different prices: he may offer to buy some good X at price 2 (units of good Y per unit of good X) or he may offer to sell some at price 1. Agent 2 learns from agent 1's offer what trading price is selected: if agent 1 offers to buy X, she must demand to sell a certain amount of this good and the market clears by uniform rationing (in particular, if she demands zero, there is no trade); if agent 1 offers to sell X, she must demand to buy a certain amount of X and the market clears by uniform rationing.

Consider the direct mechanism where agents report (convex, continuous) preferences and the corresponding offers are deduced in the usual utility maximization fashion (drawing a figure will prove useful). Show that the direct mechanism is strategy-proof and deduce a dominant strategy for each agent in the original mechanism.

Fair Division: The No Envy Test

4.1. INTRODUCTION

Chapters 4 and 5 deal with the allocation of private (and public) goods in the common ownership regime. Fair division is the embodiment of a normative problem of distributive justice formulated in the language of microeconomics. Our primary concern is the definition of reasonable tests of equity, selecting as often as possible a small subset of efficient and "just" outcomes. This is cooperation in the justice mode. Yet, in accord with the program stated in Chapter 1, we address fair division in the decentralized mode as well; see the discussion of divide and choose at the end of this chapter.

The core and the competitive equilibrium play a key role in the search for tests of fairness. The no envy test is inspired by the competitive idea, and closely related to the fair-division method "competitive equilibrium with equal incomes" (this chapter). On the other hand, the core leads to the stand alone test (Chapter 5). It turns out that the no envy approach has more bite than the stand alone test in the classical problem of dividing fairly a set of unproduced goods—the cake division problem, whereas the stand alone approach overshadows the no envy test in the cooperative production problems, such as the provision of public goods or the regulation of a natural monopoly. Accordingly, Chapter 4 studies mostly manna division problems, whereas cooperative production problems are the subject of Chapter 5.

The simplest fair-division problem involves a bundle of unproduced commodities (manna from heaven, whether a gift from Mother Nature, the spoils of warfare, or a helicopter drop from the Red Cross) to be consumed privately by a given set of agents. Normally[1] we suppose that all commodities are "goods," the consumption of which no one is ever satiated (this is the "free disposal" assumption: I am not less happy from consuming 10 kilos of steak per day than from consuming 150 grams because I can throw away whatever I do not want to eat; my preferences would be quite different if I had to eat all the steak I receive!).

---

[1] But not exclusively; see Section 4.5.

The commodities are the common property of a given society, or group of agents. How should we distribute the goods among them given the differences in their tastes for those goods? The problem goes far beyond the familiar story of dividing a (heterogeneous) pizza among children (some like the parts with more cheese, some dislike tomatoes, and so on). Indeed, many privately consumed commodities come to us as common property: fish in the extraterritorial waters; mineral riches (from the common property underground). Inventions and discoveries continuously bring new such commodities: radio frequencies, thermonuclear energy, the resources of the moon, and so on. Inheritance and the breakup of a federation (such as the Czech and Slovak republics) are other important examples where either the resources or the agents to share them are created by an exogenous "shock." At the firm level, one of the most important tasks of a manager is to allocate a certain amount of resources (fixed once for a whole fiscal year) among its constituency (this involves a whole spectrum of private goods, from office furniture to travel budget). At a more general social level, the device by which common property resources (which in some countries include most of the health and education services, and in all countries include a substantial fraction of these services) are distributed is a considerable part of the government's task, and the subject of intense political debate.

In some of the above examples, the resources to be divided are best interpreted as a production function: fish in the ocean (or mineral riches) is not manna falling from heaven, but rather the opportunity of catching fish if we spend the effort of fishing (or digging). The fish (oil) materializes as a consumable good only at the cost determined by the technological constraints of the fishing (digging) technology. Common property technologies can exhibit both decreasing marginal costs and increasing marginal costs. Conceptually, the common property of a technology is a more complicated problem than that of a heap of unproduced commodities. Thus we shall discuss it only briefly in this chapter (merely to show that the no envy concept is problematic in that case; see the brief discussion in Section 4.2), waiting until the next two chapters for a comprehensive discussion.

The arguments of justice are often confusing because of the multiple facets of redistribution. Three complicating factors follow. First, an individual agent is also a citizen with political opinions; whence in the case of distribution, an agent cares about the share he receives but also about the overall distribution of the goods (e.g., he may dislike excessive inequality even if the latter is justified to generate more surplus). This dimension of agent's preferences (caring about what other agents consume) adds the full complexity of a social choice problem: what is a fair way to aggregate the profile of opinions on the distributive policy itself into a collective opinion? (We do not discuss the aggregation of preferences in this volume. For a survey see Moulin [1988], Chapter 10.) Second, the profile of individual preferences (tastes about the different consumption vectors) is not the only parameter relevant to distribution. Another important factor is objective needs (for the elderly, the poor, the sick, and so on) as they differ from tastes, because agents cannot be held responsible for their needs.[2] Finally, the distribution agency may have specific bias about the different commodities, wishing, for example, to encourage the consumption of certain goods to the detriment of others (e.g., health food versus tobacco).

In our definition of a fair-division problem, we explicitly rule out all of the above difficulties. We assume, first of all, that each agent cares only about her own share of the pie (the distribution of commodities among all other agents is a matter of indifference to her, provided this does not affect her own allocation; this is the *no consumption externality* assumption). Second, we retain a strictly ordinal interpretation of preferences (never letting a cardinal measure of individual welfare have any influence on the choice process). Third, we take a neutral view of the different commodities and abstain from any ethical judgment while comparing different preferences: there are no "socially desirable" or "socially harmful" preferences, only agents with different preferences. Finally, we will interpret preferences as tastes, for which an agent is wholly responsible, and not as reflecting needs (for which they would not be responsible). Therefore, differences in preferences do not reflect differences in health, physical fitness, or mental ability (or, in general, in the ability of individuals to derive welfare from primary goods; see Sen [1985]). On the contrary, preferences are reversible and voluntarily selected by the agents (much as we "decide" to like opera or to become religious, and are free to change our opinions at any time).[3] The stylized model is thus characterized by ordinal and "voluntary" preferences with exogenous, involuntary property rights. Throughout this chapter and the next one, the property rights are jointly and equally held by all agents.

---

[2] So if I develop a taste for champagne, I do not have a good reason for demanding a supply of that good comparable to the average supply of beer of my more frugal neighbors; but if my health depends on a certain vitamin to be found exclusively in good champagne, I may deserve that a great deal more resources be spent to satisfy my needs. On this see, e.g., the discussion in Dworkin [1981].

[3] These assumptions surely make sense when the goods to be distributed can only have a small impact on the participants' overall long-run welfare (as in the familiar cake division stories); they are less tenable when we speak about the allocation of primary goods such as education, legal rights, and so on.

Thus we stay squarely within the "resourcist" tradition to fair division, as opposed to the welfarist tradition. The latter tradition attaches an objective meaning to the cardinal measure of welfare; it is a measure of the extent to which individual needs are satisfied. It does not hold agents responsible for their preferences. Consequently, welfarism sees the profile of individual (cardinal) welfares as the only ethically relevant object, and seeks a fair distribution (for instance, egalitarian) of these. See Fleurbaey [1993] for a clear account of the debate between the resourcists (and of the important contributions of Rawls [1971], Dworkin [1981], Cohen [1986] and Roemer [1986] to this debate).

In a nutshell, Chapters 4 and 5 push the resourcist approach to its limits, by demonstrating that several different solutions have some claim to represent the legitimate interpretation of resourcism.

In fair division, the two most important tests of equity, namely, "fair share guaranteed" and the "no envy" test, are illustrated by the time-honored method of divide and choose."[4,5] Divide and choose is just because (i) the Divider, by splitting the resources in two shares of equal value to him, is guaranteed to receive his fair share of the resources, and (ii) the Chooser is guaranteed that her share is at least as valuable to her as the other share: she cannot complain that she got the "bad" share. As a matter of strategic gaming, the Divider may not divide the goods into two piles of equal value to him (but rather of nearly equal value to the Chooser; see Section 4.8), and if his guess about the Chooser's choice is wrong, he (the Divider) may end up with less than his fair share. However, this does not matter because the Divider could have cut two shares of equal value to him (and you cannot be unjust to yourself; an argument cynically used by the lion in Aesop's fable of the Lion, the Fox and the Ass).

"Fair share guaranteed" is a very general equity test that will play a central role in this and the next chapter. If the common property goods

---

[4] "Here, for example, is how Kallistratos, the speaker in the case against Olympiodoros, describes their agreement to divide up the inheritance of Komon (to which they were admittedly not entitled): When we had sworn mutual oaths and the agreement was deposited with Androkleides, I divided the property into two shares, men of the jury. One share was made up of the house in which Komon himself lived and the slaves, sailmakers; the other share was another house and the slaves, color grinders . . . I gave the defendant Olympiodoros the choice to take which of the shares he wished and he chose the color grinders and the little house. I took the sailmakers and the other house." Finley [1951], pp. 66, 67.

[5] During the Law of the Sea conference, invoking the principle that "the seabed is part of the common heritage of mankind, a global commons in which all countries have a stake," the conference proposed the creation of an "enterprise" to mine the seabed on behalf of the LDC's. Every time that a mining company would apply for permission to mine at a particular location, it should develop two parallel sites from which the enterprise could choose one (Sebenius [1984]).

are perfectly divisible, this test simply say that everyone should receive (utilitywise) at least as much as $(1/n)$th of the total resources (where n is the number of participants). If some goods are indivisible, fair share guaranteed is also fairly easy to define; see Sections 4.3 and 4.5. The second test is "no envy": every participant should view his share as at least as valuable as anybody else's share. Both tests are simple and appealing, and both are satisfied by the solution of fair division that is most popular among economists, namely, the competitive equilibrium with equal incomes (in short, CEEI).

The CEEI translates common ownership into equal private ownership and solves the resulting exchange economy by competitive equilibrium. If, for instance, goods are perfectly divisible, the CEEI takes equal split of the cake as the initial endowment (thus distributing privately the property rights) and operates the competitive equilibrium to exhaust the cooperative opportunities above the equal-split allocation. If goods are indivisible (or if the resources are a production function), the agents receive equal fractional rights to these resources, in the traditional Arrow–Debreu fashion (e.g., agents are shareholders of the profit-maximizing firm allocating the goods). A CEEI outcome is thus efficient, just a priori in the sense of fair share guaranteed, just a posteriori in the sense of no envy (everybody chooses from the same budget set; therefore, nobody prefers someone else's share to his own), as well as stable in the sense of the core, if agents engage in direct trade of their fair shares (that is, when goods are divisible). See Lemma 4.1 for a formal statement.

Yet, despite this harvest of arguments in favor of the CEEI solution, there are also some serious ethical objections against it. The most forceful objection comes from the stand alone test (developed in Chapter 5). To put this chapter in the perspective of the book, we present in Section 4.2 the stand alone objection against no envy, in the especially limpid case where utility is transferable (via money), the resources consist of homogeneous indivisible goods, and each agent wants no more than one good. It turns out that if the common property resource is a technology with increasing marginal costs, no envy (and the CEEI solution) retain all their appeal; if, however, the marginal costs decrease, the stand alone objection is devastating and no envy must be abandoned. This important cleavage is thoroughly investigated in Chapter 5.

In Sections 4.3 to 4.5, we take a close look at the logical connections of the CEEI solution with the two tests, fair share guaranteed and no envy. Section 4.3 studies the common property version of the assignment economies (Section 3.5), where the no envy test selects no more and no less than the set of CEEI allocations. This remarkable coincidence disappears in Sections 4.4 and 4.5, where the fair-division problems involve divisible goods or a large number of (small) indivisible goods. In these models, the CEEI solution typically selects a single allocation within a large set of efficient and envy-free allocations. Its versatile applications are illustrated by the examples of Section 4.5

Section 4.6 presents the main challenge to the CEEI solution of fair division, namely, the egalitarian-equivalent (EE) solution. It distributes equally the surplus above the fair share utility, where we measure the surplus along a numeraire proportional to the resources to be divided. The EE solution has some normative advantages over the CEEI solution, and over any envy-free and efficient solution as well. In particular, it is population-monotonic: when the set of agents sharing a fixed set of resources grows, population monotonicity requires that none of the previous participants be made better off. In Section 4.7, an even more radical critique of the two tests, no envy and fair share guaranteed, is articulated around a similar resource monotonicity axiom (when the resources grow and the set of participants remains fixed, nobody should be made worse off).

Section 4.8 analyzes in some detail the strategic equilibrium of the divide and choose method and of some related fair-division games among three or more participants. Although divide and choose does not implement the CEEI solution (it does not even guarantee his fair share to the Chooser), its strategic simplicity makes it, nevertheless, a useful mechanism. No particularly simple mechanism emerges to implement the CEEI solution with some generality, whereas a simple auction mechanism, due to Crawford, implements the egalitarian-equivalent solution (Lemma 4.7).

In Appendix 4.2, we give the intuition of a convergence result about the no envy test. When the economy contains a large number of agents and when individual preferences vary in a continuous fashion over the set of all agents, then the set of efficient allocations coincides with the set of CEEI allocations. Like the Edgeworth proposition (and under similar but stronger assumptions), this result establishes the logical necessity of the competitive solution when we adopt the seemingly much weaker no envy test (or the core stability in the case of the Edgeworth proposition).

4.2. No Envy Versus Stand Along: Two Elementary Examples

In the estate division problem (Example 1.1), two agents share a single indivisible object and utility is transferable. Let u_i be the (nonnegative) value of the object to agent i (in Example 1.1, we have $u_1 = 10,000$ and $u_2 = 4000$). To fix ideas, we suppose $u_1 > u_2$, so that efficiency commands allocating the object to agent 1. The problem is to determine a fair compensation t from agent 1 to agent 2.

Start with *equal treatment of equals*. If both agents value the object equally ($u_1 = u_2$), we must split the benefit equally among the two of them: one of them gets the object and pays $\frac{1}{2}u_i$ to the other; hence the fair share guaranteed test. Says agent *i*: if agent 2 had the same utility as I, I would net my fair share of surplus $u_i/2$; in fact, agent *j*'s differs from mine and it is feasible to give more than our fair share of surplus to both of us (as $u_1 > u_1/2 + u_2/2$). Therefore, I insist upon receiving at least my fair share $u_i/2$; otherwise, I would be suffering from the difference in our preferences, whereas he would be benefiting from this difference (remember that preferences are ethically neutral). Thus the test guarantees a net benefit $u_i/2$ (or more) to agent *i*. Applying this test to both agents, we find the following bounds for *t*:

$$\frac{u_2}{2} \leq t \leq \frac{u_1}{2}. \tag{1}$$

The no envy test, on the other hand, rests on intercomparisons of the allocations, where an allocation comprises both a cash transfer and (possibly) the object. Agent 2 does not envy 1 if $t \geq u_2 - t$ (he would rather receive \$t than get the house and pay \$t), and agent 1 does not envy 2 if $u_1 - t \geq t$. Thus we find the same bounds (1) (in the estate division problem, no envy is logically equivalent to fair share guaranteed *plus* efficiency). Figure 4.1 depicts the interval *AB* of utility distributions corresponding to the envy-free allocations. The three special points *A*, *B*, and *C* can all be justified in terms of equal split of the surplus/deficit above a certain natural reference point: to select *A*, the reference point is 0 (the status quo ante when the object was not yet awarded to the agents), to select *B*, the reference point is *E* (where *each* agent consumes the object; thus, *B* is the "equal-loss" solution where each agent loses the same amount of surplus from the presence of the other), and to select *C*, it is *D* (where each agent receives her fair share; alternatively, *D* corresponds to the disagreement where the house is awarded at random, with equal probability for each agent).

The divide and choose method has a very simple interpretation in the estate division example. Agent *i* chooses a price *t* at which he offers to "buy" the house from *j*; agent *j* *either* accepts *or* buys the house from *i* at the same price *t*. This method is used to dissolve a partnership.[6] In this game, the equilibrium outcome is always envy-free (its utility

---

[6] Two partners have equal shares in a joint venture. At any time, any one of the partners can offer to buy back the other's share at a price of his choice. The partner either accepts to sell his share at that price or chooses to buy ownership of the venture at the same price. The timing of the offers is endogenous, so the method has the additional advantage of introducing no asymmetry between a Divider and a Chooser. I owe this example to J. P. Ponssard.

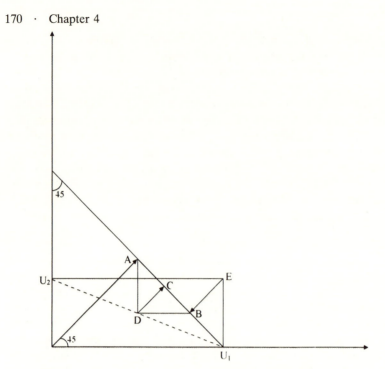

Fig. 4.1.   No envy in the estate division problem.

distribution is in the interval *AB*), where the Divider is agent 1 or agent 2. Divider who knows the Chooser's valuation can exploit this to force his preferred envy-free outcome (this general fact is analyzed in Section 4.8).

Consider next the estate division problem among *n* participants, $n \geq 3$. Agent *i*'s valuation is u_i, $u_i \geq 0$, and we assume $u_1 > u_2 \geq \cdots \geq u_n$. The object is efficiently allocated to agent 1, who in turn compensates the other agents in cash. By no envy, agents 2 and *i* (for any $i \geq 3$) must receive the same compensation *t* (least one of them prefer the other's check); therefore, agent 1 pays $(n - 1)t$. Writing that agent 2 does not envy 1 and vice versa yields the inequalities.

$$\frac{u_2}{n} \leq t \leq \frac{u_1}{n}. \tag{2}$$

Observe first that these inequalities give her fair share to each participant: agent *i*'s net benefit is at least u_i/n (for instance, agent 1 nets at least $u_1 - (n - 1) \cdot u_1/n$). Moreover, the envy-free allocations can all be interpreted as competitive allocations with equal income *t*. Indeed, pick t in the interval (2) and let the price of the house be $p = nt$. At this price, agent 1 is ready to buy and agent i, $i \geq 2$, is ready not to buy. If we split equally the proceeds of the sale, everyone receives $t, so the net allocation of agent 1 is {the house} + {pay $(n - 1)t$}.[7]

We turn now to the stand alone test. In the fair division of unproduced commodities, this is the requirement that, when we divide a vector of goods, an agent must not end up better off than if she were consuming all the goods herself. This sounds mild enough: in the estate division between Ann and Bob where $u_1 = 10,000$ and $u_2 = 4000$ (Example 1.1), the stand alone test says that Bob should not be paid more than $4000, or the value of enjoying the painting without having to "share" it with Ann. Similarly, the test says that Ann, who gets the painting by efficiency, should not receive a cash payment from Bob on top of he painting! Yet, even in this simple example, the stand alone test rules out certain efficient outcomes deemed fair by the no envy test, namely, those where Ann pays between $4001 and $5000 to Bob. Our next example shows that the stand alone and no envy tests can be outright incompatible.

Basketball Fans

Three roommates, Ann, Bob, and Charles, find in their common mailbox a ticket for an exciting basketball game (the legitimate recipient of the ticket cannot be reached). Bob and Charles are basketball fans and Bob would gladly pay $250 for it, whereas Charles could only afford to pay $200. Ann, on the other hand, does not care for sports and would immediately resell the ticket if she got it alone; the resale value (taking transaction costs into account) is $50.

Efficiency commands giving to the ticket to Bob, who in all fairness should compensate his roommates. The stand alone test places a $50 cap on the payment to Ann; for instance, Bob could give $20 to Ann and $100 to Charles. However, Ann would then envy Charles's allocation: the no envy test implies that (i) Ann and Charles receive the same amount of cash, and (ii) both receive at least $66.67 (see above). Thus in all envy-free allocations, Ann *must* receive more than what the item is worth to her.

A division of the benefits proportional to willingness to pay would

---

[7] In Section 4.3, we show that the coincidence of the no envy set and of the CEEI set goes far beyond estate division model, to the general class of fair-assignment problems.

pass the stand alone test (namely, Bob pays $25 to Ann and $100 to Charles). In this example, it is precisely the egalitarian-equivalent solution discussed in Appendix 4.3.

The interpretation of common ownership proposed by the no envy test (which, in this case, is equivalent to insisting on the CEEI solution) is equal private ownership. Thus Ann, who owns an equal share of the ticket, can legitimately extract a rent from the fact that Bob and Charles value the ticket so much (just as a seller is entitled to a share of the surplus, namely, of the difference being the buyer's valuation and his own). The stand alone test, on the other hand, says that common ownership implies a duty to share the benefit of the resources with the co-owners, hence cannot yield more benefit than the exclusive consumption of the resources.[8]

Organs for transplantation come to us as a common property resource. A donor dies and his organ is rushed to the hospital where patients in need of the organ are waiting; those patients are the common beneficiaries of the organ. Say those patients bid for the organ and the highest bidder gets it, his bid being donated to the hospital. Surely this will bring an envy-free outcome, for all the losers will get the same benefit from the manna, and they will be unwilling to match the highest bid.[9] On the other hand stand alone is violated because some of the beneficiaries of the donation are not even in need of this organ (even if the money is earmarked for the department that exclusively transplants this kind of organ, some of the patients there are not physiologically matched to this particular organ). Now consider the actual method of giving the organ to this patient who scores best on a scale combining needs, waiting time, and chances of success of the transplantation, without compensating the losers in any way (see Young [1994]). This passes the stand alone test (because those who for sure would reject the organ get no benefit), and the losers will be envious. Note that the stand alone test does not rule out all compensations to

---

[8] The key word is "consumption" as opposed to "ownership": with equal private ownership of the goods, an agent could get more benefit than from the exclusive consumption of all the goods, although she could not get more benefit than from the exclusive ownership of all the goods.

[9] In practice, they may simply be unable to match the bid of the wealthiest patient; yet, given our general assumption of responsibility for one's preferences (Section 4.1), it is irrelevant whether a patient is too poor or genuinely willing to save money. That this interpretation of preferences is untenable in the organ allocation problem is the subject of the next note.

the losers, monetary or otherwise; but they should be directed only to those patients who could have used the organ themselves.[10]

Natural resources such as the best spot on the beach are normally offered free of charge to those who come earliest; no compensation is paid to other beach users, who have the same (common) property rights to the site, in accordance with the stand alone test and in contradiction of the no envy test. On the other hand, all citizens benefit from natural resource such as fish (via the taxes paid by the fishermen), even those who do not eat fish, so this kind of common property goods falls in the realm of no envy and outside that of stand alone.[11]

Our next example shows that, when the agents own in common a technology, the no envy test combined with the efficiency requirement may lead to an unacceptable solution. Say that the technology produces q (indivisible) cars at cost $c(q)$. The demand is binary: agent i wants to consume at most one car and is willing to pay at most u_i. We still assume $u_1 \geq u_2 \geq \cdots \geq u_n$.[12] Suppose that efficiency tells us to produce q^* cars: the surplus $\sum_{i=1}^{q} u_i - c(q)$ is maximal at q^* (note that we no not make any assumption on the way marginal costs vary). In an envy-free and efficient allocation, agents $1, 2, \ldots, q^*$ consume a car and pay p, whereas agents $q^* + 1, \ldots, n$ receive t (all active agents receive the same car, so by no envy they must pay the same price). To compute p and t, assume that u_q^* and u_{q+1}^* are so close to $\partial c(q^*)$ that we can equalize all three numbers.[13] The two critical agents in applying the no envy test are q^* and $q^* + 1$:

$$q^* \text{ does not envy } q^* + 1 \Leftrightarrow t \leq u_{q^*} - p,$$

$$q^* + 1 \text{ does not envy } q^* \Leftrightarrow u_{q^*+1} - p \leq t.$$

---

[10] In the real world, most of us view the need for an organ as a feature of our preferences for which we are not responsible (it is a need, not a taste; see Section 4.1). This is the compelling ethical argument against the auctioning of organs, and in turn against the no envy test in this context. Thus organ allocation does not fall within the realm of the problems circumscribed in Section 4.1; it is used here merely as a thought-provoking device.

[11] Of course, mixed examples can also be found, as when a water-rich community charges for its water and rations some users (based on their needs, or past consumption, or whatever): rationing follows the logic of equal consumption rights; charging for the water makes every citizen the private owner of a share of the surplus generated by the common property resources (i.e., the water). As a result, neither the no envy nor the stand alone test is met!

[12] Estate division is the particular case where the cost function is $c(1) = 0$, $c(2) = +\infty$.

[13] The efficiency of production q^* tells us that the demand and marginal cost curves must cross at q^*: $\partial c(q^*) \leq u_{q^*}$, $u_{q^*+1} \leq \partial c(q^* + 1)$. We are taking the first-order approximation $\partial c(q^*) \cong u_{q^*} \cong u_{q^*+1}$.

By our first-order approximation, this means $t + p = \partial c(q^*)$. Invoking the budget-balance condition

$$q^* p - (n - q^*)t = c(q^*),$$

this completely determines p and t:

$$t = \frac{q^*}{n} \cdot (\partial c(q^*) - ac(q^*)),$$

(3)

$$p = \partial c(q^*) - t.$$

Therefore, *everyone* (active and inactive agents) receives the lump-sum transfer t and, in addition, the active agents pay the marginal cost $\partial c(q^*)$, but note that the *sign* of the transfer t depends on which of marginal or average cost is higher. If marginal costs exceed average costs, as is the case if *marginal costs are increasing*, the transfer t is positive: inactive agents are subsidized, and active agents get a lump-sum rebate on the marginal cost. On the other hand, if marginal costs are below average costs, as is the case if *marginal costs are decreasing*, then t is negative: inactive agents must *pay* in order not to be envied by the active ones. That is to say, if inactive agents did not pay, the active agents would pay more than the marginal cost $\partial c(q^*)$ and agent q^* would suffer a net loss; he would then envy any inactive agent.

If the common property technology has decreasing marginal costs, we cannot accept the combination of no envy and efficiency as a guideline for fair division: the outcome they recommend violates the principle of voluntary participation. The common property of valuable resources cannot translate into a negative net utility. Stated differently, this means that I should always have the option to leave and forfeit my rights before the division takes place.[14] Alternative interpretations of the common property of decreasing marginal cost technologies are the key theme of Chapter 5. In our simple example with indivisible goods, a reasonable solution suggests asking no money from (nor giving any

---

[14] Of course, regulated monopolies who run at a loss and then spread the loss uniformly to all taxpayers do exactly what no envy recommends, but this policy is politically tenable only if the goods offered by the state monopoly are in some sense useful to the society at large, so that even inactive agents benefit from them (think of railway services). By contrast, it does not make ethical sense to subsidize in this fashion the consumers of opera or of soccer games unless the overall social benefit of the consumption of these goods is arguably larger than the total willingness to pay of its consumers (perhaps this is how we should interpret the fact that many entertainment goods are routinely subsidized in this fashion).

money to) the inactive agents, whereas active agents pay a uniform price or the full value of the object to them, whichever is smaller.[15]

If the common property technology has increasing marginal costs, the efficient allocation (2) makes good sense: active agents pay the marginal cost and the corresponding surplus is redistributed equally to all participants. This is, of course, the competitive equilibrium with equal income further discussed in Section 5.2.

4.3. THE FAIR-ASSIGNMENT PROBLEM: NO ENVY EQUALS CEEI

A fair-assignment problem is identical to an assignment economy, except for the absence of a private ownership structure. So we have n agents, q heterogeneous indivisible houses, and money. Utilities are quasi-linear and each agent wants to consume at most one good. Formally, a problem is a triple (N, H, u) (same notation as in Section 3.5).

A standard example is job assignment. A group of n firms share a large contract with q jobs (where $q \leq n$). Each firm can perform one job at most (perhaps each job is in a different location). Each job will bring a certain revenue (known in advance); our firms are different inasmuch as the opportunity cost of performing each job varies among firms (and among jobs). Note that a random assignment (e.g., with equal probabilities of any task to any agent) is certain to miss the efficiency goal (although it is unmistakably just). On the other hand, if some jobs are clearly worse than others (that is, everyone agrees that they are less profitable), surely the firm who gets that job deserves a compensation: the question is, how much?[16]

The first and foremost justice argument is *anonymity* (or justice a priori; see Section 1.7). This says that permuting the preferences of any two agents only results in permuting their welfare levels and leaves other agents' welfare level unchanged. In particular, anonymity implies equal treatment of equals. Now to some justice arguments rooted in the structure of the fair-division model.

---

[15] Find p as the solution of equation $\sum_{i=1}^{q^*} \min\{u_i, p\} = c(q^*)$. Then an active agent i pays $\min\{u_i, p\}$. Thus the only envy is directed at active agents who pay less than the full price, but those agents are making zero profit anyway! See Section 5.5 for a general discussion of this solution.

[16] In an equivalent version of this problem, the firms receive a global payment for the contract and must share it along with the jobs: by dividing in arbitrary fashion the payments among the various jobs, we are back to the framework described above, and the set of just outcomes is unaffected; see Exercise 4.3, question (d) for an example.

The idea of fair share guaranteed is the oldest normative idea of the modern literature on fair division.[17] Consider first the assignment problem with a *single* house. If agent i's utility for the house is u_i, his fair share of the resources is simply u_i/n (where n is the number of agents). The fair share guaranteed property requires that every agent's net gain from the division of the house be at least u_i/n.

Consider next a fair-assignment problem with q house and n agents. We assume $|H| = q \leq n$ throughout this section; most of the results can be adapted to the case $q > n$ as well. See Remark 4.2 for details. One way to interpret agent i's "fair share" of the resources would be as the nth part of the value to agent i of consuming all the resources, or

$$\frac{1}{n} \max_{h \in H} \{u_i(h)\}.$$

This reasonable lower bound, however, is not the best we can do. It is always feasible to guarantee to agent i the net gain α_i, where

$$\alpha_i = \frac{1}{n} \left(\sum_{h \in H} u_i(h) \right). \tag{4}$$

Indeed, denote as $v(N)$ the gross surplus in this assignment economy: $v(N)$ is the surplus resulting from an optimal assignment of the q houses. Then for any one-to-one mapping τ from H into N, we have

$$\sum_{h \in H} u_{\tau(h)}(h) \leq v(N).$$

Summing up over all such mappings τ and taking averages, we get

$$\sum_{i \in N} \alpha_i \leq v(N),$$

establishing that the requirement "agent i should be guaranteed α_i (at least)" is feasible. We call α_i the *fair share utility* of agent i.

Remark 4.1

The lower bound α_i on agent i's final utility is precisely his fair utility in the hypothetical problem where all agents have the same utility as his

---

[17] From Steinhaus [1948] to Dubins and Spanier [1961], this property was taken as the definition of fairness.

(this follows at once from equal treatment of equals). This hypothetical unanimous economy used to compute an agent's fair share guaranteed will be used repeatedly (see Exercise 4.17 and Section 5.2). Moreover, it is easy to show that the utility level α_i is the highest lower bounded on agent i's final utility that (i) does not depend on other agent's utilities, and (ii) is feasible at all utility profiles.[18]

We illustrate the bite of the lower bound (4) in a numerical example with three agents and three houses. The utility profile is as follows:

$$
\begin{array}{c|c|c|c}
 & u_1 & u_2 & u_3 \\
\hline
h_1 & 3 & 9 & 9 \\
\hline
h_2 & 12 & 6 & 6 \\
\hline
h_3 & 9 & 6 & 3 \\
\end{array}
\qquad (5)
$$

Total surplus is 27, achieved by the uniquely optimal assignment $1 \to h_2$, $2 \to h_3$, $3 \to h_1$. Agents i's fair share utility is his average utility over all three houses:

$$\alpha_1 = 0, \qquad \alpha_2 = 7, \qquad \alpha_3 = 6.$$

Thus the fair share lower bound leaves only six units of "loose" surplus, or less than $\frac{1}{4}$ of total surplus (not unusually low in 3×3 situations, as the reader can check in Exercise 4.3). By contrast, the conservative lower bound $\max_h\{u_i(h)/3\}$ leaves 17 units of loose surplus.

Fair share guaranteed is a simple and intuitive sort of axiom: every agent receives a (decentralized) statement about his worst-case utility level. This is one measure of the risks implied by a particular solution. To require that the worst-case utility level be as high as possible can be interpreted as providing the best opportunity ex ante to every agent.

---

[18] Such a constraint places a lower bound $\varphi_i(u_i)$ on agent i's utility that does not depend upon other agents' utilities (although it depends on the number of other agents). Under anonymity, φ_i is independent of i, so $\varphi_i = \varphi$ for all i. Feasibility of the constraint φ means

$$\sum_{i=1}^{n} \varphi(u_i) \leq \max \sum_{k=1}^{q} u_{i_k}(h_k) \quad \text{for all ordering } h_1, \ldots, h_q \text{ of } H.$$

Applying this to the (unanimous) profile where $u_i = u_0$ for all i, we get

$$n\varphi(u_0) \leq \max \sum_{k=1}^{q} u_0(h_k) \Rightarrow \varphi(u_0) \leq \alpha_i(u_0).$$

The second important axiom of the fair-division literature can be interpreted as providing equal opportunity ex post to every agent.

The *no envy test* say that no agent should wish to exchange his own allocation (awarded by a particular solution) for the allocation of any other agent. Think of a referee dividing total resources into n lots, where n is, as usual, the number of agents. We wish to give the same choice opportunities to every agent, namely, the ability to pick any of the lots. This is feasible if (and only if) there is a way to assign lots to agents so that the resulting allocation passes the no envy test: everyone gets, in her view, the best lot, whence this division is self-enforcing (provided no one can propose another arrangement of lots).

The no envy test, in the formal sense just described, is a more recent concept than fair share guaranteed.[19] It was stated by Foley [1967], and became known to the economic profession through the work of Kolm [1972] and Varian [1974]. It quickly became the dominant argument of justice within microeconomic theory.[20] I see three reasons for such a success. First, no envy is indeed an intuitively appealing concept, deftly using interpersonal comparison of shares in lieu of the prohibited interpersonal comparison of welfare. Second, in fair division, the no envy test has generally much more bite than fair share guaranteed (it selects a much smaller set of efficient outcomes). Third, the no envy concept is deeply related to competitive-price ideas (see below), a particularly dear feature for economists.

While no envy has a valid claim to the preeminence among other tests of justice for the fair division of unproduced commodities (this is the main theme of this chapter), one should not forget that alternative, conflicting tests are worthy of our attention, too (Section 4.2, 4.6, and 4.7), and that for the fair division of a technology, no envy is only a secondary (and sometimes an unacceptable) requirement (Sections 4.2, 5.4, and 5.6).

Definition 4.1. In the fair-assignment problem (N, H, u), the allocation (σ, t) (where σ assigns agents to houses, including the null house, and t is a balanced vector of transfers) passes the no envy test (is envy-free) if

$$\text{for all } i, j \text{ in } N: \quad u_i(\sigma(i)) + t_i \geq u_i(\sigma(j)) + t_j. \tag{6}$$

---

[19] Although both are *implicitly* present in the discussion of divide and choose; see Section 4.1.

[20] So much so that the unfortunate usage of defining "fairness' as the combination of Pareto optimality and the no envy test was very popular for about a decade.

In the numerical example (5), we observe first that a suboptimal assignment such as $1 \to h_3$, $2 \to h_1$, $3 \to h_2$ cannot be combined with a balanced vector of transfers to produce an envy-free allocation. Suppose t is such a vector. We would have

agent 1 does not envy agent 3: $\quad 9 + t_1 \geq 12 + t_3,$

agent 2 does not envy agent 1: $\quad 9 + t_2 \geq 6 + t_1,$

agent 3 does not envy agent 2: $\quad 6 + t_3 \geq 9 + t_2,$

hence a contradiction by summing up. The general proof that no envy implies Pareto optimality in fair assignment (Theorem 4.1) simply copies the above argument.

Now let σ_* be the optimal assignment in example (5) ($\sigma_*: 1 \to h_2$, $2 \to h_3$, $3 \to h_1$). The vector of transfers t yields an envy-free allocation if and only if

1 envies neither 2 nor 3: $\quad 12 + t_1 \geq 9 + t_2, \quad 12 + t_1 \geq 3 + t_3,$

2 envies neither 3 nor 1: $\quad 6 + t_2 \geq 9 + t_3, \quad 6 + t_2 \geq 6 + t_1,$

3 envies neither 1 nor 2: $\quad 9 + t_3 \geq 6 + t_1, \quad 9 + t_3 \geq 3 + t_2.$

Taking feasibility ($t_1 + t_2 + t_3 = 0$) into account, we can reduce these conditions to a system in t_1, t_2. After deleting redundant inequalities, that system boils down to

$$t_2 \leq 3 + t_1, \quad 3 \geq t_1 + 2t_2, \quad 2t_1 + t_2 \leq 3,$$

which cuts a triangle in the (t_1, t_2) plane. This triangle is represented as ABC in Figure 4.2, along with the much larger triangle DEF cut by the fair share lower bound within efficient allocations, namely, $12 + t_1 \geq 8$, $6 + t_2 \geq 7$, and $9 + t_3 \geq 6$.

Theorem 4.1

(*i*) *In any fair-assignment problem, there exists at least one envy-free and Pareto optimal allocation.*
In any fair-assignment problem with no more objects than agents ($q \leq n$):
   (*ii*) *every envy-free allocation is Pareto optimal as well; it also meets fair share guaranteed;*
   (*iii*) *the set of envy-free allocations coincides with the set of competitive equilibrium allocations with equal income.*

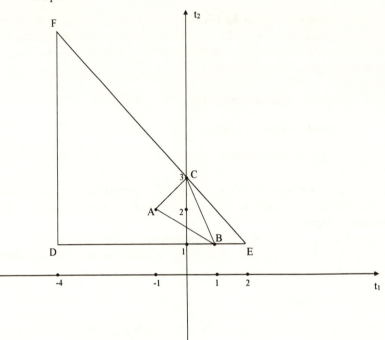

Fig. 4.2. The assignment problem (5): the no envy (*ABC*) and fair share guaranteed (*DEF*) areas.

A constructive proof of statement (i) is given in Appendix 4.1 (note that, unlike statements (ii) and (iii), statement (i) holds true for any n, q). Let us prove (ii). Assume for simplicity that $q = n$. Let (σ, t) be an envy-free allocation and $\tilde{\sigma}$ be an arbitrary assignment. For all i, in (σ, t), agent i does not envy this agent j who receives the house $\tilde{\sigma}(i)$ [or $j = \sigma^{-1}(\tilde{\sigma}(i))$: σ is invertible because $q = n$]:

$$u_i(\sigma(i)) + t_i \geq u_i(\tilde{\sigma}(i)) + t_j.$$

Summing up these inequalities for all i, upon noticing $\sum_i t_i = \sum_j t_j = 0$ (because j ranges over all N as i does), we find that the (gross) surplus of σ is not smaller than the surplus of $\tilde{\sigma}$. As $\tilde{\sigma}$ was arbitrary, this proves the optimality of assignment σ. In order to check that the allocation (σ, t) meet fair share guaranteed, we write the no envy property for an arbitrary agent i:

$$u_i(\sigma(i)) + t_i \geq u_i(\sigma(j)) + t_j \quad \text{for all } j.$$

Summing up the above inequalities for all j (including $j = 1$) and taking averages, we get

$$u_i(\sigma(i)) + t_i \geq \frac{1}{n} \sum_H u_i(h).$$

Adapting the above proof to the case $q < n$ is easy and omitted. For a discussion of the above case $q \geq n$, see Remark 4.2.

Finally, we check statement (iii) in the case $q \leq n$. Recall from Definition 3.2 that the set of competitive prices does not depend on the property right structure. If p is a competitive price and σ an optimal assignment (condition (a) and system (13)), we define a balanced vector of transfers t as follows:

$$t_i = \frac{1}{q} \sum_H p_h - p_{\sigma(i)} \quad \text{for all } i.$$

Clearly, the allocation (σ, t) is envy-free and coincides with the CEEI allocation at that price. Conversely, if σ is an optimal assignment and if (σ, t) is envy-free, we set $t^* = \max_{i \in N} t_i$ and define a price vector p as follows:

$$\sigma(i) = h \Rightarrow p_h = t^* - t_i \geq 0 \quad \text{for all } i, h. \tag{7}$$

This definition makes sense because if two agents receive the same house, that house must be the null house and by no envy we must have $t_i = t_j$. Moreover, if $\sigma(i) = h_\varnothing$, then t_i must be largest among all t_j (since all houses are goods, not bads), hence $p_{h_\varnothing} = 0$. Now the system (6) is at once equivalent to (13) in Chapter 3, so that p is a competitive price. Furthermore, (7) and the fact that t is balanced imply

$$t^* = \frac{1}{|H|} \sum_{h \in H} p_h,$$

and the allocation (σ, t) is indeed the CEEI at price p. Q.E.D.

More generally, consider a problem of *fair division with transferable utility*. That is, we have an economy with money and an arbitrary number of (divisible or indivisible) goods, where all preferences are represented by quasi-linear utilities. We want to divide fairly a set of (nonmonetary) goods (monetary compensations are possible). Therefore, just as in fair assignment, efficiency commands finding an allocation maximizing the joint surplus $v(N)$, let us say,

$$v(N) = u_1(z_1) + \cdots + u_n(z_n)$$

(where z_i is the allocation of nonmonetary goods to agent i). We assume that such an allocation exists (for instance, if all goods are divisible, existence is guaranteed when the utility functions u_i are continuous and monotonic).

Now we freeze the shares $z_1, \ldots, z_n$ and we allow only reallocations of those shares among the agents: this artificial restriction converts our problem into one of fair assignment where the n objects are (now indivisible) vectors z_i. By Theorem 4.1, there exists a set of transfers completing the division $z_1, \ldots, z_n$ into an envy-free allocation of this fair-assignment problem; but this must be an envy-free allocation in the initial economy (where the shares z_i, after defrosting, become divisible again) as well, because the no envy test only involves comparing pairs z_i, z_j. Thus we have the following.

Corollary to Theorem 4.1. In any fair-division problem with transferable utility, there always exists at least one envy-free and Pareto optimal allocation.

Remark 4.2. The Case with More Objects than Agents

In an assignment problem with q houses and n agents with *identical* utilities u_i, an optimal assignment picks the n best houses for u_i. Therefore, when $q > n$, the fair share utility is worth

$$\alpha_i(u_i) = \frac{1}{n} \max_{H_0} \{u_i(h_1) + \cdots + u_i(h_n)\}, \tag{8}$$

where the maximum bears on all subsets H_0 of n houses.

In the general case where utilities are not necessarily identical, the no envy test is defined as above (inequality (6)). This time the test does not imply Pareto optimality, because it places no condition on the houses not consumed. For instance, take two identical agents and three houses: one good house and two bad ones. Giving one bad house to each agent is envy-free but inefficient.

Even if we restrict ourselves to Pareto optimal allocations, the no envy property does not imply fair share guaranteed anymore. To see this, consider the following example:

	u_1	u_2
h_1	3	0
h_2	0	1
h_3	2	0

The optimal assignment is $1 \rightarrow h_1$, $2 \rightarrow h_2$. The corresponding allocation where agent 1 pays 1 to agent 2 is envy-free (as $3 - 1 \geq 0 + 1$), but it gives less utility to agent 1 (i.e., 2) than his fair share utility ($(3 + 2)/2 = 2.5$).

Should we conclude from these very simple examples that no envy is not an operational concept in the fair-assignment problem with $q > n$? Not so, because there is, in every such assignment problem, at least one envy-free and efficient allocation guaranteeing fair share utilities. A competitive allocation with equal incomes meets all three properties. The proof of these claims is the subject of Exercise 4.6

In the next section, we turn to the fair division of divisible goods when preferences are no longer quasi-linear, and the equivalence between no-envy and CEEI breaks down. Yet even in the transferable utility context, this equivalence disappears as soon as we drop the central assumption of the fair-assignment problem, namely, that an agent wants to consume at most one unit of indivisible good. Consider the following example with two indivisible goods a, b and quasi-linear utilities:

	u_1	u_2
a	6	4
b	4	5
a, b	12	6

Efficiency commands giving both houses to agent 1. Check that the envy-free allocations are those where agent 1 pays at least 3 and at most 6 to agent 2. However, in a competitive allocation with equal incomes, the prices p_a, p_b are such that $p_a \geq 4$ and $p_b \geq 5$ (so that agent 2 will not buy any good); hence, agent 1 must give at least \$4.50 (and at most \$6) to agent 2.

4.4. THE COMPETITIVE EQUILIBRIUM WITH EQUAL INCOMES

We use the Arrow–Debreu economies with divisible goods (and often convex preferences) to define the two equity tests and state the basic properties of the CEEI solution. The notations are the same as in Section 3.6. Throughout the rest of this chapter, we use an ordinal preference ordering R_i to represent agent i's preferences. In some examples, we revert without notice to a utility notation.

A fair-division problem $(R_1, \ldots, R_n; \omega)$ specifies the preference profile and the vector ω of resources (an element of $\mathbb{R}_+^K$) to be divided among the n participants. Equal split of ω (giving ω/n to each agent) is feasible (goods are divisible) and unquestionably fair (yet inefficient). Hence our first equity test is easy to write.

FAIR SHARE GUARANTEED (ALSO CALLED EQUAL-SPLIT LOWER BOUND)

In the fair-assignment problem $(R_1, \ldots, R_n; \omega)$, agent i does not strictly prefer the nth share of the resources to his actual allocation:

$$z_i \, R_i \left(\frac{\omega}{n} \right) \quad \text{for all } i$$

(where z_i is agent i's allocation in the proposed division). (9)

The axiom is easily implemented by giving to each agent the right to veto any allocation and claim her fair share of the resources. However, note that the equal-split lower bound does not amount to giving private ownership of ω/n to each participant: to be entitled to the welfare level resulting from consuming ω/n is not the same as owning ω/n (and being able, for instance, to trade it with other agents). Therefore, equal-split lower bound is implied by the above veto right but does not imply it.

Remark 4.3

When goods are perfectly divisible, the lower bound can be adjusted to account for unequal rights over the resources: if λ_i represents agent i's fraction of these rights (with $\sum_n \lambda_i = 1$), we would simply require that agent i does not prefer $\lambda_i \omega_i$ over z_i.

Remark 4.4

Under convex preferences, it is easy to show that the lower bound (9) is the highest feasible lower bound on each individual utility that does not depend upon other agents' preferences; see Exercise 4.17 for details. This remark is the key to defining fair share guaranteed in other contexts, in particular, when some of the goods are indivisible; an example is formula (4) for the assignment problem.

The no envy test is defined exactly as in the assignment problem:

$$no\ envy: \quad \text{for all pairs of agents } i, j: \quad z_i\ R_i\ z_j. \quad (10)$$

The next concept is a variant of the competitive equilibrium where the private ownership of the resources (represented by the initial endowments ω_i) is replaced by a distribution of shares of total income. Given a vector of weights λ_i, $i \in N$ ($\lambda_i \geq 0$ for all i, and $\Sigma_N \lambda_i = 1$), a *competitive equilibrium with λ-income shares* is a pair $(p; (z_1, \ldots, z_n))$ comprising a price and a feasible allocation, such that

$$p \cdot z_i \leq \lambda_i(p \cdot \omega) \text{ and for all } z' \in \mathbb{R}_+^K \quad \{p \cdot z' \leq \lambda_i(p \cdot \omega) \Rightarrow z_i\ R_i\ z'\}.$$

One especially interesting example is the *competitive equilibrium with equal incomes*, where each agent has a $1/n$-th share of total income ($\lambda_i = 1/n$). The competitive allocations with equal incomes is the most interesting solution ever proposed to the fair-division problem: it simply translates common ownership into equal private ownership, and computes the competitive set of the exchange economy thus created. Every allocation selected in this way passes our two tests of fairness.

Lemma 4.1. A competitive equilibrium with equal incomes (in short, CEEI) passes the no envy test and meets the equal-split lower bound. Moreover, if preferences are monotonic, it is also Pareto optimal.

To prove that a CEEI allocation generates no envy, simply observe that agent i's consumption z_i maximizes his preferences over the budget set

$$B = \left\{ z \mid p \cdot z \leq \frac{1}{n}(p \cdot \omega) \right\}.$$

As every agent has the same budget set, each vector z_i is in B and agent i cannot strictly prefer z_j over z_i, for all j.

Here is a related statement in the context of exchange economies: at a competitive equilibrium allocation (Section 3.6), no agent envies the *net trade* of another agent. That is to say, agent i does not strictly prefer $\omega_i + (z_i - \omega_i)$ over $\omega_i + (z_j - \omega_j)$, for all j. The proof is equally easy: by definition of the competitive equilibrium, the net trade $\tau_i = z_i - \omega_i$ of agent i satisfies $p \cdot \tau_i \leq 0$ (where p is the competitive price) and, moreover,

$$\text{for all } \tau_i' \in \mathbb{R}^K: \quad \{p \cdot \tau_i' \leq 0\} \Rightarrow \{(\omega_i + \tau_i) R_i (\omega_i + \tau_i')\}.$$

As agent j's net trade τ_j satisfies $p \cdot \tau_j \leq 0$ as well, the claim $(\omega_{i.} + \tau_i) R_i (\omega_i + \tau_j)$ follows.

The property of no envy on net trades is the fundamental equity property of the competitive exchange mechanism.

Theorem 4.2. In a fair-division problem with divisible goods, convex and continuous preference, at least one competitive equilibrium with λ-income shares exists as long as each share is positive. In particular, a CEEI allocation exists.

This theorem is an immediate consequence of Theorem 3.4 about the existence of a competitive equilibrium (every agent has a positive endowment of every good).

Of course, the CEEI allocation is not necessarily unique (just as an exchange economy may have several competitive allocations; see the discussion after Theorem 3.4). Yet for many simple preference profiles, uniqueness of the CEEI is guaranteed, at least utilitywise, e.g., when all agents have Leontief preferences (Exercise 4.8) or when all have linear preferences and there are only two goods (see Exercise 4.10).

In Appendix 4.2, we discuss an important converse property of Lemma 4.1 originally suggested by Varian [1974]. When agents are negligibly small and have convex preferences, it may happen that the CEEI allocation(s) is (are) the only efficient allocation(s) passing the no envy test. Varian's proposition gives sufficient conditions on the distribution of individual preferences implying this property. Similar results (due to Schmeidler and Vind [1972]) allow one to characterize the competitive equilibrium in exchange economies by the property of "no envy on net trades." See Exercise 4.4 for a very simple example.

As in the case of exchange economies, the CEEI may fail to exist if preferences are not convex. In fact, under nonconvex preferences, even the existence of any envy-free and Pareto optimal allocation is not guaranteed. Easy examples of this situation are constructed by allowing preferences to be not only nonconvex, but also nonstrictly monotonic. Choose $n = 2$, $K = 2$, $u_i(x, y) = \max\{x, y\}$, for $i = 1, 2$ and $\omega = (5.3)$. There are only two efficient distributions of ω, namely, $z_i = (5, 0)$, $z_j = (0, 3)$. Hence the equality/efficiency dilemma[21] arises: even though

---

[21] A simplistic version of which has indivisible goods *without* money: the only envy-free allocation of two identical goods among three agents is to throw them away, which is surely inefficient!

all agents have identical preferences, there is no Pareto optimal allocation where all agents enjoy the same utility. Of course, an envy-free allocation must treat equals equally (utilitywise). See Exercise 4.14 for more samples of the dilemma. See also Exercise 4.15 for an example (due to Varian) where the nonexistence of envy-free and efficient allocations is not due to an equality/efficiency dilemma (there are two goods, two agents, both preferences are strictly monotonic and continuous, and one of them is convex). Finally, Exercise 5.13 gives an example where the resources are a technology with decreasing marginal costs, individual preferences are convex, and yet no envy-free and efficient allocation exists.

Remark 4.5

Obviously, there exist some fair-division problems where the CEEI solution does not exist, whereas envy-free and efficient allocations do exist. For instance, in the transferable utility context, we know that an efficient and envy-free allocation always exists, provided preferences are continuous and monotonic but not necessarily convex (see Corollary to Theorem 4.1), whereas the CEEI often does not exist (the numerous examples in Section 2.8 can be adapted to the equal-income case). See also an example with divisible goods (and no money) in Exercise 4.13.

Remark 4.6

One should not conclude from the discussion of the last two sections that the CEEI is the only systematic (often single-valued) selection from the set of efficient and envy-free allocations. A variety of other selections have been proposed, of which the simplest example is due to Diamantaras and Thomson [1990]. Start from any allocation $(z_1, \ldots, z_n)$. For some nonnegative number λ, we have, for all $i, j, i \neq j$; $z_i R_i (\lambda z_j)$. An allocation is envy-free if and only if we can find such a number λ no smaller than 1. An efficient allocation where λ is maximal is a natural choice within the envy-free and efficient set.[22] Moreover, if every efficient allocation generates envy, we can still pick in this way a plausible efficient allocation where mutual envy is evenly distributed (but this time, the highest λ to which corresponds a feasible allocation is smaller than 1). Diamantaras and Thomson [1990] study its properties under continuous and monotonic preferences. One serious defect of this solution, however, is that it may not guarantee fair shares.

---

[22] Exercise: in Example (23), check that λ is $(2)^{1/5}$ or approximately 1.16, and compute the corresponding allocation.

4.5. THREE EXAMPLES OF THE CEEI SOLUTION

The three examples underline the versatility of the CEEI solution. In the first one, preferences are represented by linear utilities. The second one is the celebrated land (or cake) division problem where utility is additive over the various pieces of land. The third shows how the CEEI solution can be adapted in the case of nonmonotonic preferences. In all three examples, the CEEI solution selects a single allocation, whereas the set of allocations that are efficient, envy-free, and guarantee a fair share is a "large" set.

Fair Division with Linear Preferences (Hylland and Zeckhauser [1979])

Three agents share two indivisible paintings (of no commercial value) and they cannot use cash for compensations. They agree to allocate the objects at random. Each agent has a von Neumann–Morgenstern utility: he compares lotteries by comparing their expected utilities. Here is their utility profile:

	Ann	Bob	Charles
Object 1	4	2	2
Object 2	1	1	5

(11)

Randomizing the allocation of the paintings turns these objects into divisible goods: we have one unit of two goods and agents have linear preferences. For instance, if we give away each painting independently and with an equal probability ($\frac{1}{3}$) to each agent, we get the equal-split allocation $z_i = (\frac{1}{3}, \frac{1}{3})$, $i = 1, 2, 3$, generating the utilities

$$u_1 = 4 \cdot (\tfrac{1}{3}) + 1 \cdot (\tfrac{1}{3}) = \tfrac{5}{3}, \qquad u_2 = 1, \qquad u_3 = \tfrac{7}{3}.$$

This allocation is not efficient, however, as it does not take advantage of the differences in tastes. A Pareto superior allocation is to *either* give the first painting to Ann and the second to Bob *or* give the first to Bob and the second to Charles (each option with equal probability $\frac{1}{2}$). The resulting allocation yields better expected utilities, namely, $u_1 = 2$, $u_2 = \frac{3}{2}$, $u_3 = \frac{5}{2}$.

Compute first the Pareto optimal allocations. We claim that the set of efficient allocations consists of the following three subsets:

Type α:	z_1	z_2	z_3
	1	0	0
	α_1	α_2	α_3

where $\alpha_i \geq 0$, $\sum_1^3 \alpha_i = 1$,

	Z_1	Z_2	Z_3
Type β:	$1 - \beta_1$	β_1	0
	0	β_2	$1 - \beta_2$

where $0 \le \beta_1 \le 1$, $0 \le \beta_2 \le 1$,

	Z_1	Z_2	Z_3
Type γ:	γ_1	γ_2	γ_3
	0	0	1

where $\gamma_i \ge 0$, $\sum_1^3 \gamma_i = 1$.

To check the claim, observe that an allocation where Ann has a positive probability of receiving object 2, and does not get object 1 for sure, is not Pareto optimal (because there is a mutually advantageous trade with Bob). Similarly, if Charles has a positive probability of receiving object 1, and does not get object 2 for sure, the allocation is Pareto inferior. This shows that an efficient allocation must be one of the three types, α, β, or γ. The converse inclusion (all such allocations are efficient) is obvious and omitted.

The equal-split lower bound rules out type-α allocations ($u_2 \ge 1$ yields $\alpha_2 = 1$, and $u_2 \ge \frac{1}{3}$ yields $\alpha_3 > 0$) and leaves the following subsets of type-β and type-γ allocations:

type β: $\qquad 0 \le \beta_1 \le \frac{7}{12}$, $\qquad 0 \le \beta_2 \le \frac{8}{15}$, $\qquad 1 \le 2\beta_1 + \beta_2$,

type γ: $\qquad \frac{5}{12} \le \gamma_1$, $\qquad \frac{1}{2} \le \gamma_2$.

The no envy test is not feasible with type-α allocations. For type-γ allocations, it implies that $\gamma_1 = \gamma_2$ and $2\gamma_2 \ge 2\gamma_3 + 1$; hence, $\gamma_1 = \gamma_2 = \frac{1}{2}$, and the corresponding allocation—in fact, the CEEI—can be viewed as a type-β allocation. Straightforward computations show that the set of envy-free and Pareto optimal allocations is the following subset of type-β allocations:

$$2 \le 4\beta_1 + \beta_2, \qquad 2\beta_1 + 10\beta_2 \le 5, \qquad 8\beta_1 + \beta_2 \le 4.$$

This set is depicted as $\mathcal{N}$ in Figure 4.3. It is a subset of the set $\mathcal{U}$ of Pareto optimal allocations meeting the equal-split lower bound (this is always true with linear utility functions; see Exercise 4.9).

Compute now the competitive equilibrium with equal incomes. The relative price p_1/p_2 of painting 1 to painting 2 must be 2: if the relative price is higher, both Bob and Charles want to spend all their income

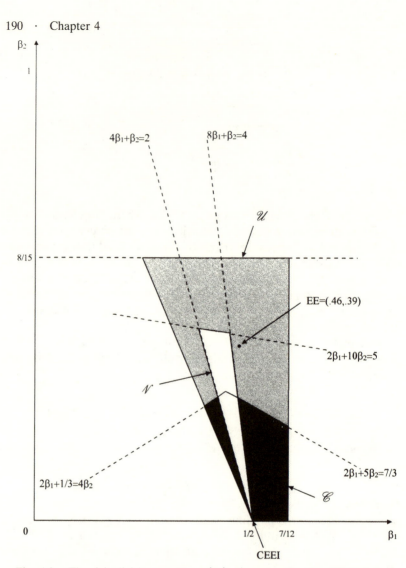

Fig. 4.3. The fair-division problem (11): the core $\mathscr{C}$, the efficient and envy-free set $\mathscr{N}$, and the CEEI.

$((p_1 + p_2)/3)$ buying as much good 2 (probability of receiving painting 2) as they possibly can, and the resulting demand $(2 \cdot (p_1 + p_2)/3p_2)$ exceeds 1; similarly, if p_1/p_2 is below 2, Ann and Bob buy all the good 1 they can, resulting in excess demand of good 1 $(2 \cdot (p_1 + p_2)/3p_1)$. Therefore, the competitive price can be taken as $p_1 = 2$, $p_2 = 1$, and the unique competitive allocation is

	Z_1	Z_2	Z_3
painting 1	1/2	1/2	0
painting 2	0	0	1

For the sake of completeness, compute the core with equal initial endowments (initially each agent owns $\frac{1}{3}$ of each good), even though this is not a normatively attractive solution to fair division (for instance, it does not treat equals equally). Consider a type-β allocation. By allocating its initial endowment ($\frac{2}{3}$ units of each good) efficiently, coalition $\{1, 3\}$ achieves the following intervals of utilities:

$$u_1 + 2u_3 = \tfrac{28}{3} \quad \text{with } 0 \le u_1 \le \tfrac{8}{3}$$

and

$$5u_1 + u_3 = \tfrac{50}{3} \quad \text{with } 0 \le u_3 \le \tfrac{10}{3}.$$

This rules out as core unstable all type-β allocations such that

$$\{2\beta_1 + 5\beta_2 > \tfrac{7}{3} \text{ and } \beta_1 \ge \tfrac{1}{3}\} \quad \text{or} \quad \{4\beta_1 + \beta_2 > \tfrac{5}{3} \text{ and } \beta_2 \ge \tfrac{1}{3}\}.$$

A similar argument can be made for coalitions $\{2, 3\}$ and $\{1, 2\}$.

Combining this with the equal-split lower bound, we find a set of type-β core allocations defined by:

$$1 \le 2\beta_1 + \beta_2, \quad 0 \le \beta_2, \quad \beta_1 \le \tfrac{7}{12}, \quad 2\beta_1 + 5\beta_2 \le \tfrac{7}{3}, \quad 4\beta_2 \le 2\beta_1 + \tfrac{1}{3}.$$

There are also a few type-γ allocations in the core (that the conscientious reader may easily compute).

Figure 4.3 illustrates the relative shapes of the core $\mathscr{C}$, the set $\mathscr{N}$ of efficient and envy-free allocations, and the CEEI (all drawn with the parameterization of type-β allocation). The core and envy-free sets overlap; the CEEI is an extreme point of the core, the envy-free set, as well as of the set cut by the equal-split lower bound. Its fairness is somewhat questionable: Bob gets only his equal-split utility, Ann gains 20% over and above her equal-split utility, whereas Charles more than doubles his equal-split utility. A plausible alternative solution to the CEEI is the egalitarian-equivalent solution computed at the end of Section 4.6. See also Example (14) for a related critique of CEEI.

Exercise 4.10 analyzes a fair-division problem similar to (11) with two goods and linear utilities.

Our second example is the fair division of a plot of land, where agents compare various pieces of land with an additive measure. This appealing model has generated a substantial literature (spanning from Steinhaus [1948], to the recent survey by Brams and Taylor [1995], and developed by mathematicians rather than economists) that pays attention not only to the familiar fairness properties (no envy, fair share guaranteed, and so on) but also to the geometrical shape of the shares (see Berliant, Dunz, and Thomson [1992] for an economist's viewpoint, and the references therein).

The Land Division Problem

A piece of land, represented as a subset Ω of $\mathbb{R}^2$, is to be divided among n agents. Feasible allocations are partitions $A_1, \ldots, A_n$ of Ω, with A_i the share of agent i. Each agent compares two subsets X, Y by comparing their "weight" according to his subjective measure of the value of these pieces; he prefers X to Y if and only if $u_i(X) > u_i(Y)$, where u_i is an *additive*, positive measure over subsets of Ω.[23] The case where the set Ω is finite is not particularly interesting,[24] so we think of Ω as a region in the plane and we assume that no agent gives a positive weight to any single point of Ω.[25] An important mathematical result, known as Lyapounov's theorem, says that when we consider all conceivable partitions $A_1, \ldots, A_n$ of Ω and compute the corresponding utility vectors $(u_1(A_1), \ldots, u_n(A_n))$, they cover a convex and compact subset of $\mathbb{R}^n$ (see Dubins and Spanier [1961]). This implies, for instance, that we can partition Ω in such a way that every agent receives her fair share:

$$u_i(A_i) = \frac{1}{n} u_i(\Omega) \quad \text{for all } i = 1, \ldots, n.^{26}$$

[23] For instance, if Ω is a finite set, he gives a certain weight to each point in Ω and the weight of X is simply the weighted sum of its elements. When Ω is an arbitrary closed subset of R^2, we rely on the apparatus of measure theory with its technical requisites, such as measurability of the subsets X and σ-additivity of the measure. These are ignored in the discussion of the example.

[24] Ω is a collection of heterogeneous indivisible goods, and we cannot use money or lotteries to smooth out the indivisibilities; accordingly, there may exist no envy-free allocation and/or no egalitarian-equivalent allocation.

[25] Formally, the measures are assumed to be nonatomic; see, e.g., Rudin [1974].

[26] Indeed, the partition giving everything to agent i yields the utility vector $(0, \ldots, 0, u_i(\Omega), 0, \ldots, 0)$; the above utility vector is the arithmetic average of these vectors for $i = 1, \ldots, n$.

Despite the absence of money (or lotteries) in the land division economy, we can define a competitive price as an additive positive measure p. A partition $(A_1, \ldots, A_n)$ is a competitive allocation with equal incomes corresponding to this price if we have, for all $i = 1, \ldots, n$,

$$p(A_i) = \frac{1}{n} p(\Omega),$$

and

$$\{\text{for all } B \subseteq \Omega: \quad u_i(B) > u_i(A_i) \Rightarrow p(B) > p(A_I)\}.$$

Efficiency of a CEEI, as well as the existence of at least one such allocation, follow from the assumption that all measures u_i are nonatomic and positive.[27]

We illustrate our two solutions in a simple numerical example where Ω is the interval $[0, 1]$ and we have only two types of agent:

$$m_1 \text{ agents of type 1:} \quad u^1(A) = \int_A (1 - x)\, dx,$$

$$m_2 \text{ agents of type 2:} \quad u^2(A) = \int_A x\, dx. \tag{12}$$

Type-1 agents like the land near 0: their utility for a piece of land decreases linearly from 0 to 1. Type-2 agents, symmetrically, like the land near 1. A division of the land is Pareto optimal if, and only if, the type-1 agents together own an interval $[0, \alpha]$ and the type-2 agents own $[\alpha, 1]$.

We compute the CEEI. With at least two agents of each type $(m_1, m_2 \geq 2)$, this allocation is unique. (Exercise 4.20 shows that this is not true when one type of agents is represented by a single agent.) Indeed, suppose a CEEI allocation gives the combined share $[0, \beta]$ to type-1 agents and $[\beta, 1]$ to type-2 agents. The price p must be proportional to u^1 on $[0, \beta]$ (and proportional to u^2 on $[\beta, 1]$); otherwise, all the type-1 agents would demand the same cheap subset of $[0, \beta]$ (namely, those parts where the ratio $p(dx)/u^2(dx)$ is lowest) and total demand of type-1 agents would not equal $[0, \beta]$. Therefore,

$$p(dx) = a \cdot (1 - x)\, dx \quad \text{for } 0 \leq x \leq \beta,$$

$$p(dx) = b \cdot x\, dx \qquad\qquad \text{for } \beta \leq x \leq 1.$$

---

[27] To prove that a CEEI is efficient is an easy exercise (observe that $u_i(B) \geq u_i(A_i)$ must imply $p(B) \geq p(A_i)$); the existence result is more difficult. See Berliant [1985] and Weller [1989].

Moreover, inequality $a(1 - \beta) < b\beta$ would imply that a type-2 agent prefers to buy a piece dx immediately to the left of β over a piece immediately to the right of β, and $a(1 - \beta) > b\beta$ would similarly imply that type-1 agents do not demand $[0, \beta]$. Thus, we can take $a = \beta$, $b = (1 - \beta)$ (since price can be normalized to our liking), and β is finally determined by equating the income of all agents. Setting $r = m_1/m_2$, we get

$$\frac{p([0, \beta])}{m_1} = \frac{p([\beta, 1])}{m_2} \Leftrightarrow \beta(2\beta - \beta^2) = r \cdot (1 - \beta)(1 - \beta^2). \quad (13)$$

One checks that this equation has a unique solution β in $[0, 1]$ for all $r \geq 0$. For instance, with $r = 2$ we get $\beta = 0.59$ and a utility about 50% higher for the type-2 agents (the rare types).

Fair Division with Single-Peaked Preferences

A single divisible item is being divided among agents whose preferences for the item are not monotonic. There are many items that we consume privately but for which our preferences are not monotonic. The baby-sitting story of Example 1.7 is one of many examples: as soon as the item cannot be freely disposed, we expect our preferences to decrease after a point. Good meals is our obvious example; another one is a share of company (when ownership brings an uncertain flow of returns) that I wish to purchase for my portfolio.

Most of the results presented in this section can be adapted to deal with nonmonotonic preferences. For the sake of illustration, it will be enough to look at a fair-division problem with a single divisible good, namely, that of Example 1.7. We have 15 units of the item and three agents. Ann's preferences increase from 0 to 15; Bob's preferences increase from 0 to 6 and decrease afterwards; Charles's preferences increase from 0 to 3 and decrease afterwards. Here the competitive allocation with equal income is the uniform rationing allocation (already discussed in Example 1.7), namely,

$$x_A = x_B = 6, \qquad x_C = 3.$$

The corresponding competitive price is $p = 1$, with an income of 6 per agent; this is *not* one-third of $p \cdot \omega = 15$. Thus, Ann and Bob spend all their incomes to buy 6 units, while Charles spends only half his income (money can be freely disposed of). When demand falls short of total resources, the price of the good becomes negative. For instance, say that Ann's peak is at 7 (her preferences increase up to 7 and decrease afterwards), Bob's peak is at 2, and Charles's peak is at 3. The competitive price is -1 and (equal) income is -4; in other words, the budget constraint requires one to consume *at least* 4 units of the good. So Bob and Charles buy exactly 4 units, while Ann buys 7 units:

$$x_A = 7, \qquad x_B = x_C = 4.$$

Remarkably, the "direct" mechanism defined by the CEEI (each agent needs only to report the peak of his/her preferences) is strategy-proof (even with respect to coalitional manipulations), a fact already noted in the discussion of Example 1.7. Moreover, the CEEI is the only strategy-proof mechanism implementing an efficient allocation and satisfying equal treatment of equals. See Exercise 4.11 for details.

Which general normative properties of the CEEI are preserved in the case of nonmonotonic preferences? Every CEEI passes the no envy test, for each agent is facing the same budget set. Another important feature of the CEEI is Pareto optimality. This is guaranteed only in the weak sense that no other allocation exists that all agents strictly prefer to the CEEI. However, a general result due to Mas-Colell [1985] gives some sufficient conditions on individual preferences for (a) the existence of at least one CEEI and (b) the Pareto optimality of all such allocations. The key condition is the convexity of preferences, augmented by the assumption that satiation points are isolated.

4.6. THE EGALITARIAN-EQUIVALENT SOLUTION

A critique of the CEEI solution based on the stand alone test is offered in Section 4.2 for fair division with transferable utility.[28] Now we give an argument against CEEI in the general case where utilities are not necessarily quasi-linear.

We have two goods denoted X and Y, $\omega = (2, 2)$, and the two agents have the following utilities:

$$u_1(x_1, y_2) = x_1 \cdot y_1, \qquad u_2(x_2, y_2) = 2x_2 + y_2. \tag{14}$$

The CEEI, with price $p = 2$ (units of Y for one unit of X), is the allocation $(x_1, y_1) = (\frac{3}{4}, \frac{3}{2})$; $(x_2, y_2) = (\frac{5}{4}, \frac{1}{2})$. See Figure 4.4. This looks unfair when we compare it to the equal-split allocation, for agent 2 only gets his equal-split utility, whereas agent 1 keeps all the "surplus" (similarly, in Example (11), Bob gets only his equal-split utility). Yet,

---

[28] Recall that in the estate division problem, the envy-free set coincides with the set of CEEI allocations: Theorem 4.1.

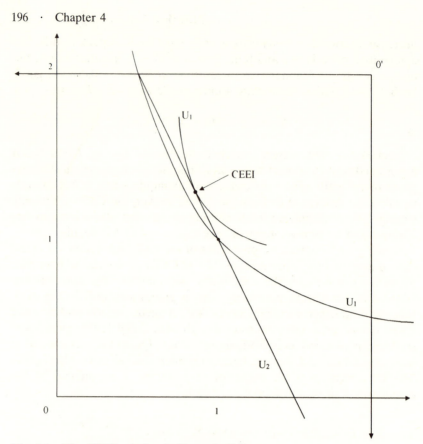

Fig. 4.4. The CEEI solution in Example (14).

there are many efficient and envy-free allocations where each agent gets a positive share of surplus (Remark 4.6 at the end of Section 4.4 gives one way for selecting one).

The idea of giving "equal" shares of the surplus above the equal-split allocation materializes as the egaliterian-equivalent solution. The difficulty is to find a way of measuring surplus in a world of ordinal preferences. Pazner and Schmeidler [1978] were the first to propose to measure agent i's surplus at the allocation z_i as a share of total resources that she views as just equivalent to z_i. An *ω-egalitarian-equivalent allocation* in the fair-division problem $(R_1, \ldots, R_n; \omega)$ is an allocation $(z_i, i = 1, \ldots, n)$, such that

$$z_i \, I_i \, (\lambda \cdot \omega) \quad \text{for all } i = 1, \ldots, n \tag{15}$$

(recall that I_i denotes indifference). The important point is that λ does not depend upon i.

The *ω-egalitarian-equivalent solution* looks for the highest number λ^* such that a corresponding egalitarian-equivalent feasible allocation exists, and picks one of those allocations. Even though we may have more than one feasible allocation satisfying the system (15) for the maximal value λ^*, they are by definition welfare-equivalent: every agent is indifferent between any two of those allocations.

In Example (14), for instance, the ω-egalitarian-equivalent (ω-EE) solution picks $\lambda = 6/(2\sqrt{2} + 3)$ and $x_1 = 3\sqrt{2}/(2\sqrt{2} + 3)$, $y_1 = 6\sqrt{2}/(2\sqrt{2} + 3)$.[29] At the end of this Section, we compute the ω-EE solution in the Examples (11) and (12), but fist, some general properties of this solution.

Lemma 4.2. Given is a fair-division problem $(R_1, \ldots, R_n; \omega)$ with divisible goods, where the individual preferences are strictly monotonic (strictly increasing in every good) and continuous. Then the ω-egalitarian-equivalent solution is well defined and all corresponding allocations are Pareto optimal.

Proof. Denote by I the set of real numbers λ such that there exists a feasible division $(z, \ldots, z_n)$ of ω, such that

$$z_i \, R_i \, \lambda \cdot \omega \quad \text{for all } i. \tag{16}$$

Then I is a closed subinterval of $[0, 1]$ (by continuity of preferences) and we denote its upper bound by λ^*. Let z^* be a corresponding allocation (by (16)), and suppose that z^* is Pareto inferior to another allocation z. Assume, for instance, $z_1 \, P_1 \, z_1^*$, $z_i \, R_i \, z_i^*$, for all $i \geq 2$. Then we can reduce z_1 to z_1' (by decreasing at least one nonzero component of z_1) while preserving the relation $z_1' \, P_1 \, z_1^*$. If we redistribute (evenly) the vector $z_1 - z_1'$ to every other agent, we get a new allocation z' where every agent i, $i \geq 2$, is strictly better off than at z (because of the strict monotonicity of preferences). Therefore, we have $z_i' \, P_i \, (\lambda^* \cdot \omega)$ for all i, so that we can raise λ^* a little to λ and preserve the relations (16); this is a contradiction of the definition of λ^*.

We have shown that z^* is Pareto optimal. It remains to show that, in fact, we have $z_i^* \, I_i \, (\lambda^* \cdot \omega)$, for all i. Suppose $z_1^* \, P_1 \, (\lambda^* \cdot \omega)$ and apply the same trick (reduce z_1^* and redistribute the "savings" evenly) to derive, again a contradiction of the definition of λ^*.   Q.E.D.

---

[29] Check that this yields a Pareto optimal allocation satisfying (15).

Thus the ω-egalitarian-equivalent solution fares better than the competitive equilibrium with equal incomes on two accounts. First, it picks a unique utility vector in the Pareto set (whereas multiple utility vectors —even a continuum of them—in the competitive set with equal incomes are not uncommon). Second, this utility vector is well defined even for nonconvex preferences (in contrast, see Theorem 4.2 and the discussion following it).

On the other hand, egalitarian-equivalence is rooted in the monotonicity of preferences (allowing us to calibrate individual welfare levels along the half-line defined by the vector ω), an assumption that is not essential to the logic of the competitive equilibrium, as shown at the end of Section 4.5.[30] Moreover, it cannot easily be adapted to accommodate indivisible goods except in the case of quasi-linear preferences; see Appendix 4.3.

We now compare the two competing interpretations of common ownership from the normative angle.

Lemma 4.3.   Assumptions as in Lemma 4.2.

(*i*) *The ω-egalitarian-equivalent solution (just like the competitive one) meets the equal-split lower bound.*

(*ii*) *These two solutions have the equal-treatment property (two identical agents end up with the same utility level).*

(*iii*) *The ω-egalitarian-equivalent solution, unlike the competitive one, is population-monotonic: if more agents show up to share the same bundle of resources ω, none of the previous agents is made better off.*

(*iv*) *The ω-egalitarian-equivalent solution, unlike the competitive one, may have the domination property: the allocation z_i to a certain agent i may be larger than the allocation z_j to agent j in every good ($z_i^k > z_j^k$ for all $k = 1, \ldots, K$). Note that domination may occur only with three or more agents.*

Thus both approaches meet the minimal criterion of justice a priori (equal treatment of equals) and the intuitive equal-split lower bound. They differ in two interesting criteria comparing allocations across agents within a fixed population, or across populations for a fixed agent. The ω-EE solution may generate an amplified form of envy: as I receive less of every good than you, not only do I envy you, but everyone can see that you get more in every single good! This argument is hard to resist and will probably suffice to reject the ω-EE allocation at those preference profiles where the domination property occurs. On the other

---

[30] In the last example of Section 4.5 (i.e., the baby-sitting story Example 1.7), one checks that the only ω-egalitarian-equivalent allocation is equal split, a Pareto inferior outcome.

hand, the CEEI allocation is not convincing when it gives all the surplus above the equal-split allocation to a single agent (as in Example (14)). As shown in Exercise 4.21 (question (c)), this will always be the case when the n agents split into $(n - 1)$ agents with identical preferences and one "eccentric" agent, and n is large.

A related property of the ω-EE solution (discussed in Moulin [1991]) looks at the upper bound on the utility of a given agent when the utility of other agents varies: if we fix the preference ordering R_1, the resources ω, and the number n of agents, agent 1 can never achieve a utility level above that of $(K/n)\omega$, where K is the number of goods. The CEEI solution, on the other hand, could give her a much higher utility level; see Exercise 4.21 for details.

We turn to statement (iii), uncovering a serious defect of the CEEI solution. When new agents show up to share the same resources, it may happen that the CEEI solution increases (in every good) the share of certain agents: with more mouths to feed, one child gets a bigger share of the cake! This is ruled out by the population monotonicity axiom (due to Thomson [1983]), an amplified form of the stand alone test.[31] Therefore, this critique of the CEEI (namely, it is not population-monotonic) is essentially the same as the critique of no envy by the stand alone axiom for fair division with transferable utility (Section 4.2). It can be shown, for instance, that in fair division *without* money, there is no population-monotonic solution selecting an efficient and envy-free allocation at all profiles of convex preferences and for all sizes of the population (see Moulin [1991]). I conjecture that the same impossibility statement holds true when we replace no envy by the (weaker) no domination property.

Proof of Lemma 4.3

Statement (i): As goods are divisible, the allocation $z_i = \omega/n$ for all i is feasible; therefore, the number λ^* in the definition of the ω-EE solution is at least $1/n$, and the claim follows at once from the monotonicity of preferences.

Statement (ii) is obvious.

Statement (iii): Fix a problem $(R_1, \ldots, R_n; \omega)$ with corresponding EE parameter λ^*, and call z^* an allocation in the ω-EE solution. Now remove agent n: the allocation $(z_1^*, \ldots, z_{n-1}^*)$ is feasible in problem $(R_1, \ldots, R_{n-1}; \omega)$ (by free disposal of the goods; alternatively, a topo-

---

[31] Say that a solution is population-monotonic, and fix a certain fair-division problem with n agents. When we remove the agents one at a time until only agent i is left (keeping the resources ω constant), agent i's welfare will be nondecreasing at every step.

logical argument shows that we can redistribute z_n^* among agents $1, \ldots, n-1$ and maintain the egalitarian-equivalence property with a parameter not smaller than λ^*). Therefore, the EE parameter in problem $(R_1, \ldots, R_{n-1}; \omega)$ is at least λ^*, implying that none of the agents $1, \ldots, (n-1)$ is worse off in the ω-EE solution of this problem that in the ω-EE solution of the n-agents problem.

Statement (iii) (continued): That the CEEI solution fails population monotonicity is demonstrated by an example (inspired by Chichilnisky and Thomson [1987]). We start with a four-agent problem with two goods. Global resources are $\omega = (24, 24)$ and the preferences are represented by the following utilities:

$$u_1(x_1, y_1) = \min\{2x_1 + 8, y_1\},$$

$$u_i(x_i, y_i) = \min\{18x_i + 100, 25y_i + 132\} \quad \text{for } i = 2, 3, 4.$$

The competitive price with equal incomes is $p = (3, 2)$, with associated CEEI:

$$z_1 = (x_1, y_1) = (2, 12), \qquad z_i = (x_i, y_i) = (\tfrac{22}{3}, 4) \quad \text{for } i = 2, 3, 4.$$

Next, remove agent 4 and compute the new competitive price $p' = (2, 7)$, with CEEI

$$z_1' = (1, 10), \qquad z_i' = (\tfrac{23}{2}, 7) \quad \text{for } i = 2, 3.$$

Thus agent 1's allocation is strictly worse in every good for the three-agent problem. See Figure 4.5. (*Exercise:* Compute the EE solutions in these two problems.)

Statement (iv): We give an example of domination in a three-agent and two-good problem. Global resources are $\omega = (12, 12)$ and the three utilities are as follows:

$$u_1(x_1, y_1) = \min\{3x_1, 5y_1\},$$

$$u_2(x_2, y_2) = x_2 + y_2,$$

$$u_3(x_3, y_3) = \min\{5x_3, y_3\}.$$

The EE parameter is $\lambda^* = \tfrac{5}{12}$, with corresponding ω-EE allocation

$$z_1 = (5, 3), \qquad z_2 = (6, 4), \qquad z_3 = (1, 5).$$

To check this, observe first that z is feasible and is egalitarian-

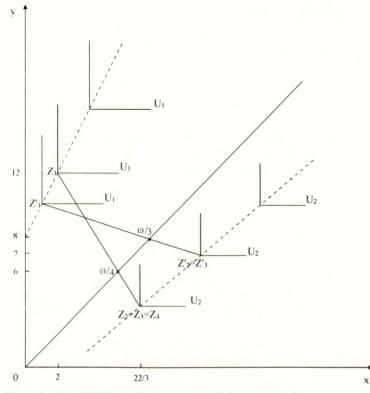

Fig. 4.5.   The CEEI solution is not population-monotonic.

equivalent $(u_i(z_i) = u_i(5, 5)$ for all i). By Lemma 4.2, it is enough then to show that z is a Pareto optimal allocation. This is obvious given the shape of agents 1 and 3's preferences (represented by Leontief utility functions): any allocation z' such that $u_i(z_i') \geq u_i(z_i)$ for $i = 1$ and 3 must have $z_i' \geq z_i$ for $i = 1$ and 3, whence by feasibility, $z_2' \leq z_2$.

The point of the example is that agent 1 receives strictly less than agent 2 in each good.   Q.E.D.

A common feature to both the competitive and the egalitarian-equivalent approaches is their versatility. They can be adapted to other contexts beyond the canonical fair-division problem with divisible commodities. This has been amply established in Sections 4.3 to 4.5 for the CEEI solution. Appendix 4.3 explains how the ω-EE solution can be adapted to fair division with transferable utility. For the time being, we compute it in two examples from the previous section.

ω-EE Solution in Example (11)

By definition, an egalitarian-equivalent allocation $z_i = (x_i, y_i), i = 1, 2, 3$ (where x_i is the probability that agent i gets object 1, and y_i the probability that he gets object 2), is such that

$$4x_1 + y_1 = 5\lambda, \qquad 2x_2 + y_2 = 3\lambda, \qquad 2x_3 + 5x_3 = 7\lambda.$$

We look for the highest λ such that the above system has a feasible solution. By Lemma 4.2, it is enough to look for a Pareto optimal allocation. For a type-β efficient allocation, the system reads

$$4 - 4\beta_1 = 5\lambda, \qquad 2\beta_1 + \beta_2 = 3\lambda, \qquad 5 - 5\beta_2 = 7\lambda,$$

with unique solution

$$\beta_1 = \tfrac{21}{46}, \qquad \beta_2 = \tfrac{9}{23}, \qquad \lambda^* = \tfrac{10}{23}.$$

This is the unique egalitarian-equivalent solution: as illustrated in Figure 4.3, it is just outside the envy-free set and occupies a more central position than the CEEI solution within the set of efficient allocations meeting the equal-split lower bound.

ω-EE Solution in the Land Division Example (12)

In this context, an egalitarian-equivalent allocation is a partition of Ω such that there is a number λ, $0 \le \lambda \le 1$, with

$$u_i(A_i) = \lambda u_i(\Omega) \quad \text{for all } i = 1, \dots, n. \tag{17}$$

The ω-EE solution picks an ω-EE allocation where the parameter λ is as high as possible.[32] Obviously, λ is as least $1/n$; hence, every agent gets at least his fair share of the land.

The joint utility of type-1 agents for $[0, \alpha]$ is $\int_0^\alpha (1 - x)\, dx$ and that of type-2 agents for $[\alpha, 1]$ is $\int_\alpha^1 x\, dx$. Therefore, (17) implies

$$\frac{1}{m_1}\left(\alpha - \frac{\alpha^2}{2}\right) = \lambda \cdot \frac{1}{2}, \qquad \frac{1}{m_2}\left(\frac{1}{2} - \frac{\alpha^2}{2}\right) = \lambda \cdot \frac{1}{2}.$$

---

[32] The compactness property of the set of utility vectors $(u_1(A_1), \dots, u_n(A_n))$ guarantees that the ω-EE solution is well defined.

So after setting $m_1/m_2 = r$ and assuming $r \geq 1$, we get

$$\alpha = \frac{r}{1 + \sqrt{r^2 - r + 1}}.$$ (18)

For instance, $r = 2$ gives $\alpha = 0.73$ and $\lambda = 0.46/m_2 = 0.92/m_1$.

Comparing α (given by (18)) with β (given by (13)) amounts to comparing the ω-EE and the CEEI solutions. For instance, when r goes to infinity, one checks that $1 - \beta$ goes to zero much more slowly than $1 - \alpha$ (the former as $1/\sqrt{2r}$, the latter as $1/2r$).

4.7. RESOURCE MONOTONICITY

Consider two fair-division problems $(R_1, \ldots, R_n; \omega)$ and $(R_1, \ldots, R_n; \omega')$ that differ only in the global resources to be divided. If the resources grow from ω to ω' (that is, for every commodity k, $\omega^k \leq \omega'^k$), the second problem is undoubtedly a better problem, for there is more manna to share among the same population (and preferences are monotonically increasing in all commodities). A natural requirement is that everyone should benefit from growth: no one should end up worse off from the division of ω' than from the division of ω. This is the resource monotonicity[33] axiom:

for all i, if agent i receives z_i in $(R_1, \ldots, R_n; \omega)$ and

z_1' in $(R_1, \ldots, R_n; \omega')$, then $\{\omega \leq \omega' \Rightarrow z_i' \, R_i \, z_i\}$.

First, we construct an efficient and resource-monotonic solution for all such problems. We pick an arbitrary vector e in $\mathbb{R}_+^K$ (called the vector of numeraire) and we call *e-egalitarian-equivalent* an allocation $z_1, \ldots, z_n$ such that, for some positive number λ, we have

$$z_i \, I_i \, (\lambda e) \quad \text{for all } i = 1, \ldots, n.$$ (19)

The e-EE solution picks an allocation where the parameter λ is as high as possible. Under the same assumptions as Lemma 4.2, such an allocation exists and is Pareto optimal. (Exercise: prove this claim.) Obviously, the e-EE solution is resource-monotonic because any e-EE allocation in the problem $(R; \omega)$ is also an e-EE allocation in the problem $(R; \omega')$ (by free disposal of the goods). The key is that the vector of numeraire e does not depend on the resources ω to be

[33] The axiom was introduced by Chun and Thomson [1988].

divided. By contrast, the EE solution defined along the vector ω (Section 4.6) is *not* resource-monotonic (see below). On the other hand, like the ω-EE solution, the e-EE solution is population-monotonic (see Lemma 4.3; the proof is identical).

The e-EE solution is not, however, a serious solution of the fair-division problem for two reasons. First of all, it does not guarantee the equal-split lower bound. Say $n = 2$ and $k = 2$, and take the first good (money) as numeraire ($e = (1, 0)$). The two agents have linear preferences represented by

$$u_1(x_1, y_1) = x_1 + 10y_1, \qquad u_2(x_2, y_2) = x_2 + y_2.$$

With total resources $\omega = (11, 11)$, the e-EE solution picks the efficient allocation $z_1^* = (0, 2)$, $z_2^* = (9, 11)$, because each agent views this as equivalent to a gift of 20 units of money. However, agent 2 gets about one-third of his equal-split utility![34]

The second defect of the e-EE solution is, in my opinion, even worse. In order to compute its allocation, we must take into account the shape of individual preferences at irrelevant consumption vectors, namely, vectors that could never be awarded in the particular problem at hand because of the feasibility constraints. In the example, agent i views z_i^* as equivalent to a fictitious 20 units of money that do not even exist in the pile to be divided! The ω-EE solution, on the other hand, makes a gift that everyone perceives as equivalent to a fraction λ^* of the *actual* resources (λ^* is at most 1); hence, i can be computed once we know individual preferences over all conceivable shares z (namely, the rectangle $0 \leq z \leq \omega$).

Resource monotonicity, as it turns out, is the culprit for the first defect of the e-EE solution.

Lemma 4.4. (Moulin and Thomson [1988]). In the fair-division problem with divisible goods and convex preferences,

(i) there is no efficient and resource-monotonic solution satisfying the equal-split lower bound,

(ii) every efficient and resource-monotonic solution must have the domination property at some preference profiles (in particular, it must fail the no envy test).

---

[34] Note that furthermore, z_2^* dominates z_1^*; this can happen as well with the ω-EE solution (see Lemma 4.3), but only with three or more agents. In this particular example, the ω-EE solution picks $z_1 = (0, 7.8)$, $z_2 = (11, 3.2)$.

Furthermore, I conjecture that the only efficient and resource-monotonic solutions which are computed only from the restriction of the preference profile to the rectangle $[0, \omega]$ are dictatorial (a certain agent gets all the manna, always).

Proof of Statement (i). The example involves two goods and two agents with the following utility functions:

$$u_1(z_1) = \min\left(\frac{x_1}{2}, y_1\right), \qquad u_2(z_2) = \min\left(x_2, \frac{y_2}{2}\right).$$

Say that S is a solution satisfying all three axioms: efficiency, resource monotonicity, and equal-split lower bound. If ω are the resources to divide, $S_i(\omega)$, $i = 1, 2$, denotes agent i's final utility after division. By the equal-split lower bound, $S_1(12, 6) \geq 3$, so by feasibility, $S_2(12, 6) \leq \frac{3}{2}$; hence, by resource monotonicity, $S_2(6, 6) \leq \frac{3}{2}$. A symmetrical argument shows $S_1(6, 6) \leq \frac{3}{2}$, but by dividing $(6, 6)$ into $z_1 = (4, 2)$, $z_2 = (2, 4)$, each agent reaches the utility level 2, so we contradict the efficiency of S. (Exercise: draw a figure.)

The proof of statement (ii) is omitted for brevity.   Q.E.D.

Note finally that in the context of fair division with transferable utility, we have the same incompatibilities.[35] Moreover, there is no efficient and resource-monotonic solution meeting the stand alone test. See Exercise 4.23.

However, in the fair-assignment problem, a certain application of the Shapely value is together efficient, resource monotonic, and passes both the stand alone test and (a version of) the equal-split lower bound; see Example 7.11 in Section 7.7.

4.8. DIVIDE AND CHOOSE, MOVING KNIVES AND AUCTIONS

Several games implement the no envy and (at least for one player) the fair share guaranteed tests. The simplest games are divide and choose, and moving knife. Both admit a number of interesting yet more complicated variants.

---

[35] See Exercise 4.22 for a proof than an efficient and resource-monotonic solution generates envy; this is true even in the fair-assignment problem. That no efficient and resource-monotonic solution can meet the equal-split lower bound is proven with the same example as above (where utilities are now interpreted as willingness to pay for a non-monetary allocation).

The divide and choose method (D & C) is a very general fair-division game among two agents. It can be used to split a vector ω of divisible goods (the Divider cuts two shares z_1 and $z_2 = \omega - z_1$, the Chooser picks one of these shares, and the remaining share is for the Divider) as well as a piece of land, or a finite set of indivisible goods when monetary transfers are feasible (the Divider makes two piles of indivisible goods and attaches a price t to one pile, $-t$ to the other pile; the Chooser either picks the first pile and pays t to the Divider, or picks the second pile and receives t from the Divider).

One obvious limitation of D & C is the asymmetric role given to the players. Indeed, the equilibrium of the D & C game reflects this inequality, when the Divider is aware of the Chooser's preferences.

Lemma 4.5. *Assume the preferences are continuous and strictly monotonic, and that goods are divisible (alternatively, assume that agents have unbounded reserves of money and that preferences are continuous and strictly increasing in money).*[36] *When the Divider is completely informed of the Chooser's preferences, and the Chooser uses an undominated strategy, the Nash equilibrium allocation is the best envy-free allocation from the Divider's point of view. The Divider gets at least her fair share utility but the Chooser does not. The Divider has a strategic advantage: both agents prefer to play the game as Divider rather than Chooser. Normally, the equilibrium allocation is not Pareto optimal, and any (strictly) Pareto improving move from this allocation makes the Chooser envious.*

Proof. The Chooser picks his preferred share in the second stage of the game (by our assumption that the Chooser uses an undominated strategy; notice that requiring the property of subgame perfection[37] for the equilibrium would lead to exactly the same outcome). Hence, (z_1^*, z_2^*) is an equilibrium allocation if and only if z_1^* is a solution of the program

$$\max\{u_1(z_1) \mid u_2(z_2) \geq u_2(z_1)\}. \tag{20}$$

By the continuity and monotonicity assumptions, this program has the same optimal solutions as the following:

$$\max\{u_1(z_1) \mid z_2 \, I_2 \, z_1\} \tag{21}$$

---

[36] Note that we do not assume here that preferences are quasi-linear, but simply that there is enough money to compensate for any feasible amount of non-monetary goods.

[37] See, e.g., Myerson [1991].

(indeed, if $z_2 \, P_2 \, z_1$, the Divider can trim z_2 a little to the benefit of z_1 without changing the preference relation for the Chooser). Existence of at least one optimal solution z_1^* follows from the continuity assumptions.

Suppose the Divider envies the Chooser at the equilibrium allocation: $u_1(z_2^*) > u_1(z_1^*)$. Then the Divider can augment the share z_1^* to z_1, little enough so as to preserve $u_1(z_2) > u_1(z_1)$ but enough to make the Chooser choose the share z_1 (because $u_2(z_1^*) = u_2(z_2^*)$), resulting in z_2 for the Divider, namely, a better share for her (as $u_1(z_2) > u_1(z_1) > u_1(z_1^*)$). We conclude that both programs (20) and (21) have the same set of optimal solutions as the following:

$$\max\{u_1(z_1) \mid (z_1, z_2) \text{ is envy-free}\}. \qquad (22)$$

This proves the first statement of the theorem. To check that the Divider gets her fair share utility, observe that splitting ω in two equals shares is one option open to the Divider. The fact that the Divider has a strategic advantage follows at once from (22) (note that if any agent is indifferent between the role of Divider or that of Chooser, then the other agent is also indifferent between the two roles).

Two examples below will show why the D & C equilibrium outcome is normally not Pareto optimal. We check now that a Pareto improvement of the equilibrium allocation makes the Chooser envious. Consider a division (z_1, z_2) strictly Pareto superior to (z_1^*, z_2^*) and where the Chooser is not envious. This contradicts the optimality of z_1^* for program (20), because z_1 would satisfy the constraint $u_2(z_2) \geq u_2(z_1)$ and bring a higher utility to the Divider.

The first example has two goods, $\omega = (15, 15)$, and two agents with the following utilities:

$$u_1(x_1, y_1) = x_1^2 y_1, \qquad u_2(x_2, y_2) = x_2 \cdot y_2. \qquad (23)$$

The set of envy-free allocations is depicted in Figure 4.6. It consists of two lenses meeting at $\omega/2$. The straight line $x_1 + y_1 = 15$ is the boundary of the envy-free set where agent 2 is indifferent between z_1 and z_2, namely, $x_1 \cdot y_1 = (15 - x_1) \cdot (15 - y_1)$. When agent 1 divides, she picks her preferred allocation on that line, namely, $z_1^* = (10, 5)$. Notice that the Chooser does not get his equal-split utility: $u_2(5, 10) < u_2(7.5, 7.5)$. Moreover, the equilibrium is not efficient.

In our next example, utility is transferable via money and the two agents share one unit of divisible (nonmonetary) good. Agent 1 cuts two shares (x_1, t_1) and $(1 - x_1, -t_1)$, where $0 \leq x_1 \leq 1$ and t_1 is an arbitrary real number. The Chooser (agent 2) is indifferent between the two

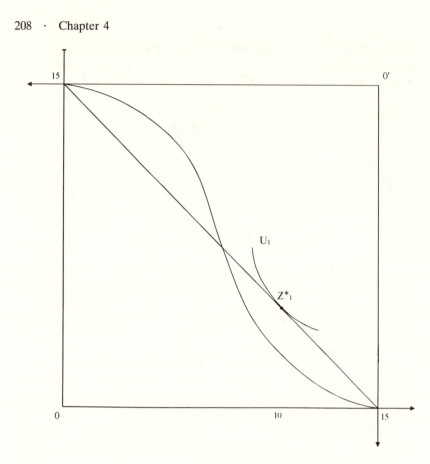

Fig. 4.6. Agent 1 divides and agent 2 chooses in Example (23).

shares if

$$u_2(x_1) + t_1 = u_2(1 - x_1) - t_1.$$

Therefore, the Divider solves the program:

$$\max_{0 \le x_1 \le 1} u_1(x_1) + \tfrac{1}{2}(u_2(1 - x_1) - u_2(x_1))$$

$$\Leftrightarrow \max_{0 \le x_1 \le 1} \tfrac{1}{2}(u_1(x_1) + u_2(1 - x_1)) + \tfrac{1}{2}(u_1(x_1) - u_2(x_1)).$$

Thus the divider is not maximizing the correct objective function $u_1(x_1) + u_2(1 - x_1)$ (in order to bring an efficient division), but only the half sum of the correct objective and of the difference $(u_1 - u_2)$. Exercise 4.24 offers some more numerical examples of this problem.

Q.E.D.

In some interesting cases, the D & C mechanism will, however, yield an efficient equilibrium division. One such case is the fair-assignment problem (Section 4.3), because the no envy property implies Pareto optimality (theorem 4.1). Moreover, the equilibrium division is the best CEEI allocation from the Divider's point of view (see the estate division example discussed in Section 4.2).

Another case is when utilities are additive, as with land division (Example (12)) or Example (11). Then the no envy test is equivalent to the equal-split lower bound $(u_1(z_1) \geq u_1(\omega - z_1) = u_1(\omega) - u_1(z_1) \Leftrightarrow u_1(z_1) \geq u_1(\omega/2)$; see Exercise 4.9), whence the equilibrium allocation maximizes the Divider's utility over the set of allocations meeting the equal-split lower bound. Therefore, it is an efficient allocation where the Divider keeps all the cooperative surplus above the equal-split utilities (and the Chooser get precisely his equal-split utility).

The *moving knife* game, like Divide and Choose, is applicable to any fair-division problem with either divisible goods (and continuous preferences) or indivisible goods when monetary transfers are feasible. Just as in divide and choose, what matters is the possibility to continuously increase or decrease a share without causing a jump in any player's preferences. For instance, in the case of land division, a utility function represented by a nonatomic measure (see note 26) meets the continuity requirement. Unlike D & C, the moving knife method is defined for an arbitrary number of players.

A knife moves continuously across the cake (the resources ω). At time t, the shares $\omega(t)$ and $r(t)$ (remainder) are cut with $\omega(0) = \varnothing$, $r(0) = \omega$, and $\omega(1) = \omega$, $r(1) = \varnothing$. We assume that $\omega(t)$ and $r(t)$ are both continuous in t. At any time t, any one of the players can say stop. If the fist player calling stop does so at time t_1, he gets the share $\omega(t_1)$ and exits. The remainder $r(t_1)$ is then divided in the same fashion among the other players.

The moving knife method treats all players anonymously (no player has a systematic strategic advantage). On the other hand, the choice of the cutting pattern greatly influences the outcome, as one sees easily in the land division game of Example (12). Assume there is only one agent of each type $(m_1 = m_2 = 1)$. If the knife moves from left to right on $\Omega = [0, 1]$ (so that $\omega(t) = [0, t]$), then the informed agent 1 gets all the surplus (above the fair share utility for agent 2) by stopping the knife shortly before agent 2 is indifferent between $[0, t]$ and $[t, 1]$, namely, at $t^* = \sqrt{2}/2$. Similarly, if the knife moves from right to left $(\omega(t) =$

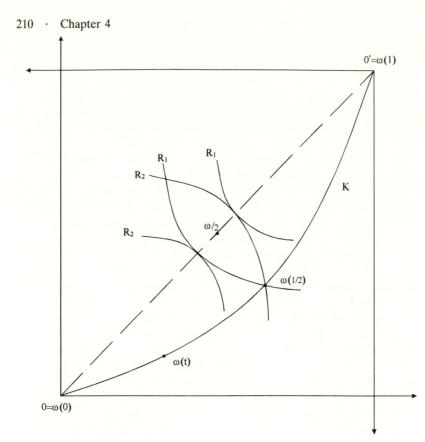

Fig. 4.7. Proof of Lemma 4.6.

$[1 - t, 1])$, then agent 2 captures all the surplus by stopping the knife at $t^* = \sqrt{2}/2$ (so this time, agent 2 gets $[0.3, 1]$ as opposed to $[0.7, 1]$ when the knife moves from the left).

Lemma 4.6. *Assume that preferences are continuous and strictly monotonic and that goods are divisible. In the moving knife mechanism, it may happen that no player can guarantee his/her fair share utility. On the other hand, the equilibrium outcome (among completely informed players using undominated strategies) is envy-free. In general, the equilibrium outcome is not Pareto optimal.*

Proof. Consider the two-person, two-good economy depicted in Figure 4.7. Both agents have identical preferences (e.g., represented by $u_i(x, y) = x \cdot y$) and the knife moves along the path K. Thus no division of the resources proposed by the knife is Pareto optimal (except when one player gets everything). The equilibrium is at $t = \frac{1}{2}$, where both players are indifferent to switching shares (hence the equilibrium is envy-free), but both are below their equal-split utility. Note that, given that player 2 will stop the knife at $t = \frac{1}{2}$ unless player 1 stops it first, player 1 has no way of achieving his equal-split utility. This proves the first and third statements of Lemma 4.6.   Q.E.D.

The moving knife mechanism amplifies the defects of divide and choose: all players may end up below their fair share utility (as opposed to only the Chooser, in divide and choose)[38]; furthermore, the inefficiency of the equilibrium outcome is imposed by the choice of the slicing pattern (even a direct agreement of both players could not restore efficiency in the example of Figure 4.7).

The moving knife mechanism is, therefore, not a serious solution of the fair-division problem. In some cases, however, there is a natural cutting path and the resulting moving knife mechanism is quite appealing. For an example in the fair-assignment problem, see Exercise 4.28.

The main advantage of divide and choose and the moving knife mechanisms is strategic simplicity: each participant acts once and only once before the division is completed. The price we pay for this simplicity has just been described. We turn now to more complex mechanisms where the agents act more than once in a multistage pattern. This opens up many possibilities. For instance, consider the modification of D & C where the Chooser has the option to reject the allocation proposed by the Divider and enforce equal-split instead. Then one serious defect of D & C disappears, namely, the fact that the Chooser is not guaranteed his fair share (of course, this variant is only defined when the goods are divisible). Other multistage variants of D & C are discussed in Exercise 4.25. In these, the asymmetry between Divider and Chooser is greatly diminished by a division of the "cake" in many small crumbs over which players alternate between the role of Divider and of Chooser.

Several multistage generalizations of D & C with three or more participants have been proposed. One such celebrated method is attributed to Knaster by Steinhaus [1948] (from whom the quotation is borrowed):

«The partners being ranged $1, 2, 3, \ldots, n$, 1 cuts from the cake an arbitrary part. 2 has now the right, but is not obliged, to diminish the slice cut off.

---

[38] If, however, preferences are represented by additive utility functions, a player guarantees a fair share by stopping the knife whenever it reaches a share worth $(1/n)$th of the total resources. Exercise: prove this claim.

Whatever he does, 3 has the right (without obligation) to diminish still the already diminished (or not diminished) slice, and so on up to n. The rule obliges the "last diminisher" to take as his part the slice he was the last to touch. This partner thus disposed of, the remaining $n - 1$ persons start the same game with the remainder of the cake. After the number of participants has been reduced to two, they apply the classical rule (one divides while the other chooses) for halving the remainder.»

To feel the strategic complexity of this method, the reader may ponder the following open problem: is the equilibrium of the Knaster mechanism[39] an envy-free allocation? The answer to this question is not known even in the case of additive utilities.

In fact, the literature on the generalization of D & C to three or more agents does not address strategic manipulations by the participants; instead, it assumes a sincere behavior by participants who are not informed of preferences other than their own. The task is one of informational decentralization: to define a simple algorithm to reach an envy-free division in a (small) number of steps where each participant uses information from his own preferences only. Successful algorithms have been found based either on a sequence of divide or choose decisions[40] or on "shaving and padding" steps. See Exercises 4.26 and 4.27 for examples. An excellent survey of this literature is Brams and Taylor [1995].

To summarize, divide and choose and its relatives offer easy methods to implement an envy-free allocation, but not necessarily an efficient one (unless preferences are represented by additive utilities). In particular, they do not yield an easy implementation of the CEEI solution. By contrast, the ω-egalitarian-equivalent solution is easily implemented by a three-stage mechanism where the role of dividing the resources is allocated by auction between all participants.

Definition 4.2. Crawford's auction mechanism (Crawford [1979]). Case of two agents, and divisible goods.

STAGE 1: Each player bids a number λ_i, $0 \leq \lambda_i \leq 1$. The highest bidder, say player 1, becomes the Divider in stage 2 (possible ties are broken in an arbitrary fashion).

STAGE 2: The Divider proposes an allocation (z_1, z_2) of the resources ω. Her proposal is that agent 1 gets z_1 and agent 2 gets z_2.

STAGE 3: The remaining player chooses one of two options: either he accepts the allocation proposed in Stage 2 (so that he gets z_2); or he claims $\lambda_1 \cdot \omega$ (and player 1 get $(1 - \lambda_1) \cdot \omega$).

---

[39] Among fully informed agents who play a subgame perfect equilibrium.

[40] Those mechanisms, however, require additive utilities.

Lemma 4.7. Assume preferences (over divisible goods) are continuous and strictly monotonic. Among fully informed agents who use a subgame perfect equilibrium, the auction mechanism implements an ω-egalitarian-equivalent allocation.

The generalization of Crawford's mechanism to three or more agents is intuitively clear: Stage 1 selects a Divider who proposes a single allocation in Stage 2. In Stage 3, the other agents successively approve or reject the proposal (the order in which they talk is determined by the Stage 1 bidding, higher bidders speaking first). The first one who rejects gets $\lambda_1 \cdot \omega$, and the mechanism tries to give $\lambda_1 \cdot \omega$ to every other non-Divider agent as well. This may not be feasible, though, and some rationing is called for. See Demange [1984] for details.

APPENDIX TO CHAPTER 4

A4.1. Proof of Statement (i) in Theorem 4.1

We are given a fair-assignment problem with q objects and n agents, $q \le n$. We prove constructively the existence of an envy-free allocation. The argument is taken from Aragones [1992]. Let σ be an optimal (i.e., joint utility-maximizing) assignment, and denote

$$q_{ij} = u_i(\sigma(i)) - u_i(\sigma(j)) \quad \text{for all } i, j \text{ in } N,$$

$$\text{and} \quad r_i = \min\{q_{ii_1} + q_{i_1 i_2} + q_{i_2 i_3} + \cdots + q_{i_{t-1} i_t}\}, \tag{24}$$

where the minimum bears on all finite sequences $(i, i_1, \ldots, i_{t-1}, i_t)$ (with possible repetitions) starting from i. Note that r_i cannot be $-\infty$, because along a cycle $(i_0, i_1, \ldots, i_T = i_0)$, we have

$$\sum_0^{T-1} q_{i_t i_{t+1}} = \sum_0^{T-1} u_{i_t}(\sigma(i_t)) - \sum_0^{T-1} u_{i_t}(\sigma(i_{t+1})) \ge 0.$$

This inequality follows from the optimality of σ. Hence, r_i is well-defined and nonpositive (because we can take the sequence (i, i)). Consider the vector of transfers t:

$$t_i = \frac{1}{n}\left(\sum_N r_j\right) - r_i.$$

It is balanced: $\sum_N t_i = 0$. We claim that (σ, t) is an envy-free allocation. Indeed, the inequality $u_i(\sigma(i)) + t_i \ge u_i(\sigma(j)) + t_j$ reads $q_{ij} + r_j \ge r_i$. To show the latter, pick a sequence $j, j_1, \ldots, j_t$ such that $r_j = q_{jj_1} +$

$q_{j_1j_2} + \cdots + q_{j_{t-1}j_t}$. Then $q_{ij} + r_j$ is the sum corresponding to the sequence $i, j, j_1, \ldots, j_t$, hence it cannot be below r_i.

Aragones [1992] also shows that this construction picks, within the envy-free set, that allocation where the largest out-of-pocket payment (the largest negative t_i) is as small as possible. For instance, in estate division (Section 4.2) it chooses $t = 2$, in Exercise 4.2 it chooses $t = (n - p)/n \cdot u_{p+1}$, and in Example (5) it chooses $t_1 = 0$, $t_2 = 1.5$, $t_3 = -1.5$.

A4.2. The Varian Proposition (Varian [1974])

In a fair-division problem with monotonic preferences, a competitive allocation with equal incomes is both efficient and envy-free. In the important special case of the fair-assignment problem where the number of houses is not larger than the number of agents every envy-free (hence efficient) allocation is, conversely, a competitive allocation with equal incomes (Theorem 4.1).

The Varian proposition gives some sufficient conditions guaranteeing the equality of the competitive set with equal incomes and of the set of envy-free and efficient allocations. Once again, each agent must have convex preferences and be negligibly small (when all participants have equal rights over the resources, the latter means that we have a very large—e.g., infinite—number of agents), yet these assumptions alone are not enough. We must also rule out any "gap" in the preferences represented in the economy; thus, if there is an agent with preferences R and one with preferences R', we must be able to find a "path" of agents whose preferences vary continuously from R to R'.

Here is a heuristic proof of the result. Consider an economy with two goods and a continuum of agents each endowed with a Cobb–Douglas utility function:

$$u_\alpha(a, b) = a^\alpha \cdot b^{1-\alpha}.$$

Assume that the parameter α varies in some interval of $[0, 1]$ (for the general proof, the key feature is that the preferences of our agents form a connected subset in the topological space of preferences). Consider a Pareto optimal allocation z: it is supported by some price π for exchanging goods A and B; hence, the allocation z_α to the type-α agent takes this form:

$$z_\alpha = \lambda_\alpha \cdot (\alpha, (1 - \alpha) \cdot \pi),$$

where λ_α is some positive number. Express the no envy property:

$$\lambda_\alpha \cdot \alpha^\alpha (1 - \alpha)^{1-\alpha} \geq \lambda_{\alpha'} \cdot \alpha'^\alpha (1 - \alpha')^{1-\alpha}.$$

Note that the function $x^\alpha(1 - x)^{1-\alpha}$ reaches its maximum at $x = \alpha$, and so for α' close to α, the first-degree approximation of $\alpha'^\alpha(1 - \alpha')^{1-\alpha}$ is just $\alpha^\alpha(1 - \alpha)^{1-\alpha}$.

This means $\lambda_\alpha \geq \lambda_{\alpha'}$ for α' close to α, implying that λ_α is constant with respect to α. The reader can now check that for λ independent of α, the allocation is precisely a competitive allocation with equal incomes.

After Varian's original result (Varian [1974], Champsaur and Laroque [1981]), Zhou [1992] obtains a similar result where the assumption of a connected set of types is dispensed with, but where the no envy test is strengthened to require that no agent envies the averaged allocation of any coalition of other agents (see also Schmeidler and Vind [1972] and Kolm [1973]).

A4.3. The Egalitarian-Equivalent Solution in Fair Division with Money

In the canonical representation of preferences by utility functions $u_i(z_i) + t_i$, we interpret, as usual, t_i as the monetary transfer and z_i as the allocation of nonmonetary goods (which can be divisible or not). The normalization $u_i(0) = 0$ guarantees uniqueness of the function u_i. We speak of the value $u_i(z_i)$ (in dollars) to agent i of the share z_i. Therefore, given global resources ω (made of divisible and/or indivisible goods), the consumption of (z_i, t_i) is equivalent for agent i to a share λ of the stand alone surplus if

$$u_i(z_i) + t_i = \lambda u_i(\omega).$$

An ω-egalitarian-equivalent allocation (in the quasi-linear sense) is an allocation (z_i, t_i), $i = 1, \ldots, n$, such that the coefficient λ is the same for all agents. The ω-egalitarian-equivalent solution picks an ω-egalitarian-equivalent allocation such that the common share λ^* is as high as feasible. Given monotonic preferences (and the normalization $u_i(0) = 0$), the number λ^* is computed as follows. Denote by $v(N)$ the maximal joint surplus from dividing ω:

$$v(N) = \max\left\{ \sum_{i \in N} u_i(z_i) \right\} \tag{25}$$

(where the maximum bears over all feasible divisions of ω among N).

Then λ^* is defined by (we rule out the uninteresting case where nobody cares for the resources: $u_i(\omega) = 0$ for all i):

$$v(N) = \lambda^* \cdot \left\{ \sum_{i \in N} u_i(\omega) \right\}, \qquad (26)$$

and an ω-egalitarian-equivalent solution consists of a surplus-maximizing division of ω as z_i^*, $i \in N$ (a solution of (25)) and the vector of transfers defined by

$$u_i(z_i^*) + t_i^* = \lambda^* \cdot u_i(\omega). \qquad (27)$$

Note that λ^* must be between $1/n$ and 1; this follows at once from the inequalities

$$\frac{1}{N} \sum_N u_i(\omega) \le v(N) \le \sum_N u_i(\omega).$$

(Exercise: prove these inequalities for monotonic utility functions.)

Therefore the ω-egalitarian-equivalent solution meets the *stand alone* test: $u_i(z_i) + t_i \le u_i(\omega)$; and it guarantees to each agent a fair share of the benefits: $u_i(z_i) + t_i \ge u_i(\omega)/n$.[41]

In the basketball story (Section 4.2), the ω-egalitarian-equivalent solution requires Bob to pay \$25 to Ann and \$100 to Charles (computed by solving $250 = 250\lambda^* + 200\lambda^* + 50\lambda^*$). In the most general fair-assignment problem (Section 4.3), this solution still divides joint surplus in proportion to the agents' utilities for their best object. It guarantees to everyone his fair share of his best house, but not necessarily his fair share utility (4). This is proven by the following example:

u_1	u_2	u_3
2	2	1
0	1	1
0	0	1

,

where the ω-egalitarian-equivalent solution gives to agent 3 one-fifth of the joint surplus 4, whereas her fair share utility is 1. This defect can be fixed by measuring benefit shares above and beyond the fair share utilities (instead of the zero utilities). In other words, we define the

---

[41] Note two differences between this lower bound and the equal-split lower bound $u_i(\omega/n)$: the latter is defined for divisible goods only, but preferences need not be quasi-linear; the former is defined for quasi-linear preferences only, but goods need not be divisible.

number μ^* between 0 and 1 by

$$v(N) = \mu^* \cdot \left(\sum_{i \in N} u_i(\omega) \right) + (1 - \mu^*) \cdot \left(\sum_{i \in N} \alpha_i(u_i) \right),$$

and the modified ω-egalitarian-equivalent solution by $u_i(z_i) + t_i = \mu^* u_i(\omega) + (1 - \mu^*)\alpha_i(u_i)$.

We check the properties listed in Lemma 4.3. The ω-EE solution defined by (26) and (27) satisfies properties (ii) and (iv) of Lemma 4.3 and a lower bound analog to equal split, namely, $u_i(\omega)/n$: everyone is guaranteed a "fair share" of the resources. On the other hand, it fails property (iii). For instance, consider a fair-assignment problem with one house and two identical agents for whom the value of the house is 10; hence, they both enjoy a net surplus of 5 at the ω-EE solution. Now add a third agent who values the house at 30: in the ω-EE solution of the three-person problem, our first two agents enjoy a surplus of 6 (check that (26) implies $\lambda = \frac{3}{4}$); hence, they have benefited from the addition of a third player. Thus population monotonicity is violated. On the other hand, whenever a new agent enters, either all the previous agents benefit, or they all suffer. This weaker property, known as population solidarity, is already incompatible with no envy; see Exercise 4.22.

It turns out that in the fair-assignment problem, there is an interesting solution meeting population monotonicity (that solution follows an application of the Shapley value formula, quite a different approach from egalitarian-equivalence). However, in the more general (quasilinear) fair-division problem with money (described in formulas (26) and (27)), there is in general no efficient and population-monotonic solution, even with convex preferences.

We prove the last statement. For any coalition T of agents, we denote by $v(T)$ the (gross) surplus generated by coalition T from consuming ω:

$$v(T) = \max \left\{ \sum_T u_i(z_i) \,|\, z_i \geq 0 \quad \text{all } i \in T, \quad \sum_T z_i = \omega \right\}.$$

Suppose there is an efficient and population-monotonic solution for a given ω and a certain maximal population $N = \{1, 2, 3, 4\}$. We denote by $S_i(T)$ agent i's net surplus (as computed by the solution) when the resources ω are divided among the population T. The monotonicity assumption implies

$$S_i(\{13\}) \geq S_i(\{134\}) \quad \text{for } i = 1, 3.$$

Summing up and rearranging, we get

$$S_4(\{134\}) \geq v(\{134\}) - v(\{13\}).$$

Combining this with a similar inequality for $S_3(\{134\})$, we deduce

$$S_1(N) \leq v(13) + v(14) - v(134).$$

Combining the above inequality with similar statements for $S_i(N)$, $i = 2, 3, 4$, we get

$$v(N) + \sum_1^4 v(N \setminus i) \leq 2(v(13) + v(14) + v(23) + v(24)). \quad (28)$$

Next, consider the following fair-division problem with concave utility functions. We have four goods, $\omega = (6, 6, 6, 6)$, and utility profile:

$$u_1(a, b, c, d) = \min\left(a, \frac{b}{5}, c, \frac{d}{5}\right), \quad u_2(a, b, c, d) = \min\left(\frac{a}{5}, b, \frac{c}{5}, d\right),$$

$$u_3(a, b, c, d) = \min\left(a, \frac{b}{5}, \frac{c}{5}, d\right), \quad u_2(a, b, c, d) = \min\left(\frac{a}{5}, b, c, \frac{d}{5}\right).$$

It is an easy matter to check that inequality (28) is not satisfied in this particular problem.

EXERCISES ON CHAPTER 4

Exercise 4.1. A General Property of Efficient Allocations

In an efficient allocation of private goods, show that there is at least one agent whom nobody envies and at least one agent who envies no one.

Exercise 4.2. Dividing p Identical Objects (Fair Assignment)

The n agents divide p identical objects $p \leq n$. We order them so that agent i's utility u_i for an object decreases with i: $u_1 \geq u_2 \geq \cdots \geq u_n$. A (Pareto) optimal assignment gives a house to the first p agents and yields the surplus $v(N) = u_1 + \cdots + u_p$. Assume $u_p > u_{p+1}$ so that the optimal assignment is unique.

(a) Show that an allocation is envy-free if and only if (i) it is Pareto optimal, and (ii) all agents $1, \ldots, p$ pay t and all agents $p + 1, \ldots, n$

receive t', where

$$pt = (n - p)t' \quad \text{and} \quad \frac{n - p}{n} u_{p+1} \leq t \leq \frac{n - p}{n} u_p.$$

Show that in an envy-free allocation, each agent receives at least his fair share utility (4); and that agents $1, \ldots, p - 1$ and $p + 2, \ldots, n$ receive strictly more.

Exercise 4.3. Computing the Envy-Free Allocations in Some Fair-Assignment Problems

The first three examples have three agents and three houses.

(a)

	u_1	u_2	u_3
h_1	3	2	2
h_2	1	4	3
h_3	2	6	4

Show that we have 12 units of surplus and that fair share guaranteed leaves only three units of loose surplus. Show that the vector of surpluses (x_1, x_2, x_3) corresponds to an envy-free allocation if and only if we have

$$x_1 + x_2 + x_3 = 12, \quad 10 \leq 2x_1 + x_2 \leq 13, \quad 13 \leq x_1 + 2x_2 \leq 14.$$

In particular, agent 2 gets at least $4\frac{1}{3}$, which is $\frac{1}{3}$ above her fair share utility, and at most 6. On the other hand, the share of agent 1 can vary by as much as $2\frac{1}{3}$ within the envy-free set.

(b)

	u_1	u_2	u_3
h_1	2	6	7
h_2	4	3	4
h_3	4	2	1

Check that we have two optimal assignments. Accordingly, the set of envy-free allocations is one-dimensional. Specifically, show that a surplus vector (x_1, x_2, x_3) corresponds to an envy-free allocation if and only if

$$x_1 = 13 - 2x_2, \qquad x_3 = x_2 + 1, \quad \text{and} \quad 3\tfrac{2}{3} \le x_2 \le 4.$$

(c)

	u_1	u_2	u_3
h_1	2	6	7
h_2	5	3	4
h_3	4	2	2

We have three optimal assignments, and the envy-free set is unique utility-wise; find the corresponding surplus distribution.

(d) Fair assignment of tasks: an example. The three jobs can be performed by any one of three companies, and each firm must perform one and only one task. Completing the three tasks brings a gross revenue of $2000, and the cost to firm i of performing task j is given by the following table:

	Firm 1	Firm 2	Firm 3
task 1	700	600	400
task 2	500	500	600
task 3	700	1000	600

Show that this problem is equivalent to a fair assignment of indivisible bads. Find the optimal assignment and the set of envy-free distributions of the net revenue. Show that the envy-free allocation closest to equal split of the gross revenue (also the one selected by the algorithm described in Appendix 4.1) is

	task	revenue
Firm 1	3	800
Firm 2	2	600
Firm 3	1	600

What is the envy-free allocation closest to equal net revenue for all agents?

Exercise 4.4.   No Envy on Net Trades in the Böhm–Bawerk Market

Consider a Böhm–Bawerk market with n sellers and m buyers and utilities

$$v_1 \leq v_2 \leq \cdots \leq v_n, \qquad u_1 \geq u_2 \geq \cdots \geq u_m.$$

Suppose that some trade takes place in some competitive equilibrium ($u_1 \geq v_1$), that not all buyers and not all sellers are active in equilibrium ($u_n < v_n$ if $n \leq m$; $u_m < v_m$ if $m \leq n$). Show that any allocation where (i) some trade occurs, and (ii) no one envies the net trade of any other agent, must be a competitive allocation.

Show that the statement does not hold in a Böhm–Bawerk market where all sellers (resp. all buyers) are active in equilibrium.

Exercise 4.5.   Fair Assignment with Multiplicative Utilities

We consider a problem with n agents and n objects and utility functions of the following form:

$$u_i(h) = \alpha_i \cdot \beta_h \quad \text{for all } i, h, 1 \leq i, h \leq n,$$

where α_i and β_h are two sequences of nonnegative numbers. Many examples fall in this framework: say an object represents a job, β_h representss the number of hours needed for job h, and α_i is agent i's hourly wage.

Assume that agents and houses are ranked by decreasing parameter α_i and β_h, respectively:

$$\alpha_1 \geq \alpha_2 \geq \cdots \geq \alpha_n \geq 0, \qquad \beta_1 > \beta_2 > \cdots \beta_n > 0.$$

Show that the assignment of house i to agent i is optimal. Is it the only one? (You may use Exercise 3.8.)

Show that a vector of transfers $(t_i)_{i=1,\ldots,n}$ yields an envy-free allocation (with the allocation $\sigma(i) = i$) if and only if

$$\alpha_{i+1} \leq \frac{t_{i+1} - t_i}{\beta_i - \beta_{i+1}} \leq \alpha_i \quad \text{for all } i = 1, \ldots, n-1.$$

In particular, $t_1 \leq t_2 \leq \cdots \leq t_n$. Show that the envy-free allocation with the smallest range of transfers $t_n - t_1$ is the solution of the following system:

$$t_{i+1} - t_i = \alpha_{i+1} \cdot (\beta_i - \beta_{i+1}) \quad \text{for } i = 1, \ldots, n-1.$$

Exercise 4.6. Proof of Remark 4.2

We have n agents and q houses, with $q > n$. A competitive equilibrium with equal income is a pair (p, σ) where p is a nonnegative price vector, and σ is an assignment (a one-to-one mapping from N into H), such that

$$p_h = 0 \quad \text{for all } h \notin \sigma(N),$$

$$\text{for all } i \in N: \quad u_i(\sigma(i)) - p_{\sigma(i)} \geq u_i(h) - p_h \quad \text{for all } h \in H.$$

The associated competitive equilibrium allocation uses the vector of transfers

$$t_i = \frac{1}{n}\left(\sum_H p_h\right) - p_{\sigma(i)} \quad \text{for all } i \in N.$$

(a) Obviously, a CEEI must be envy-free. Show that it is Pareto optimal as well.

(b) Show that a CEEI meets the fair share lower bound ((4)). (*Hint:* If $H(i)$ is the set of the best n houses for agent i, write the competitive inequality for all h in $H(i)$.)

(c) To prove the existence of a CEEI, consider a set M of $(q - n)$ dummy agents with $u_j(h) = 0$ for all $h \in H$ and all $j \in M$. Let σ be an optimal assignment. Construct the number q_{ij} and r_i given by (24) in the augmented problem $(M \cup N, H)$. Show that

$$\text{for all } j \in M: \quad r_j = \min_{i \in N}\{r_i\} = r.$$

Define the price vector p as follows:

$$p_{\sigma(i)} = r_i - r \geq 0 \quad \text{for all } i \in N,$$

$$p_h = 0 \qquad \qquad \text{for all } h \in H \setminus \sigma(N).$$

From Appendix 2.1, we know that (σ, t^*) is an envy-free allocation (in $M \cup N$) if we set

$$t_i^* = \frac{1}{q}\left(\sum_{M \cup N} r_j\right) - r_i \quad \text{for all } i \in M \cup N.$$

Deduce that the restriction of σ to N and p forms a competitive price with equal incomes.

Exercise 4.7. Fair-Assignment Problems where the Fair Share Lower Bound is Tight

Consider a fair-assignment problem with n agents and n houses where $u_i(h)$ denotes the utility of agent i for house h (with i and h running from 1 to n). Show that the following properties are equivalent:

(i) The vector of fair share utilities $\alpha_i(u_i)$ (see formula (4)) is efficient (that is, $\sum_i \alpha_i(u_i)$ equals maximal joint surplus).

(ii) The utilities are additively decomposable: there exist two nonnegative sequences a_i, b_h, such that

$$u_i(h) = a_i + b_h \quad \text{for all } i, h, 1 \le i, h \le n.$$

Exercise 4.8. Economies with Leontief Preferences (Section 3.6)

(a) We have two goods, and the resources to be divided are 30 units of each good. The preferences of the three agents are represented by the following utility functions:

$$u_i(x_i, y_i) = \min\left\{x_i, \frac{y_i}{2}\right\} \quad i = 1, 2,$$

$$u_2(x_3, y_3) = \min\left\{\frac{x_3}{5}, y_3\right\}.$$

Show that the following allocation is in the core from equal split:

$$(x_1, x_2, x_3) = (5, 8\tfrac{1}{3}, 16\tfrac{2}{3}), \qquad (y_1, y_2, y_3) = (10, 16\tfrac{2}{3}, 3\tfrac{1}{3}).$$

Yet the two equal agents (1 and 2) are treated very unequally!

(b) Show the set of efficient and envy-free allocations yields, *in the utility space*, the following two intervals:

$$(10\lambda, 10\lambda, 6 - 4\lambda) \quad \text{for } \tfrac{1}{4} \le \lambda \le \tfrac{2}{3} \quad \text{and}$$

$$(10\lambda, 10\lambda, 30 - 40\lambda) \quad \text{for } \tfrac{2}{3} \le \lambda \le \tfrac{5}{7}.$$

Give an example of an efficient and envy-free allocation that violates the equal-split lower bound.

(c) Compute the CEEI allocation.

(d) Now we have K goods and a vector of resource $\omega \in \mathbb{R}^K_+$ ($\omega_k > 0$

for all k). Agent i's utility function writes

$$u_i(z_i) = \min_{k=1,\ldots,K} \left\{ \frac{z_i^k}{a_i^k} \right\},$$

where a_i^k is fixed and positive for all i, k.

Show that if (p, z) is a competitive equilibrium with an equal income of 1 for every agent, we have

$$\sum_{i=1}^{n} \frac{1}{(p \cdot a_i)} a_i^k \leq \omega_i^k \quad \text{for all } k = 1, \ldots, K,$$

with equality if and only if p^k is positive (the notation $x \cdot y$ is the inner product of $\mathbb{R}^K$). Deduce that all CEEI allocations yield the same utility profile. (*Hint:* Assume first $n \leq k$, and that the matrix $[a_i^k]_{i,k}$ is of rank n.)

Exercise 4.9.   Relations between No Envy and Equal Split Lower Bound

Throughout the exercise, n agents divide a vector of goods ω in $\mathbb{R}_+^K$, and their preferences are monotonic and continuous.

(a) Assume each individual preference is *linear*, namely, it is represented by a utility function such as

$$u_i(z_i) = \sum_k \alpha_i^k \cdot z_i^k,$$

where the coefficients α_i^k are fixed and nonnegative (and not all equal to zero). Show that the no envy property implies the equal-split lower bound. If we have only two agents ($n = 2$), show that no envy is equivalent to the equal-split lower bound. If $n = 2$, show that every envy-free and efficient allocation is a *constrainted* competitive equilibrium with equal incomes (as defined in Remark 3.4). However, show that the CEEI is unique.

(b) Assume $n = 2$ and both agents have convex preferences. Show that the equal-split lower bound implies the no envy property. With at least three agents ($n \geq 3$) with convex preferences, show that even among efficient allocations, neither of the two properties (no envy, equal-split lower bound) implies the other.

Exercise 4.10. An Example with Linear Preferences

We have two goods and the resources to be divided are $\omega = (X, Y)$ $(X > 0, Y > 0)$. Agents are of two types: n_k agents are of type k and have utilities

$$u_k(x, y) = a_k x + y \qquad k = 1, 2.$$

We assume $a_1 > a_2 > 0$.

(a) Show that an allocation is Pareto optimal if and only if

$$\{y_i = 0 \text{ for each agent of type 1}\}$$

and/or

$$\{x_j = 0 \text{ for each agent of type 2}\}.$$

(b) Show that the CEEI allocation is unique and given by the following formulas:

- if $a_1 \geq \dfrac{Y}{X} \cdot \dfrac{n_1}{n_2} \geq a_2$, then {
$$x_i = \frac{X}{n_1}; \quad y_i = 0 \text{ for type-1 agents},$$
$$x_i = 0; \quad y_i = \frac{Y}{n_2} \text{ for type-2 agents},$$

- if $a_2 \geq \dfrac{Y}{X} \cdot \dfrac{n_1}{n_2}$, then {
$$x_i = \frac{1}{n}\left(X + \frac{Y}{a_2}\right); \quad y_i = 0 \text{ for type-1 agents},$$
$$x_i = \frac{1}{n}\left(X - \frac{n_1}{n_2}\frac{Y}{a_2}\right); \quad y_i = \frac{Y}{n_2} \text{ for type-2 agents},$$

- if $\dfrac{Y}{X} \cdot \dfrac{n_1}{n_2} \geq a_1$, then {
$$x_i = \frac{X}{n_1}; \quad y_i = \frac{1}{n}\left(Y - \frac{n_2}{n_1}a_1 X\right) \text{ for type-1 agents},$$
$$x_i = 0; \quad y_i = \frac{1}{n}(a_1 X + Y) \text{ for type-2 agents}.$$

(c) Compute all efficient and envy-free allocations (from Exercise 4.9 we know that they all meet the equal-split lower bound).

Exercise 4.11. Fair Division with Single-Peaked Preferences

A single divisible item must be divided among n agents. Agent i's preferences over the various quantities of the item are single-peaked with their peak at a_i. This means that when her consumption of the good increases up to a_i, her welfare is strictly increasing; when the consumption increases from a_i up, her welfare is strictly decreasing.

Given a profile of peaks $(a_1, \ldots, a_n)$ and a quantity ω of the item to distribute, define the *uniform rationing allocation* as follows:

if $\sum_{i=1}^{n} a_i \geq \omega$, then $x_i = \inf\{a_i, r\}$, where r solves $\sum_i \inf\{a_i, r\} = \omega$,

if $\sum_{i=1}^{n} a_i \leq \omega$, then $x_i = \sup\{a_i, r\}$, where r solves $\sum_i \sup\{a_i, r\} = \omega$.

(a) Describe the set of efficient allocations.

(b) Check that this defines a unique allocation and that this allocation can be interpreted as the (unique) CEEI allocation.

(c) Check that the direct revelation mechanism defined by the uniform rationing rule is strategy-proof (even with respect to coalitional deviations). Sprumont [1991] and Ching [1993] show that uniform rationing is the only strategy-proof mechanism satisfying equal treatment of equals and Pareto optimality.

(d) In the familiar example $\omega = 15$, $n = 3$, $a_1 = 15$, $a_2 = 6$, $a_3 = 3$ (Section 4.5 and Example 1.7), compute the core from equal split. Specifically, show that the core contains a unique allocation if agent 2 does not prefer 7 over 5 (if $u_2(7) \leq u_2(5)$), and is empty otherwise.

Exercise 4.12. Where No Envy Implies Efficiency (Berliant, Dunz, and Thomson [1992])

The n agents must divide the interval $[0, 1]$ into n subintervals (no other kind of splitting is allowed; for instance, they divide a period of time, like in the baby-sitting story of Example 1.7). We assume that individual preferences over subintervals of $[0, 1]$ are strictly increasing with respect to inclusion (all prefer $[0.2, 0.4]$ over $[0.3, 0.4]$, but $[0.2, 0, 3]$ may be preferred over $[0.5, 0.8]$).

(a) Show that an envy-free division of $[0, 1]$ must be efficient.

(b) Show that an envy-free division exists for $n = 2$ and $n = 3$, if preferences can be represented by continuous utility functions.

Exercise 4.13. Where No CEEI Allocation Exists, but an Envy-Free and Efficient Allocation does Exist

We have two goods and the vector of resources is $\omega = (3, 3)$. The three agents have identical preferences, represented by the utility functions

$$u_i(a_i, b_i) = \max\left\{ a_i + \frac{b_i}{7}, \frac{a_i}{7} + b_i \right\}.$$

The Pareto optimal allocations are computed by distinguishing agents of type α (with $\{a_i + b_i/7 \geq a_i/7 + b_i\}$ and of type β (with $a_i + b_i/7 < a_i/7 + b_i$). Say that agents 1 and 2 are of type α and agent 3 is of type β:

$$u_1 = a_1 + \frac{b_1}{7}, \qquad u_2 = a_2 + \frac{b_2}{7},$$

$$u_3 = 3 - (b_1 + b_2) + \tfrac{1}{7}(3 - (a_1 + a_2)).$$

Check that the equal-split utility level is $\frac{8}{7}$, and that any two agents, by exchanging their equal-split endowments, can secure the utilities $(2, 2)$.

Check next that the core from equal split is empty. Consider a Pareto optimal allocation as above (agents 1 and 2 of type α, agent 3 of type β). If $a_1 + a_2 < 3$, so that $a_3 > 0$, show that we must have $b_1 = b_2 = 0$, and deduce $u_1 + u_2 < 3$ so that coalition $\{1, 2\}$ can object. If $a_1 + a_2 = 3$, the corresponding subset of the Pareto utility frontier has the equation $u_1 + u_2 + u_3/7 = 3\frac{3}{7}$. Deduce a contradiction again, by invoking the equal-split lower bound and the objectives of two-person coalitions.

Finally, show the existence of an envy-free and efficient allocation.

Exercise 4.14. *Two Examples of the Equality / Efficiency Dilemma*

(a) In the first example, all three identical agents have nonconvex preferences over two goods represented as follows:

$$u_i(x_i, y_i) = \max\{x_i, y_i\}.$$

This corresponds to mutually exclusive goods: an agent needs not consume any good Y (resp. X) while consuming good X (resp. Y). An example is a vacation in two distinct locations at the same time.

Our three agents share three units of good X and five units of good Y. Show that there is no efficient allocation where equals are treated equally (an example of the equality/efficiency dilemma), hence no efficient and envy-free allocation.

A "fair" allocation in this problem takes the form $z_1 = (3, 0)$, $z_2 = (0, \frac{5}{2})$, $z_3 = (0, \frac{5}{2})$ (up to a permutation of the agents). Indeed, this allocation is "as close as can be" to equal utility (we say that these three allocations are selected by the leximin ordering of welfares). Show that this allocation is in the core from equal split.

(b) With the same preferences as in (a), the agents must divide three units of good X and three units of good Y. Show that we still have an equality/efficiency dilemma. Moreover, the "fair" allocations are no longer in the core from equal split.

Exercise 4.15. Where Preferences are Strictly monotonic and yet no Efficient and Envy-Free Allocation Exists (Varian [1974])

Recall that preferences are strictly monotonic if they strictly increase whenever the consumption of a single good increases. For instance, the preferences in the previous exercise are monotonic but not strictly monotonic.

(a) Show that with continuous and strictly monotonic preferences there is no equality/efficiency dilemma. Given a fair-division problem $(R_1, \ldots, R_n; \omega)$, pick continuous utility functions $u_1, \ldots, u_n$ to represent these preferences. Assume that $u_i(0) = 0$ for all i and show the existence of (at least) one efficient allocation where everyone enjoys the same utility level. (*Hint:* Suppose such an allocation does not exist. Consider an allocation where all agents get equal utility λ; as this allocation is not Pareto optimal, show the existence of a Pareto superior equal-utility allocation.)

(b) Consider the two-person, two-good fair-division problem with $\omega = (12, 12)$ and preferences R_1, R_2 represented in Figure 4.8. Agent 2 has strictly monotonic and (strictly) convex preferences and agent 1 has strictly monotonic and convex preferences (represented by the utility function $u_1(x_1, y_1) = \max\{4x_1 + 8y_1, 4x_1 + 3y_1 + 35\}$). The contract curve (the set of efficient allocations) consists of two disconnected intervals OB and $O'A$, where $A = (4, 8)$ and $B = (9, 3)$. The two critical indifference curves (one for each agent) are tangent both at A and at B.

Check that agent 1 envies agent 2 at B as well as at any allocation on the interval OB, and that agent 2 envies agent 1 at any allocation on the interval AO'. Deduce that no efficient and envy-free allocation exists in this economy.

In this example, Vohra [1992] calls *essentially envy-free* the pair of allocations corresponding to the points A and B: they yield the same utility vector, so if the players do not know which of these allocations will be used, they are not envious "for sure." He shows that with strictly monotonic and continuous preferences, the existence of at least one efficient and essentially envy-free allocation is guaranteed.

Exercise 4.16. A Case where Envy-Free and Efficient Allocations Exist (Varian [1974])

Consider a fair-division problem $(u_1, \ldots, u_n; \omega)$ where u_i is continuous and strictly monotonic for all i and, moreover, any two Pareto optimal

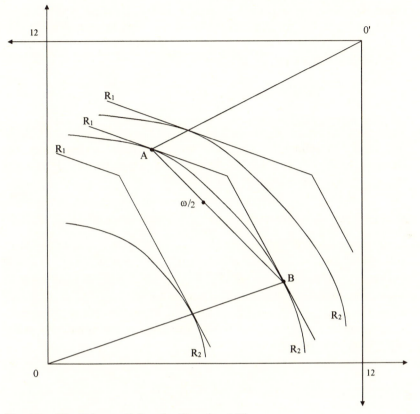

Fig. 4.8. The Edgeworth box for Exercise 4.15.

allocations are distinguished by at least one agent:

$$\{z \neq z', z \text{ and } z' \text{ Pareto optimal}\} \Rightarrow \{u_i(z) \neq u_i(z') \text{ for at least one } i\}.$$

Note that the example in Exercise 4.15 does not satisfy the above assumption. We also assume, without loss of generality, that utility functions have been normalized as $u_i(0) = 0$, $u_i(\omega) = 1$.

(a) Show that the following mapping is an homeomorphism (continuous bijection) from the set P of Pareto optimal allocations into the unit simplex Σ of R^n:

$$z = (z_1, \dots, z_n) \in P \to f(z) \in \Sigma: f_i(z) = \frac{u_i(z_i)}{\sum_1^n u_j(z_j)}.$$

(*Hint:* To show that f is onto Σ, pick λ in Σ and define $A_i(t) = \{z_i \in \mathbb{R}_+^K / u_i(z_i) \geq \lambda_i \cdot t\}$ for all $t \geq 0$. Then use the largest t such that $\Sigma_j A_j(t)$ contains ω.

(b) Define the subset M_i of P as follows:

$$M_i = \{z \in P / u_j(z_j) \geq u_j(z_i) \quad \text{for all } j = 1, \dots, n\}.$$

Show that the union of the sets M_i equals P (use Exercise 4.1). Show that $f(M_i)$ contains the face $u_i = 0$ of Σ. Then deduce from the Knaster-Kuratowski and Mazurkievitcz Lemma (see Border [1985]) that the sets $f(M_i)$ must intersect and conclude.

Exercise 4.17.   Interpreting the Equal-Split Lower Bound (Moulin [1990b])

Consider a fair-division problem $(u_1, \dots, u_n; \omega)$ with continuous convex and monotonic utility functions. In the fair-division problem $(u_i, u_i, \dots, u_i; \omega)$ where all agents have identical utilities u_i, denote by $\lambda \cdot \vec{e}$ the highest feasible utility vector (where $\vec{e} = (1, \dots, 1)$). We write $\lambda = \text{una}(u_i)$ and call this number the unanimity utility of agent i.

(a) Show that if agent i has convex preferences, the unanimity utility is the equal-split utility:

$$\text{una}(u_i) = u_i\left(\frac{\omega}{n}\right).$$

If all agents have convex preferences, show that the equal-split lower bound is the highest feasible lower bound of the fair-division problem.

To show this, consider a mapping ϕ from the utility functions into utility levels. Say that ϕ is a *feasible lower bound* in the fair division of ω if, for all profile of (convex, monotonic) preferences, the vector $(\phi(u_1), \dots, \phi(u_n))$ is feasible in the economy $(u_1, \dots, u_n; \omega)$. Then show the inequality:

$$\phi(u_i) \leq u_i\left(\frac{\omega}{n}\right) \quad \text{for all } u_i.$$

(b) If the preferences of agent i are not convex, show that the unanimity utility can be strictly higher than equal-split utility. Use the example

$$u_i(x_i, y_i) = \max\{2x_i + y_i, x_i + 2y_i\}, \qquad i = 1, 2.$$

Show that if some preferences are not convex, the vector of unanimity utilities may not be feasible (hence cannot be used as a lower bound). Use the example $n = 2$, $\omega = (1, 1)$, and

$$u_1(x_1, y_1) = \max\{2x_1 + y_1, x_1 + 2y_1\},$$

$$u_2(x_2, y_2) = \min\{2x_2 + y_2, x_2 + 2y_2\}.$$

Exercise 4.18. An Example of the ω-EE Solution Adapted from an Example of Young ([1994], p. 131)

Ann and Bob have been left equal shares of their parents' estate. The property consists of $200,000 in cash and 100 acres of land. Bob has money and a sentimental attachment to the land. Ann has little income and no interest in the land except for its resale value. Bob would be indifferent between receiving all of the land or a payment of $300,000. Ann is indifferent between receiving all of the land or a payment of $100,000, which is the amount that she believes it would bring in a sale. We shall assume that both have fixed trade-off rates between money and land, that is, Ann values each acre of land at $1000, while Bob values each acre of land at $3000.

(a) Assume first that Bob is not allowed to buy back some land from Ann out of his own wealth; in other words, he must receive a non-negative share of the land and of the money in the estate. Compute the efficient and envy-free allocations. Compute the ω-egalitarian-equivalent solution.

(b) Now Bob can use his own (unlimited) wealth to compensate Ann. Answer the same two questions as above.

(c) What division of the estate would result if the divide and choose mechanism is played with Ann as Divider? Use successively the context of question (a) and that of question (b) to answer this question.

Exercise 4.19. The ω-EE Solution in some Fair-Assignment Problems

(a) In the problem of Example (5), compute the ω-EE solution and check that it meets the equal-split lower bound. Show that it generates envy. What is the "closest" envy-free allocation?

(b) In the following two-houses, two-agents problem, compute the ω-EE solution and check that it fails the equal-split lower bound (hence it generates envy):

	u_1	u_2
h_1	4	6
h_2	2	5

.

Exercise 4.20. Land Division: Example (12) Continued

(a) Prove the statement made about efficient allocations: a division is Pareto optimal if and only if the type-1 agents own $[0, \alpha]$ and the type-2 agents own $[\alpha, 1]$. (*Hint:* Restrict the feasible allocations to those divisions where the type-1 agents (as well as the type-2 agents) own a finite collection of intervals.)

(b) In the case $m_1 = m_2 = 1$ (one agent of each type), consider the division where agent 1 gets $[0, \sqrt{2}/2]$ and agent 2 gets $[\sqrt{2}/2, 1]$. Check that agent 2 receives exactly her fair share and agent 1 gets all the surplus above his fair share. Show that this allocation is competitive for the following price:

$$p(A) = \int_A x \, dx.$$

Show that, symmetrically, the division where agent 1 gets $[0, 1 - (\sqrt{2}/2)]$ is competitive. Show that every efficient division where each agent gets at least his or her fair share is a competitive allocation. (*Hint:* Use a convex combination of u_1 and u_2 for competitive price.)

(c) What are the competitive allocations in the case $m_1 = 2$, $m_2 = 1$ (two agents of type 1, one agent of type 2)?

Exercise 4.21. Upper Bound on Utility in the CEEI and the ω-EE Solution (Moulin [1991])

(a) Given are the preference profile $R_1, \ldots, R_n$ and the resources ω in $\mathbb{R}_+^K$, $\omega_k > 0$, for all k, satisfying the assumptions of Lemma 4.2. Show that at any ω-EE allocation z^*, we have

$$\left(\frac{K}{n} \cdot \omega\right) R_i \, z_i^*.$$

(*Hint:* Denote by n_k the number of agents i such that $z_i^k > (K/n)\omega^k$ and show that $\sum_{k=1}^{K} n_k < n$.)

(b) By contrast, show a fair division where $K = 2$, n is arbitrary, and the CEEI allocation $\bar{z}$ is such that

$$\bar{z}_1 \, R_1 \left(\left(\frac{n-1}{n} \right) \cdot \omega \right).$$

(c) Fix two convex preferences R_1, R_2 and a vector ω_0 in $\mathbb{R}_+^K$. Consider the fair-division problem with n agents, profile $(R_1, R_2, \ldots, R_2)$, and resources $\omega = n \cdot \omega_0$. Show that if equal split is not Pareto optimal in the two-agent problem $(R_1, R_2; 2\omega_0)$, then as n grows large, at the CEEI allocation, each of the $(n-1)$ R_2-agents receives a utility level arbitrarily close to that of ω_0. Therefore the "eccentric" R_1-agent gets all the surplus above equal split. Compare with the limit (as n grows large) of the ω-EE allocation of agent 1.

Exercise 4.22.   Monotonicity Axioms in the Fair-Assignment Problem

Recall from Appendix 4.3 that in the quasi-linear context, the ω-egalitarian-equivalent solution satisfies population solidarity: when a new agent enters, it cannot be true that one of the previous agents strictly benefits while another one strictly suffers.

(a) Show that in the fair-assignment problem, no selection of the envy-free set can meet population solidarity. Use the following problem with four agents and two houses:

	u_1	u_2	u_3	u_4
h_1	1	6	8	8
h_2	1	6	8	8

(b) Show that in the fair-assignment problem, no selection of the envy-free set can be resource-monotonic. Use the following (example) with three agents and three houses (due to Alkan [1989]):

	u_1	u_2	u_3
h_1	1	6	6
h_2	1	6	6
h_3	1	6	6

Exercise 4.23.   On Resource Monotonicity in Fair Division with Transferable Utility

In a fair-division problem with transferable utility (see Corollary to Theorem 4.1), we show that there is no efficient solution satisfying resource monotonicity and the stand alone upper bound.

Let S be a solution satisfying all three properties. Show that for any two agents and any two vectors ω, ω', we must have

$$v(12; \omega) - u_2(\omega) \le S_1(\omega) \le S_1(\omega \vee \omega'),$$

$$v(12; \omega') - u_1(\omega') \le S_2(\omega') \le S_2(\omega \vee \omega'),$$

where $v(N; \omega)$ is the surplus that coalition N can generate by allocating (and consuming) the resources ω, and $\omega \vee \omega'$ is the coordinatewise supremum. Deduce the following necessary inequality:

$$v(12; \omega) + v(12; \omega') \le u_1(\omega) + u_2(\omega') + v(12; \omega \vee \omega').$$

Then use the following three-goods economy with convex preferences to derive a contradiction:

$$\omega = (1, 1, 0), \qquad \omega' = (1, 0, 1),$$

$$u_1(x_1, y_1, w_1) = \min\{x_1, w_1\}, \qquad u_2(x_2, y_2, w_2) = \min\{x_2, y_2\}.$$

Exercise 4.24.   Inefficiency of the Divide and Choose Equilibrium

(a) In the following three examples, we have two indivisible identical houses and two agents with quasi-linear utilities who can consume up to two goods. In each case, compute the D & C equilibrium outcome when agent 1 divides and check that this outcome is not efficient. What about the D & C equilibrium when 2 divides?

	u_1	u_2		u_1	u_2		u_1	u_2
$\partial u(1)$	4	6	,	4	6	,	5	2	.
$\partial u(2)$	5	2		4	3		2	6

(b) We have one unit of divisible good and money. The preferences of the two agents are quasi-linear:

$$u_1(x_1, t_1) = \sqrt{x_1} + t_1, \qquad u_2(x_2, t_2) = x_2 + t_2,$$

where x_i is agent i's share of the nonmonetary good. Show that if agent 1 divides and agent 2 chooses, the resulting allocation is Pareto optimal and guarantees his fair share to each agent. Show that if agent 2 divides and agent 1 chooses, the outcome is inefficient (and yields a surplus loss of approximately 16% of total surplus).

Exercise 4.25. A Multistage Variant of Divide and Choose

We have two goods X, Y and two agents with utilities

$$u_1(x_1, y_1) = 2x_1 + y_1, \qquad u_2(x_2, y_2) = x_2 + 2y_2.$$

(a) For a given vector of resources $\omega = (\bar{x}, \bar{y})$, compute the equilibrium allocation of D & C when agent 1 divides (denoted $z(\omega; 1)$) and of D & C when agent 2 divides (denoted $z(\omega; 2)$). For instance, check that $z(\omega; 1)$ is as follows:

if $\bar{x} \geq 2\bar{y}$:
$$(x_1, y_1) = \left(\frac{\bar{x}}{2} + \bar{y}, 0 \right), \qquad (x_2, y_2) = \left(\frac{\bar{x}}{2} - \bar{y}, \bar{y} \right),$$

if $\bar{x} \leq 2\bar{y}$:
$$(x_1, y_1) = \left(\bar{x}, \frac{\bar{y}}{2} - \frac{\bar{x}}{4} \right), \qquad (x_2, y_2) = \left(0, \frac{\bar{y}}{2} + \frac{\bar{x}}{4} \right).$$

(b) Consider the following mechanism. First, agent 1 divides ω in two shares ω^1, ω^2. Next, agent 2 chooses one of two options: *either* {to play $\Delta(\omega^1, 1)$ and $\Delta(\omega^2, 2)$} *or* {to play $\Delta(\omega^1, 2)$ and $\Delta(\omega^2, 1)$} (where $\Delta(\omega, i)$ is divide and choose where agent i divides).

Compute the equilibrium of this game (when agents are completely informed of their mutual preferences) and of the symmetrical game where the roles of 1 and 2 are exchanged. Show that the advantage of the first Divider is smaller than in the regular divide and choose method.

(c) Call $\Gamma(\omega, i)$ the mechanism described in question (b) where agent i gets to divide first. Then define a new mechanism as follows. First, agent i divides ω in two shares ω^1, ω^2. Next, agent j chooses either {to play $\Gamma(\omega^1, i)$ and $\Gamma(\omega^2, j)$} *or* {to play $\Gamma(\omega^1, j)$ and $\Gamma(\omega^2, i)$}. Compute the equilibrium allocation and show that the advantage to the first divider is further reduced.

(d) What happens when we iterate the above construction?

Exercise 4.26. An Extension of Divide and Choose

We have three agents and a "pie" Ω. The resources could be divisible goods as in Section 4.4, or they could be a piece of land (as in Example

(12)) that can be divided continuously.[42] For the sake of interpretation, we think of Ω as a cake. Agent 1 cuts the cake in three shares A, B, C. Next agent 2 picks one of these shares, say A, and cuts a wedge W from A: she may choose to cut nothing at all (then $W = \varnothing$). Next, agent 3 picks one of the three pieces $A' = A \setminus W$, B, or C. Two cases arise: if agent 3 did not pick A', then A' goes to agent 2 and the last piece goes to agent 1; if, on the other hand, agent 3 did pick A', then 2 chooses one of B or C and agent 1 gets the last piece (hence C or B).

So far we have distributed $\Omega \setminus W$, and there is nothing more to do if W is empty. If W is nonempty, then whomever from agents 2 or 3 did *not* get A' cuts W in three pieces. Next, the other agent 3 or 2 picks a piece of W; next, agent 1 takes a piece and the last piece goes to the Divider of W.

(a) Assume that utilities over pieces of cake are additive.[43] Check that if we have an envy-free division of $\Omega \setminus W$ among our three agents, and an envy-free division of W, then the combined division of Ω is envy-free.

(b) Show that every agent has a strategy in the above game guaranteeing that he will not envy any other share. For agent 1, this strategy consists of cutting Ω in three equal pieces, and of choosing (one of) the best available piece of W when called to choose one. For agent 2, this strategy is to choose (one of) the best pieces among A, B, C and to cut a wedge W so that $A' = A \setminus W$ is equal to her next best piece, say B (so she takes $W = \varnothing$ whenever there are two equally good first choices in A, B, C); then if called to cut W, to cut it in three equal pieces; and if no called to cut W, to pick (one of) her best piece among the partition of W. Finally, agent 3's strategy is simply to pick his best piece whenever he has to choose, and cut the wedge in three equal pieces if at all.

Brams and Taylor [1995] offer a generalization of this procedure to an arbitrary number of agents.

Exercise 4.27. Shaving and Padding

Assumptions are as in the previous exercise, except that we do not need to assume additive utilities. Any profile of continuous and strictly

---

[42] For any agent and any integer k, any subset of the original land can be divided into k shares of equal worth.

[43] They are represented by nonatomic measures over Ω.

monotonic preferences will work. The shaving/padding mechanism resembles a moving knife game where the position of the knife is endogenous.

Consider a problem with three agents. Agent 1 cuts Ω in three pieces A, B, C. Next, agents 2 and 3 each choose a piece. If they choose different pieces (say, 2 picks C and 3 picks A), they get their choice, agent 1 gets the last piece, and the game is over. If they both choose A, say, then agent 1 starts shaving A at a continuous pace, at the same time padding B and C in any way he likes with the piece he took from A. Thus, at any point in time, we have a division of Ω as A', B', C', where $A' \subseteq A$ and $B \subseteq B'$, $C \subseteq C'$. The shaving/padding continues until one of players 2 or 3, say 3, switches from his choice of A' to the choice of B', or C', say B'. He then receives B', agent 2 gets A', and agent 3 gets C'.

Show that each player has a strategy guaranteeing that he will not envy any other share. A generalization of this mechanism to four players is possible, but it is not known whether it can be extended to an arbitrary number of players (see Brams and Taylor [1995]).

Exercise 4.28. Moving a Knife for Fair Assignment

Consider a fair-assignment problem with n agents and p houses, $p \le n$. The mechanism resembles a descending auction where a price p is decreasing at constant speed starting from a high figure (higher than what any house is worth to any agent). Agents can stop the price p at any time. If the price reaches zero without anyone stopping it, the houses are allocated without transfer of money, by letting the agents choose sequentially following the fixed ordering $1, 2, \ldots, n$. If an agent, say agent 3, stops the price at the level p, he gets to choose the house he likes best and he pays $p/(n-1)$ to every other agent. The descending auction then continues among the remaining $(n-1)$ agents (with $(p-1)$ houses left) in the same way, starting from price p. Possible ties when several agents stop the price at the same time are resolved by means of the fixed ordering.

(a) Suppose we have a single house ($p = 1$). Show that this mechanism yields the same equilibrium outcome as the second-price auction (Section 4.2).

(b) Suppose $n = p = 2$, and show that the equilibrium outcome must be an envy-free allocation. In fact, it is the very selection of the envy-free set described in Appendix 2.1.

(c) With three or more houses to divide, this mechanism does not always implement an envy-free allocation. Consider the following example with $n = p = 3$:

	u_1	u_2	u_3
h_1	6	4	0
h_2	2	1	0
h_3	4	0	10

Show that in the equilibrium among completely informed agents, agent i gets h_i, for $i = 1, 2, 3$, with the following transfers: $t_1 = -2$, $t_2 = t_3 = +1$. Hence agent 1 envies agent 3.

(d) In the example (5), show that the equilibrium assignment is efficient, and even envy-free. The transfers are $t_1 = t_2 = 1$, $t_3 = -2$. Check that this envy-free allocation is different from the one constructed in Appendix 2.1.

Fair Division: The Stand Alone Test

5.1. Models of Cooperative Production

A given group of agents use jointly a certain technology namely, a production process transforming certain inputs (provided by the agents) into certain outputs.

This formulation is general enough to include many different models, where the output may be a private or a public good and the marginal cost of production may increase or decrease.

A first series of examples are from common property resources such as fisheries, forests, pastures, mines, and other increasing marginal cost technologies. The agents contribute the fishing (or cutting, or digging) input and collect the fish (or wood, or minerals) output. For instance, in Example 1.4, the technology is the pasture transforming lean cows into fat cows; or in a cooperative production (Israelsen [1980]), workers contribute some input and collect a share of total output produced.

In a natural monopoly problem (see Section 2.6), a certain good, such as power supply or a road network, is produced under decreasing marginal costs, and is assigned to a regulated monopoly (see the arguments leading to regulation in Sections 2.6 and 2.7). In the common property resources and natural monopoly models, the general question is: how do we cooperatively arrange production among a set of agents with given preferences? Who should contribute how much input and consume how much output?

Public goods offer another rich family of cooperative production models. A pure public good is a commodity consumed without rivalry (my consumption of the good does not reduce yours) and without exclusion (the same amount of good is available to all agents). Flowers and art objects in public places (see Examples 1.2 and 1.5), radio programs, laws, and armies are typical examples or pure public goods. Just as in the case of decreasing marginal cost technologies producing private goods, efficiency commands producing the public good with a single machine (avoiding wasteful duplications). The question is: how much public good should be produced and how should its cost be divided?

Public bads (such as air, noise, or litter pollution) raise a symmetrical problem, where the agents enjoy privately the benefits from polluting and consume without rivalry or exclusion the bad resulting from the sum of individual polluting actions.

In all four models above, cooperative production entails externalities. In public good (resp. bad) models, this externality is immediately visible, because when I increase my contribution to the public good (bad), my fellow agents at once benefit (suffer) from the increase of public good (bad). In the private good model, the externality comes indirectly via the returns to scale: if agent 1 asks more output, the corresponding increase in total output affects total returns (improving them if marginal costs decrease and vice versa), and in turn affects the cooperative opportunities of the rest of the agents.

In this chapter and in Chapter 6, we apply our tri-modal cooperative methodology to these four cooperative production models. We find it very difficult, indeed often impossible, to fulfill the cooperative program stated in Chapter 1. In other words, we cannot identify a cooperative solution that is together fair, stable by direct agreements, and is the equilibrium outcome of a convincing decentralized mechanism. To be sure, several interesting solutions can be designed, but none clearly stands out. In this chapter, we concentrate on the two cooperative modes based on the efficiency postulate, namely, core stability and justice. We note that core stability often cuts a large subset of efficient outcomes and must be supplemented by fairness criteria (Sections 5.4, 5.7, 5.8). Several conflicting interpretations of fairness coexist, each one guided by its own normative principle. As will be clear in Chapter 6, there are no simple decentralized mechanisms to implement the first-best solutions proposed in this chapter. Yet the simple mechanisms discussed in Chapter 6 are crucially related to the normative tests of equity (such as the no envy or the stand alone tests) discussed below.

Throughout this chapter, the agents are endowed with equal property rights over the production opportunities. The interpretation of equal property rights over a certain technology, represented by a production set Y, will take two forms: either access to Y is the indivisible property of the agents as a whole (the common property regime) or each agent "owns" a copy of Y, namely can use the technology Y independently of the other agents (the equal private ownership regime).

In the *common property regime*, there is a single copy of the technology: for instance, a lake or a forest is a technology transforming fishing or wood-cutting effort into fish or wood (Ostrom [1991] reviews many other examples). The technology may have increasing marginal costs (decreasing returns), e.g., the fishing grounds, or decreasing marginal costs (increasing returns), e.g., the supply of electricity. The important point is that it cannot be reproduced (or could be at a prohibitive cost).

In the *equal private ownership regime*, each agent owns a copy of the same technology Z: by pooling their resources together (coordinating input contributions and output shares), they have access jointly to the technology $n \cdot Z$.[1]

The crucial difference is that in the common property regime our agents, individually or in coalitions, have no alternative but to cooperate as a whole, hence the stability by direct agreements has no bite (all coalitions are powerless except for the grand coalition); whereas in the equal private ownership regime the threat of direct agreements within all coalitions cuts a smaller set of stable outcomes within the efficiency frontier (e.g, the stand alone core in the provision of public good problem or in the case of a private good produced under decreasing marginal costs). However, if the two regimes differ in the positive mode of cooperation by direct agreements, they look alike in the normative mode of justice. In particular, stability arguments such as the stand alone test convey an important normative principle of no subsidization in the common property regime.

Keeping in mind these two alternative interpretations, we look successively at the production of private goods under increasing marginal costs (Sections 5.2 and 5.3), then under decreasing marginal costs (Sections 5.4 and 5.5) before turning to public goods (Sections 5.6 to 5.8) and public bads (Section 5.9). In order to relate these different models to one another, we note that the provision of a pure public good is formally equivalent to the cooperative production of a private good when the output must be divided in equal shares among all agents, no matter how much output is produced and how its cost is distributed. Similarly, the production of public bad amounts to the production of a private good of which the cost must be divided equally.[2]

There is a strong analogy between cooperative production and fair division (Chapter 4): in both cases, all agents have equal property rights over a set of resources that they can freely allocate among themselves. In both models (cooperative production and fair division), we seek to strike a balance between efficiency and fairness: efficiency forces differ-

---

[1] In the private good case. The public good (bad) case yields a different operation (see Section 5.6). Notice also that the production set $n \cdot Z$ coincides with Z in some cases (explained below) so that equal private ownership can be interpreted as free access of each agent to the common technology; in other cases, the set $n \cdot Z$ is bigger than Z, so that cooperation creates some genuine production possibilities.

[2] A different but also plausible structural constraint requires that input and output shares be in the same proportions for all agents (same average cost for all agents); we discuss it in the next chapter; Sections 6.4 and 6.5

ent final consumptions for agents with different preferences, but exactly how much inequality is required by equity?

As it turns out, the cooperative production of private goods under increasing marginal costs (decreasing returns to scale) is essentially a variant of fair division and entertains the same key solution, namely, the competitive equilibrium with equal incomes (CEEI); see Section 5.2. The CEEI solution is envy-free and is in the core of the equal private ownership regime (where each agent owns an increasing marginal cost machine and cooperation allows them to combine several machines). However, in the common property regime, the CEEI solution can be criticized by virtue of the stand alone test. The test requires that no agent should enjoy a greater utility than when she is free to use the technology unencumbered by the presence of other users (Section 5.3).

The stand alone test underlines the fact that under increasing marginal costs, production externalities are "negative," because the more other agents use the technology, the worse the overall returns (so everyone wishes he had fewer partners consuming less output). In fair division, similarly, the more other agents eat, the smaller my share. As in fair division, the challenger of the CEEI solution based on the stand alone test is an egalitarian-equivalent solution (Section 4.6).

Next, consider the production of a private good (or a public good) under decreasing marginal costs (or arbitrary marginal costs in the public good case). Here the production externalities are positive: the more other agents use the technology, the better for me. The same observation applies to the provision of a public good (one agent will do us no harm because of nonrival consumption and he will pay part of the cost), where production externalities are positive as well. It should come as no surprise that the problems with positive externalities are radically different from those with negative externalities. The main difference is that under positive externalities, the stand alone test is compelling and no envy must be dismissed, whereas under negative externalities, both tests (stand alone, no envy) make good sense and lead to conflicting interpretations of fair division.

In Section 5.4, we discuss the model where one private good is produced under decreasing marginal costs. Under equal private ownership, each agent has free access to the technology Y and cooperation gives rise to the same set of feasible outcomes as if a single copy of the technology existed. The stability property derived from free access is the stand alone core (just as in the public good context—see below). The stand alone core is nonempty in the one-input, one-output case.[3] It is

---

[3] However, recall from Section 2.8 that the possibility of an empty core promptly arises when we allow subadditive cost functions of which the marginal cost is nonmonotonic. Also, the presence of multiple inputs or outputs raises the possibility of an empty core.

not compatible with the no envy test (hence with any competitive-like outcome such as would result from a multipart tariff).

The stand alone core is fairly large within the Pareto efficiency frontier and leaves much room to the normative task of defining a single-valued solution. In the common property regime, the stand alone core test has considerable normative appeal: its violation means that a group of consumers is subsidizing the complement coalition. The stand alone core (normative) property also implies that the cost share of a given coalition of consumers is at most the cost of serving the demands of that coalition (its stand alone cost) and at least the additional cost of serving this coalition when the complement coalition is already served (its incremental cost). These are simple and intuitive guidelines for cost-sharing (Faulhaber [1975], Baumol [1986]). Two single-valued selections of the stand alone core following the egalitarian-equivalence logic are discussed in Section 5.5.

Sections 5.6 to 5.8 are devoted to the provision of a public good. In a free access regime (equal private ownership of the common technology) the stand alone core expresses stability with respect to direct agreements and is nonempty.[4] As above, the stand alone core has considerable appeal even in the common property regime. As above, it covers a large subset of efficient outcomes. The normative choice of a single-valued core outcome is an important and much-debated question. We discuss the three main ideas (they are mutually incompatible) in Sections 5.7 and 5.8.

The most popular idea goes back to the time-honoured principle of "pay according to benefit". The modern formulation of this idea (due to Lindahl [1919] and Kaneko [1977]) suggests sharing the cost of the public good in proportion to the marginal benefits of individual agents. The second idea is, once again, egalitarian-equivalence, and the third idea uses the unanimity utility level of agent i as an upper bound on agent i's actual utility. Here, as in the fair-division problems of Chapter 4, the unanimity utility of agent i is the unambiguously fair utility level in the hypothetical economy where all other agents share agent i's preferences..

Section 5.9 discusses the (surprisingly) simpler public bad production problem, where neither the stand alone test nor the no envy test plays any role, and egalitarian-equivalence still yields a reasonable solution.

The models discussed in this chapter and the next are limited by at least three restrictive assumptions. First is the equality of property

---

[4] Here again the assumption that a single public good is produced is important.

rights ruling out any technological advantage and/or difference of skills between the two agents. Of course, when each agent owns a (perhaps) different technology, the discussion of the core in Chapter 2 can be adapted. However, the normative arguments such as egalitarian-equivalence cannot.

Second is the impossibility of direct trade of output for input, or of output for a third good such as money. This materializes in the assumption that agents own initially no output and that output (and, sometimes, input) is not transferable. A celebrated example by Scarf (see Exercise 5.12) shows that the core may be empty if we drop this assumption.

Third is that our assumptions on individual preferences are the same as in Chapter 4. Preferences are ordinal and noninterpersonally comparable. Moreover, each agent is responsible for her preferences (differences in preferences are ethically neutral) that reflect her tastes and not her needs. For instance, in a model where input is labor and output is corn, we shall not call John "lazy" and Peter "industrious" if John's marginal rate of substitution between leisure and corn is higher than Peter's,[5] nor will we interpret this difference in preferences as the expression of John's "handicap."[6] Instead, we simply say that our agents have different rates of arbitrage between labor and corn.

Accordingly, our approach to fair cooperative production develops the resource-equality viewpoint (and not the welfare-equality viewpoint; see, in particular, Remark 5.1) in the context of production, just as Chapter 4 did for the fair division of unproduced commodities.

5.2. Increasing Marginal Costs: The CEEI Solution

The discussion in this section is very similar to that of Chapter 4. The prominent solution is a variant of the competitive equilibrium with equal incomes, and its justification is two-fold. On the positive side, it is a selection of the core under equal private ownership, and on the normative side, it is envy-free.

We assume first that each agent owns a copy of the same technology, transforming a single input into a single output and described by the cost function γ. We maintain the following assumption in this section:

$$\gamma \text{ is increasing, convex, and } \gamma(0) = 0. \tag{1}$$

---

[5] If at any allocation (x, y), John is willing to provide less additional work than Peter in exchange for one additional unit of corn.

[6] On such alternative interpretations, see Remark 5.1 at the end of Section 5.2.

In a feasible allocation, each agent i contributes some amount of input x_i and receives a share of output y_i. Throughout the entire Chapters 5 and 6, individual preferences decrease (at least do not increase) in the input contribution x_i, and increase (at least do not decrease) in the output share y_i. Suppose each agent owns a copy of the technology γ independently of the other agents: agent i chooses an allocation (x_i, y_i), where $x_i = \gamma(y_i)$ (presumably this choice results from maximizing her preferences, but we do not need to be specific at this stage). Because γ is convex, we can expect an opportunity to cooperate: average cost will be smaller if each one of the n identical "machines" produces the average output:

$$\sum_i \gamma(y_i) \geq n\gamma\left(\frac{\Sigma_i \, y_i}{n}\right),$$

with a strict inequality unless all output levels y_i are equal (an unlikely event, if preferences differ across agents), or unless γ is linear on the interval spanned by the numbers y_i. If the above inequality is strict, coordination of the production activities is a logical consequence of the efficiency postulate.[7] Similarly, every coalition of agents has the opportunity to cooperate, which yields a notion of core illustrated by our first example.

Example 5.1.   Indivisible Goods

The technology γ can produce up to 3 indivisible units of output, at the following cost:

$$\gamma(1) = 3, \qquad \gamma(2) = 10, \qquad \gamma(3) = 21, \qquad \gamma(4) = +\infty$$

(note that incremental costs $\gamma(k + 1) - \gamma(k)$ increase in k). Four agents own a copy of γ and are willing to consume up to 3 units of output. Their utilities are quasi-linear, and described by the following table, where $u_i(y)$ represents, as usual (see Section 2.2), agent i's

---

[7] Of course, if γ is linear over its entire domain (constant marginal cost), there are no cooperative opportunities and the core will consist of the allocation where each agent runs his own machine independently of others.

willingness to pay for y units of output:

y	u_1	u_2	u_3	u_4
1	6	8	10	12
2	10	14	18	22
3	12	18	24	30

(note that preferences are convex, i.e., incremental utilities $u_i(y + 1) -$ $u_i(y)$ decrease in y). To describe the cooperative opportunities, we compute, as usual (see Chapters 2 and 3), the net surplus feasible for any coalition. First of all, individual agents can run the technology by themselves, yielding the surplus

$$v(1) = 3, \quad v(2) = 5, \quad v(3) = 8, \quad v(4) = 12$$

(agents 1 and 2 produce 1 unit each, whereas agents 3 and 4 produce 2 units each). Assume next our agents want to efficiently coordinate their production plans: the optimal organization of production maximizes

$$\sum_1^4 u_i(y_i) - \sum_1^4 \gamma(t_i),$$

where the integers y_i, t_i are chosen arbitrarily except for the feasibility constraint $\sum_1^4 y_i = \sum_1^4 t_i$. This program is easily solved in two stages. Compute first the cooperative cost function, namely, the cost function resulting from the efficient combination of the four copies of technology γ:

y	1	2	3	4	5	6	7	8	9	10	⋯
∂c	3	3	3	3	7	7	7	7	11	11	⋯

It remains to maximize the surplus $\sum_1^4 u_i(y_i) - c(\sum_1^4 y_i)$. Hence, 6 units must be produced and $y_1 = 0$, $y_2 = 1$, $y_3 = 2$, $y_4 = 3$, for a total surplus $v(N) = 30$.

In this example, it turns out that there is a unique core allocation, with the following monetary transfers:

$$t_1 = 4, \quad t_2 = -3, \quad t_3 = -10, \quad t_4 = -17,$$

and surplus shares

$$\alpha_1 = 4, \qquad \alpha_2 = 5, \qquad \alpha_3 = 8, \qquad \alpha_4 = 13.$$

Notice that we allow agent 1 to be subsidized by his fellow agents, in line with the quasi-linear assumption (see Chapter 2).[8]

To prove the claim, compute successively

$$v(14) = 17 \Rightarrow v(14) + v(2) + v(3) = v(N)$$

$$\Rightarrow \alpha_1 + \alpha_4 = 17, \quad \alpha_2 = 5, \quad \alpha_3 = 8,$$

$$\left.\begin{array}{c} v(13) = 12 \\ v(24) = 18 \end{array}\right\} \Rightarrow v(13) + v(24) = v(N)$$

$$\Rightarrow \alpha_1 + \alpha_3 = 12, \quad \alpha_2 + \alpha_4 = 18.$$

The unique core allocation, on the other hand, is also the competitive equilibrium with price $p = 7$ and equal income 4. Indeed, the price 7 yields the demands $\delta_1 = 0$, $\delta_2 = 1$, $\delta_3 = 2$, $\delta_4 = 3$, and the supply of 6 units by the (competitive) firm using the technology c. Of course, at that price, the firm may supply 4, 5, 6, or 7 units for an optimal profit of 16. This profit is then distributed equally among the four agents.

In general, the competitive equilibrium with equal incomes (CEEI) solution is the competitive equilibrium where each agent owns an equal share of the firm, and where agents and firm react competitively to the price signal. This definition parallels that of Chapter 4 (Section 4.4) and gives one more example of the versatility of this concept (already illustrated in Section 4.5).

To give the formal definition, we assume for convenience that input and output are divisible. Given n copies of the technology γ (satisfying (1)), the cooperative opportunities are described by the cost function c:

$$c(y) = n\gamma\left(\frac{y}{n}\right). \tag{2}$$

(Exercise: prove this claim.)

Throughout this chapter, it will be convenient to use the utility notations instead of the ordinal description of preferences (used in some of Chapters 3 and 4).

---

[8] In some of the examples of this chapter we allow likewise transfers of input across agents; in some other examples we do not. Ruling out the transfers may complicate the analysis, as in the present example where we would lose the convenience of the quasi-linear framework.

Definition 5.1. Let the preferences of agent i be represented by the utility function $u_i(y_i, x_i)$ increasing in output share y_i and decreasing in input contribution x_i. A CEEI allocation is a list $(p, r; (y_i, x_i), i = 1, \ldots, n)$, such that

(i) the allocation $((y_1, x_1), \ldots, (y_n, x_n))$ is feasible: $\sum_1^n x_i = c(\sum_1^n y_i)$,

(ii) for all i, the allocation (y_i, x_i) maximizes u_i over the budget set

$$py_i' \leq x_i' + r, \qquad y_i' \geq 0,$$

(iii) the aggregate production $(y, x) = \sum_1^n (y_i, x_i)$ maximizes the firm's profit $n \cdot r$:

$$n \cdot r = p \cdot y - x = \max_{y' \geq 0} p \cdot y' - c(y').$$

We always assume nonnegative output shares $y_i \geq 0$ for all i. This holds true if no agent has any initial endowment of output, or if output is not tradable. This assumption plays little role in the increasing marginal cost context, but does play an important role in the core analysis under decreasing marginal costs; see the comments after Theorem 5.1. Depending on the context, we allow input contributions x_i of an arbitrary sign (when the input is transferable, say, money; e.g., Example 5.1) or we only allow nonnegative contributions $x_i \geq 0$ (when the input is nontransferable, say, labor; e.g., Example 5.2). In both cases, the total input $\sum_1^n x_i$ must be nonnegative for feasibility.

Existence of a CEEI allocation is guaranteed under convexity and continuity of preferences (in the models with x_i nonnegative or of an arbitrary sign), provided every agent has a finite endowment of input. The proof is similar to that of Lemma 2.3.

The discussion of the CEEI solution parallels that of the fair-division problem in Chapter 4. The allocation where each agent runs her own machine γ (independently of other agents) corresponds to the equal-split allocation in Chapter 4. The core under equal private ownership of γ corresponds to the core from equal split. In particular, a CEEI allocation belongs to the core (the proof of Lemma 4.1 is easily adapted).

The two main normative properties of the CEEI solution are the unanimity lower bound and the no envy test. The unanimity lower bound requires that the utility of any agent be at least what he can reach by using a machine γ for his own sake; here, if c is the common property technology, the cost function γ is defined by (2). The no envy test says that no agent prefers the pair (output share, input contribution) of another agent to his own pair. The CEEI solution is envy-free almost by definition, because each agent gets one of his favorite allocations from the same budget set (described in statement (ii) of Definition 5.1).

A few examples of the CEEI solution occupy the rest of this section. Consider the especially simple example where the cost function c is linear ($c(y) = c \cdot y$, constant marginal costs). In this case, the function γ is *identical* to c: as returns to scale are constant, there is no disadvantage or advantage to replacing several identical machines by a single machine. The unanimity utility of agent i obtains by maximizing u_i over the set: $c \cdot y_i \le x_i$. Denote by (y_i, x_i) a solution to this program, and consider the list $((y_1, x_1), \ldots, (y_n, x_n))$: this is a feasible allocation and equals the CEEI allocation with $p = c$, $r = 0$ (the firm's profit is zero at all levels of production). One checks easily that there is no other CEEI allocation (except for possibly multiple solutions to the individual utility maximization over the budget set). In the case of constant marginal cost, the decentralized use of the technology produces an efficient outcome (there are no production externalities whatsoever); this outcome is the only feasible allocation meeting the unanimity lower bound, hence the only one in the core from equal split and the unique CEEI allocation.[9] (Note that this allocation is not, in general, the only efficient and envy-free allocation. Exercise: prove this claim.)

The simplest case of nonconstant marginal costs is the subject of our next example.

Example 5.2.   Zero-One Marginal Costs

The cost and production function are as follows:

$$c(y) = \alpha \cdot (y - a)_+, \qquad y \ge 0 \Leftrightarrow f(x) = \frac{x}{\alpha} + a, \qquad x \ge 0$$

(where we denote $(z)_+ = \max(z, 0)$).

We have a units of output for free, and additional units cost α. We think of the combination of two problems, namely, the fair division of a units of output and the joint use of a constant marginal cost technology. The production externalities result from the interdependency of these two simple production processes.

Compute the "equal-split" technology γ defined by (2):

$$\gamma(y) = \alpha \cdot \left(y - \frac{a}{n} \right)_+ .$$

---

[9] This outcome is also selected by each one of the two egalitarian-equivalent solutions discussed in Sections 5.3 and 5.5, and by the stand alone upper bound test. Also, the decentralized mechanisms of Chapter 6 all have this allocation as their unique equilibrium outcome.

The first a/n units are free, each additional unit costs α. Let agent i maximize her utility over the allocations afforded by γ at the allocation (y_i, x_i). Because utilities increase in output, we have $y_i \geq a/n$ and $x_i = \alpha(y_i - a/n)$. We claim that the resulting allocation $((y_1, x_1), \ldots, (y_n, x_n))$ is Pareto optimal *if we assume that input is non-transferable* ($x_i' \geq 0$ for all i at any feasible allocation).[10] As in the case of constant marginal costs ($a = 0$), the program defining the unanimity utility is also the maximization of utility over the competitive budget set: $\alpha \cdot y \leq x + \alpha \cdot a/n$; hence, the allocation $((y_1, x_1), \ldots, (y_n, x_n))$ is precisely the CEEI solution. Both the unanimity lower bound and the CEEI concept make identical recommendations in this problem: split the free output equally and let everyone use freely the technology to buy additional output.

This remarkable coincidence of the two concepts does not hold anymore if input is transferable across agents. In that case, it may not be efficient to give some free output to an agent i who does not care for the output (who has a high marginal rate of substitution between input and output); one should instead give the output to other agents and compensate agent i with some input. Consequently, the vector of unanimity utilities is no longer efficient, and the CEEI solution improves upon it. See Figure 5.1 for a two-person example and Exercise 5.7 for a systematic discussion.

In our next example, marginal costs increase at every level of production and we have a typical configuration: a simple CEEI allocation among many allocations in the core with private ownership of γ and many envy-free and efficient allocations. The situation in Examples 5.1 and 5.2, with the core containing only the CEEI allocation, is exceptional.

Example 5.3. A Simple Production Economy

There are two goods, labor and corn, and a single field plowed by both agents. If x is the total amount of hours worked in the field, the

---

[10] Suppose it is not and let $((y_1', x_1')), \ldots, (y_n', x_n'))$ be a Pareto superior allocation. With the help of a figure, one checks that

$$y_i' \geq \frac{a}{n}, \qquad x_i' \leq \alpha \cdot \left(y_i' - \frac{a}{n}\right) \qquad \text{for all } i,$$

and the right-hand inequality must be strict if agent i strictly prefers (y_i', x_i') to (y_i, x_i). Summing up the right-hand inequalities yields a contradiction.

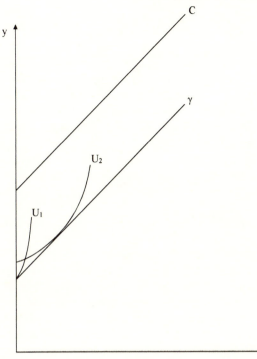
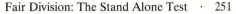

(a)

Fig. 5.1. (a) The CEEI in Example 5.2: when input is not transferable; (b) or transferable (Fig. 5.1b).

quantity of corn produced is

$$f(x) = 6 \cdot \sqrt{2x} \quad \text{all } x \geq 0 \qquad (3)$$

Thus average and marginal product decrease at the rate $1/\sqrt{x}$.

Two agents with respectively High and Low marginal rate of substitution between labor and corn own this field in common:

$$u_H(y, x) = y - x, \qquad u_L(y, x) = y - 3x.$$

Agent High is willing to work three times as much as agent Low for one unit of corn. Recall that we do not call Low lazy or handicapped.

Labor is not transferable among agents, so we consider only allocations where $x_i \geq 0$ for $i = H, L$. Because utilities are linear, efficiency plus voluntary participation (nonnegative net utility) for Low commands

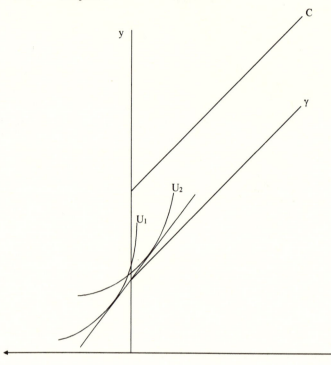

(b)

Fig. 5.1. Continued.

having High working in the field at the level $x_H^* = 18$ (the level of input where marginal product equals marginal disutility), whereas Low does not work at all. The 36 units of output can then be shared arbitrarily between the two agents (that is, every such division yields a Pareto efficient allocation). The price of the CEEI allocation must be $p = 1$, because $p > 1$ generates zero supply of work, and $p < 1$ generates too much supply (High wants to work as much as possible, and we assume that he is endowed with more than 18 hours of labor). The firm's profit is $1.36 - 18 = 18$, and the CEEI solution recommends an equal split of this surplus. Thus Low receives 9 units of corn (and does not work), whereas High gets 27 units of corn and works 18 hours.

The no envy test cuts a fairly large subset of efficient allocations. Let $x_H = 18$, $x_L = 0$, and y_H, y_L be such that $y_H + y_L = 36$. Then no envy requires

$$y_H - 18 \geq y_L, \quad y_L \geq y_H - 3.18, \quad \Leftrightarrow 27 \leq y_H \leq 36, \quad 0 \leq y_L \leq 9.$$

Characteristically, the CEEI allocation picks the worst envy-free and efficient allocation from the point of view of High (see a similar pattern in the first example of Section 4.5).

Now apply the other test of equity, namely, the equal-split lower bound. The argument is the same as in fair division. The equal-split utility of High obtains by considering an economy with two High agents. The highest feasible equal-utility vector is $(9, 9)$. Similarly, in an economy with two Low agents, 12 units of corn are produced and the efficient equal utility vector is $(3, 3)$. Thus the equal-split lower bound picks the following efficient allocations:

$$27 \leq y_H \leq 33, \qquad 3 \leq y_L \leq 9. \tag{4}$$

Once again, the CEEI allocation is the worst allocation for High within these bounds.

Remark 5.1

If the agents are not held responsible for their preferences, the whole discussion of no envy changes dramatically. The typical (and much discussed) example has two (or more) agents with different innate talents for producing the input necessary to the cooperative production. For instance, one hour of talented John's time produces more usable input than one hour of untalented Bessie's time. If Bessie is not held responsible for her lack of talent (due, perhaps, to a physical handicap), we ought to apply the no envy test to the hours worked by each (counting one hour of Bessie's time as equivalent to one hour of John's time, because this is what matters to their eventual welfare). This interpretation of no envy, however, is often incompatible with efficiency, even if the marginal cost of production is constant and preferences are convex. A simple example (due to Pazner and Schmeidler [1978]) follows.

Two agents share the linear technology $y = x_1 + x_2/10$. They provide different inputs: one hour of work by agent 1 is ten times more productive than one hour of work by agent 2. However, if agents are not responsible for their productivity, no envy means that agent i does not want to exchange his vector (output share, hours worked) for the vector (output share, hours worked) of agent j. Note that input (labor) is not transferable across agents. The utilities are linear:

$$u_1 = \tfrac{11}{10} y_1 - x_1, \qquad u_2 = 2y_2 - x_2.$$

Agents are initially endowed with one unit of labor each. Let (z_1, z_2) be a Pareto optimal allocation. We must have $x_1 = 1$; otherwise, agent 1 can work ϵ more, receive ϵ more output, and be better off (while agent 2 is unaffected). Moreover, we cannot have both $x_2 > 0$ and $y_2 > 0$. Otherwise, agent 2 will work ϵ less, receive $\epsilon/10$ less output, and be better off (while agent 1 is unaffected). Now distinguish two cases. First, suppose $y_2 = 0$. Then we have $z_1 = (1, y_1)$, $z_2 = (x_2, 0)$, with $y_1 \geq 1$ and $0 \leq x_2 \leq 1$; therefore, agent 2 envies agent 1. Second, suppose $x_2 = 0$. Then we have $z_1 = (1, y_1)$, $z_2 = (0, y_2)$, with $y_1 + y_2 = 1$. The no envy test requires

$$\text{agent 1 not envious:} \quad \tfrac{11}{10}y_1 - 1 \geq \tfrac{11}{10}y_2,$$

$$\text{agent 2 not envious:} \quad 2y_2 \geq 2y_1 - 1;$$

hence a contradiction, and the example is complete.

5.3. INCREASING MARGINAL COSTS: STAND ALONE TEST AND EGALITARIAN-EQUIVALENCE

In the common property regime, the CEEI solution can be criticized by means of the stand alone test, introduced in Section 4.2. Consider Example 5.3, where Low gets a free ride, namely, 9 units of corn for no work. Under the efficiency postulate, we cannot avoid a free ride: because of the linearity of preferences, efficiency prevents Low from working (see Exercise 5.3 for a general statement); and surely there are good reasons to give some surplus to Low. After all, he would work in the field if he were alone (or were the agent with the highest marginal rate of substitution). His unanimity utility is a plausible lower bound on the corn he should receive (see (4)). However, 9 units of free corn is arguably too much. Indeed, Low could not generate more than 6 units of surplus (by working two hours) if he were working the field by himself. Giving Low anything above 6 units of free corn allows him to benefit from the presence of High, a paradoxical outcome in the presence of negative production externalities (decreasing returns to scale).

The stand alone test (in the increasing marginal cost case) places an upper bound on each agent's utility, namely, that utility she enjoys when she has access to the technology unencumbered by the presence of other users. In the production economy of Example 5.3, this test rules out the CEEI solution, but in some examples it does not: in Example 5.1, the stand alone test places the bounds $\alpha_1 \leq 4$, $\alpha_2 \leq 9$, $\alpha_3 \leq 15$, $\alpha_4 \leq 21$; hence, the CEEI passes the test. The same is true in Example 5.2 (thanks to the assumption that the input good is not transferable; if it is, the CEEI may fail the stand alone test, as shown in Exercise 5.7, question (b)).

The tension between the stand alone test and the competitive solution runs deeper: in some very simple economies, the stand alone and the no envy tests are mutually incompatible (among efficient outcomes).[11]

Example 5.4.   A Three-Person Variant of Example 5.3

The technology for producing corn is the same ((3)) and the three agents, called High, Medium, and Low, still have linear utilities

$$u_H(y, x) = y - x, \qquad u_M(y, x) = 3y - 4x, \qquad u_L(y, x) = y - 6x.$$

As in Example 5.3, efficiency commands letting High work alone for 18 hours. The 36 units of corn are divided as $y_H + y_M + y_L = 36$. No envy requires $y_M = y_L$ (since Medium and Low do not work) and, moreover, Medium should not envy High:

$$3y_M \geq 3y_H - 4.18.$$

Hence,

$$3y_M \geq 3 \cdot (36 - 2y_M) - 72 \Leftrightarrow y_M \geq 4.$$

Thus Low receives at least 4 units of corn "for free," which exceeds her net benefit from working the field by herself:

$$\max_{y \geq 0} 6 \cdot \sqrt{2x} - 6x = 3.$$

In this example, the CEEI allocation is $z_H = (24, 18)$, $z_M = z_L = (6, 0)$. See Figure 5.2.

Further examples of the same tension between the no envy and stand alone tests in the case of decreasing returns are in Exercises 5.3 and 5.6. Note that the two tests are compatible if we allow for inefficient (Pareto inferior) allocations; an example is the serial equilibrium allocation discussed in Section 6.6.

A similar tension exists between the core under equal private ownership (each agent owns a copy of the technology γ) and the stand alone test (no agent should be better off than by having free access to the technology c, where c and γ are related by (2)). Exercise 5.5 gives an

---

[11] Note that both tests are compatible in Example 5.3. Together they cut the subset $0 \leq y_L \leq 6$. If we add the equal-split lower bound, this set shrinks to $3 \leq y_L \leq 6$.

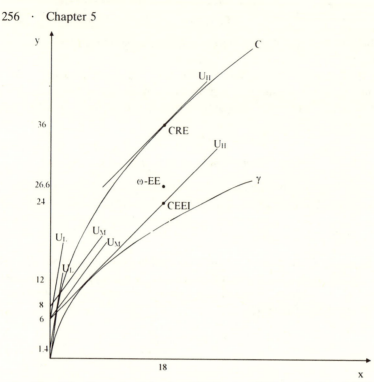

Fig. 5.2. Example 5.4: the CEEI, the CRE, and the ω-EE solutions.

example where no allocation in the core from equal split meets the stand alone test.

Therefore, the stand alone test contradicts the stability of direct agreement in the equal private ownership regime. It has no normative appeal in that regime, where the outstanding solution concept is the competitive equilibrium with equal incomes. On the other hand, in the common property regime, the stand alone test leads to an interpretation of common ownership alternative to equal private ownership. This is the view that agents have an equal right to consume the common property resources (the technology c) and that each agent must bear a share of the burden of negative externalities in consumption. See the discussion in Section 4.2.

Just as in the fair-division problem, the simplest solution embodying this notion of common property is (an adaptation of) the egalitarian-equivalent solution of Pazner and Schmeidler. Consider an arbitrary concave production function f_1. The unencumbered usage of f_1 allows one to choose an allocation (y, x), such that

$$y = f_1(x).$$

Under "equal split" of f_1, each agent owns a copy of the technology $f_{1/n}$:

$$y = f_{1/n}(x) = \frac{f(nx)}{n}$$

(note that by cooperatively running n copies of $f_{1/n}$, the aggregate production function is just f_1). When the n agents are endowed with utilities $(u_1, \ldots, u_n)$, we define the equal-split utility of agent i, also called his *unanimity utility*:

$$una(u_i) = \max_{x \geq 0} u_i(f_{1/n}(x), x) = \max_{x \geq 0} u_i\left(\frac{f(nx)}{n}, x\right)$$

$$= \max_{y \geq 0} u_i\left(y, \frac{c(ny)}{n}\right). \tag{5}$$

(Indeed, in an economy where all agents share the same utility function u_i representing convex preferences, the highest feasible equal utility vector gives $una(u_i)$ to all.) Define also agent i's stand alone utility as follows:

$$sa(u_i) = \max_{x \geq 0} u_i(f(x), x) = \max_{y \geq 0} u_i(y, c(y)),$$

and, in general, his λ-equivalent utility as

$$\phi_\lambda(u_i) = \max_{x \geq 0} u_i\left(\lambda \cdot f\left(\frac{x}{\lambda}\right), x\right) = \max_{y \geq 0} u_i\left(y, \lambda c\left(\frac{y}{\lambda}\right)\right). \tag{6}$$

Note that the vector of unanimity utilities $(\phi_{1/n}(u_1), \ldots, \phi_{1/n}(u_n))$ is feasible in the "true" economy where one copy of f is owned in common by all agents. Indeed, if (y_i, x_i) is feasible for the technology $f_{1/n}$ for all $i = 1, \ldots, n$ then $(\Sigma_i y_i, \Sigma_i x_i)$ is feasible for the technology f (as $y_i = f(nx_i)/n$ for all i implies $\Sigma_i y_i \leq f(\Sigma_i x_i)$ by the concavity of f). On the other hand, the vector of stand alone utilities $(\phi_1(u_1), \ldots, \phi_1(u_n))$ is either not feasible or barely feasible (that is, Pareto optimal). To see this, suppose that a feasible allocation $((y_i, x_i), i = 1, \ldots, n)$ exists, such that $u_i(y_i, x_i) \geq sa(u_i)$ for all i, with at least one strict inequality, and derive a contradiction. By the definition of sa, we must have $y_i \geq f(x_i)$ for all i, with at least one strict inequality. Hence $\Sigma_i\, y_i > \Sigma_i\, f(x_i)$, which by feasibility implies $f(\Sigma_i\, x_i) > \Sigma_i\, f(x_i)$. Yet f is subadditive (it is concave and $f(0) = 0$), hence the desired contradiction.

It is easy to check (and left to the reader as an exercise) that when marginal costs are constant (linear cost and production functions), both vectors of unanimity utilities and of stand alone utilities coincide and are Pareto optimal. In such an economy, the compelling cooperative outcome is to let each agent pay the real cost of his output share. No other efficient outcome passes the stand alone test, or meets the equal-split lower bound (this outcome is also the CEEI solution and the ω-EE solution).

On the other hand, when marginal costs do increase in the relevant range of output level, the vector of unanimity utilities is Pareto optimal if and only if all agents solve the program at the same allocation (for instance, if all share the same utility). However, the vector of stand alone utilities is *never* feasible; hence the stand alone test, contrary to the equal-split lower bound, never determines a single efficient outcome. It can also be shown that, in general, the equal-split lower bound bites deeper in the Pareto frontier than the stand alone test (see Watts [1991]).

Note that the concavity of f implies that the quantity $\lambda f(n/\lambda)$ does not decrease in λ. (Exercise: prove this claim.) We are now ready to define the ω-egalitarian-equivalent solution.

Lemma 5.1. Suppose the production function is concave (or, equivalently, the cost function is convex and $c(0) = 0$). Suppose the utility functions u_i are continuous (and, as usual, nondecreasing in y_i and nonincreasing in x_i). Then there is a number λ^*, $1/n \leq \lambda^* \leq 1$, such that the utility vector $(\phi_{\lambda^*}(u_1), \ldots, \phi_{\lambda^*}(u_n))$ is Pareto optimal. The corresponding allocation(s) is called the ω-egalitarian-equivalent allocation of the cooperative production problem $(f; u_1, \ldots, u_n)$. It gives to each agent at least her unanimity utility and at most her stand alone utility.

The straightforward proof of Lemma 5.1 exactly parallels that of Lemma 4.2. Just as in the fair division of unproduced commodities, the ω-egalitarian-equivalent (ω-EE) solution is a robust concept that does not require convexity of preferences (unlike the CEEI solution) and yields a unique utility vector (unlike the CEEI solution; see Exercise 5.6).

We compute the ω-EE solution in the four above examples. In Example 5.2, the vector of unanimity utilities is Pareto optimal;

hence, the critical value is $\lambda^* = 1/n$ and the ω-EE solution picks the allocation(s) achieving this utility, namely, the CEEI allocation(s)! The equality of these two solutions is, of course, exceptional.

In Examples 5.3 and 5.4, we take advantage of the quasi-linear form of utilities to compute the ω-EE solution. In Example 5.3, the maximal joint surplus is 18 (agent 1 works 18 hours) and efficiency is compatible with any division of the 18 units of surplus among the two agents. One computes easily

$$\phi_\lambda(u_H) = \max_{x \geq 0} \{6 \cdot \sqrt{2\lambda x} - x\} = 18\lambda, \qquad \phi_\lambda(u_L) = 6\lambda.$$

The critical value of λ solves $\phi_\lambda(u_H) + \phi_\lambda(u_L) = 18$; hence $\lambda^* = \frac{3}{4}$. Therefore, surplus is divided as $\pi_H = 18 \cdot \lambda^* = 13.5$, $\pi_L = 6 \cdot \lambda^* = 4.5$, and output is divided as

$$y_H = 31.5, \qquad y_L = 4.5.$$

This solution is very plausible in this particular example. It does meet the no envy and stand alone tests, and the unanimity lower bound as well.

In Example 5.4, we change the utility representation of agent M's preferences to $\tilde{u}_M(y, x) = y - (4/3)x$ so as to measure surplus with the output as numeraire (recall that input is not transferable, so it cannot be taken for numeraire). Total surplus is 18 as before and, moreover,

$$\phi_\lambda(\tilde{u}_M) = \tfrac{135}{8}\lambda, \qquad \phi_\lambda(u_L) = 3\lambda,$$

so that $\lambda^* = 0.475$. The corresponding shares of output and of surplus are

$$y_H = 26.6, \qquad y_M = 8.0, \qquad y_L = 1.4,$$

$$\pi_H = 8.6, \qquad \pi_M = 8.0, \qquad \pi_L = 1.4.$$

The ω-EE allocation is depicted in Figure 5.2. Note that agent L envies agent M. Moreover, it is not in the core from equal split: it is not acceptable to agents High and Low if each one owns the technology g, $g(x) = f(3x)/3$ (recall that when everyone owns g, the joint opportunities are precisely those afforded by f). Indeed, by combining two copies of g, the coalition (H, L) has access to the technology $h(x) = 2g(x/2) = (2/3)f(3x/2) = 4\sqrt{3x}$. The resulting joint surplus is

$$u_H + u_L = \max_{x \geq 0} 4\sqrt{3x} - x = 12,$$

whereas the ω-EE solution only awards them $8.6 + 1.4 = 10$ units of net surplus. By contrast, recall that the CEEI solution gives 6 units of surplus to each agent.

In Example 5.1, computing the ω-EE solution is slightly more involved. Indeed, the cost function c is defined for integer values of y, and must be extended to all nonnegative values before we can define the function $\gamma_\lambda(y) = \lambda c(y/\lambda)$. We use the linear interpolation:

$$
\begin{aligned}
c(y) &= 3y & &\text{if } 0 \le y \le 4, \\
&= 7y - 16 & &\text{if } 4 \le y \le 8, \\
&= 11y - 48 & &\text{if } 8 \le y \le 12.
\end{aligned}
$$

Tedious but straightforward computations give

$$
\begin{aligned}
\gamma_\lambda(1) &= 3 & &\text{if } \tfrac{1}{4} \le \lambda \le 1, \\
\gamma_\lambda(2) &= 14 - 16\lambda & &\text{if } \tfrac{1}{4} \le \lambda \le \tfrac{1}{2}, \\
&= 6 & &\text{if } \tfrac{1}{2} \le \lambda \le 1, \\
\gamma_\lambda(3) &= 33 - 48\lambda & &\text{if } \tfrac{1}{4} \le \lambda \le \tfrac{3}{8}, \\
&= 21 - 16\lambda & &\text{if } \tfrac{3}{8} \le \lambda \le \tfrac{3}{4}, \\
&= 9 & &\text{if } \tfrac{3}{4} \le \lambda \le 1.
\end{aligned}
$$

Hence the functions $\phi_\lambda(u_i)$ on the interval $\tfrac{1}{4} \le \lambda \le \tfrac{1}{2}$:

$$
\begin{aligned}
\phi_\lambda(u_1) &= 3 & &\text{if } \tfrac{1}{4} \le \lambda \le \tfrac{7}{16}, \\
&= 16\lambda - 4 & &\text{if } \tfrac{7}{16} \le \lambda \le \tfrac{1}{2}, \\
\phi_\lambda(u_2) &= 5 & &\text{if } \tfrac{1}{4} \le \lambda \le \tfrac{5}{16}, \\
&= 16\lambda & &\text{if } \tfrac{5}{16} \le \lambda \le \tfrac{1}{2}, \\
\phi_\lambda(u_3) &= 4 + 16\lambda & &\text{if } \tfrac{1}{4} \le \lambda \le \tfrac{1}{2}, \\
\phi_\lambda(u_4) &= 8 + 16\lambda & &\text{if } \tfrac{1}{4} \le \lambda \le \tfrac{11}{32}, \\
&= 48\lambda - 3 & &\text{if } \tfrac{11}{32} \le \lambda \le \tfrac{1}{2}.
\end{aligned}
$$

Total surplus is $v(N) = 30$, so the critical value is $\lambda^* = \tfrac{5}{16}$, with the corresponding surplus shares

$$
\alpha_1 = 3, \qquad \alpha_2 = 5, \qquad \alpha_3 = 9, \qquad \alpha_4 = 13.
$$

Comparing with CEEI, agent 1 gets one unit less and agent 3 gets one unit more.

Exercises 5.3, 5.7, and 5.9 illustrate further the difference between the two solutions, CEEI and ω-EE, in the increasing marginal cost case.

Finally, we note that both solutions, as well as the core from equal split, are easily extended to cooperative production involving an arbitrary number of inputs and of outputs. The technical details are much the same as in Chapters 3 and 4.

In fair division, we criticized the ω-EE solution because it may imply that an agent gets to consume more of every good than another agent (the domination property). The same argument applies in the cooperative production context; see, for instance, the ω-EE solution in Examples 5.2 and 5.3. Therefore, we may look for an efficient solution meeting the stand alone test and avoiding domination. One such solution is the *proportional solution* (Roemer and Silvestre [1993]), picking an efficient outcome where all the effective costs y_i/x_i coincide (hence are equal to the actual average cost). Exercise 5.14 explains that this solution is well defined in the one-input, one-output case if preferences are convex. Yet its ethical consequences are often unpalatable; in Examples 5.2 and 5.3, for instance, it gives the whole surplus to the efficient agent 1 and none to the others.

5.4. DECREASING MARGINAL COSTS: THE STAND ALONE CORE

In this section, we first dismiss the no envy test under decreasing marginal costs; then we extend the stand alone test to include coalitions and discuss the stand alone core.

In private ownership economies, the competitive idea breaks down under decreasing marginal costs; the core, on the other hand, often exists but tends to be large (hence nondiscriminatory); see Sections 2.6 and 2.7. These results apply in particular to the equal private ownership context. Let us repeat the essential argument.

Suppose that each agent owns a copy of the same technology c. Assume that $c(0) = 0$ and marginal cost decreases. How do we adapt the competitive equilibrium with equal incomes solution? Surely the firm cannot react competitively to the price signal, but it can set the price so as to generate the efficient demand of output. Because marginal costs decrease, the revenue thus generated will not cover costs; hence, the budget must be balanced by levying a uniform tax (e.g., in input contributions) on all agents. Thus we speak of a competitive equilibrium with equal deficits. The problem with this solution is that an inactive agent (who does not buy any output at the efficient price) is taxed all the same and therefore ends up suffering a net loss: voluntary participation is violated. A simple technology with a fixed cost and constant marginal costs makes this point very clearly.

Example 5.5. Fixed Cost and Constant Marginal Costs

To produce at any positive level, one must pay a fixed cost of b units of input, and constant marginal cost β thereafter:

$$\{c(y) = \beta y + b \text{ for } y > 0, \quad c(0) = 0\}$$

$$\Leftrightarrow \left\{ f(x) = \frac{1}{\beta} \cdot (x - b)_+, \quad x \geq 0 \right\}.$$

(Note the analogy with the technology in Example 5.2.) The CEEI solution (whether we assume that input is transferable or not) simply asks every agent to pay $(1/n)$th of the fixed cost b, and allows everyone to buy any amount of output he or she wishes at price β. In Definition 5.1, this means $p = \beta$ and $r = -b/n$. Obviously, an agent who does not want to buy any output at price β (or wants to buy only a little) suffers a net loss with this solution. See Figure 5.3 for a two-person example (where voluntary participation is violated for agent 1).

An even more striking example is the simple model with indivisible units of output and binary demands, introduced in Section 4.2 and further developed as Example 5.7 in the next section. That example also shows that using two-part tariffs or any nonlinear price schedule will not help; indeed, any efficient and envy-free outcome must violate voluntary participation.

In the common property regime, the no envy test has some normative appeal, but as we just saw, the combination of efficiency and of no envy may result in the unacceptable outcome where the net result of ownership in the valuable technology c is to decrease some agents' utility! Therefore we must reject on positive grounds the no envy test (and any kind of competitive solution using a possibly nonlinear price schedule) in the equal private ownership regime, and reject it on normative grounds in the common property regime.

There is also a third, technical reason to reject no envy in the decreasing marginal cost context. Even when agents have convex preferences, there may exist no efficient outcome passing the no envy test. A beautiful example of this difficulty (due to Rajiv Vohra) is the subject of Exercise 5.13: it involves only two agents and a very simple cost function with only two different values of marginal cost. The example is analogous to a two-person fair-division example (Exercise 4.15) where one agent has nonconvex preferences and no efficient and envy-free allocation exists. The only nonconvexity is in the production function itself.

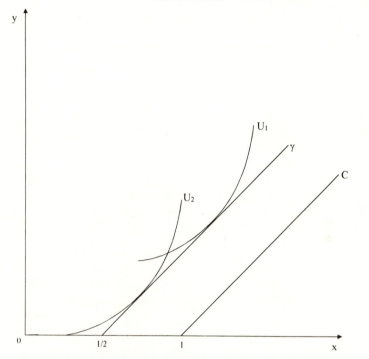

Fig. 5.3. The CEEI solution violates voluntary participation for agent 1 (Example 5.5).

We turn to core analysis. Suppose each one of n agents owns a copy of the same machine with subadditive cost function c. As discussed in Section 2.6, the efficient organization of production consists of using only one machine (property (12) in Lemma 2.4 is precisely the subadditivity of costs when all technologies are identical). Thus cooperative opportunities arise not from spreading production across several machines, but from gathering all the production on a single machine so as to benefit from higher returns to scale.[12] The relevant concept of stability by direct agreements is the stand alone core.

Definition 5.2. Let $(u_1, \ldots, u_n)$ be a profile of utility functions such that $u_i(y_i, x_i)$ is nondecreasing in y_i and decreasing in x_i. A coalition S, $S \subseteq N$, has a stand alone objection against the allocation $z = ((y_i, x_i);$ $i = 1, \ldots, n)$ if there exists an S-allocation $z'_S = ((y'_i, x'_i); i \in S)$ such that

(i) z'_S is feasible for S standing alone: $\Sigma_{i \in S}\, x'_i = c(\Sigma_{i \in S}\, y'_i)$,

---

[12] Under increasing marginal costs, we have the opposite pattern: spreading production over several machines is the reason they want to cooperate.

(ii) for all $i \in S$, $u_i(y_i', x_i') \geq u_i(y_i, x_i)$ with a strict inequality for at least one agent i.

The *stand alone core* is the set of feasible allocations against which there is no stand alone objection.

Theorem 5.1

(*i*) *If a single output is produced from a single input, if the cost function c is concave* (*decreasing marginal cost*); *if preferences are continuous, nondecreasing in* y_i *and decreasing in* x_i (*but not necessarily convex*); *then the stand alone core is nonempty.*

(*ii*) *The stand alone core may be empty when a single output is produced under subadditive cost preferences,* **or** *when several outputs are produced under concave costs,* **or** *when several inputs are used in a convex production function. All three statements hold with convex individual preferences.*

Proof. Statement (i) follows from Lemma 2.5 (in the particular case where all cost functions coincide) if we assume quasi-linear utilities. See also Lemma 5.3. The argument is constructive: a certain solution similar to the egalitarian-equivalent solution is shown to be in the stand alone core. That solution is normatively interesting in its own right; see Section 5.5.

Faulhaber's example in Section 2.8 proves the first part of statement (ii) in the context of a single output produced in indivisible units and transferable utility. There the cost function is subadditive and the core is empty. We prove the second statement in the same context.

A House Construction Story

Three agents want to build a house (each one wants at most one house). There are three possible locations A, B, C, hence three different goods (such as "a house at A"). Agents have similar utilities: a house is worth 12K to any of them if its location is "right," and nothing at all if its location is "wrong." The utilities are as follows:

agent

location	1	2	3
A	12	12	0
B	0	12	12
C	12	0	12

There are economies of scale in building several houses in the same location, specifically

$$c_t(1) = 10, \qquad c_t(2) = 16, \quad \text{where } t = A, B, \text{ or } C$$

(marginal cost decreases). There are no externalities (either positive or negative) between the production of houses in any two locations: constructive costs in various locations simply add up. Each agent has access to this cost function (the local construction market is competitive).

To show that the stand alone core is empty, we compute the net surplus of any coalition standing alone. A single agent gets $v(i) = 12 - 10 = 2$. A two-person coalition benefits from the decreasing marginal costs by building a house for each member in their common acceptable location; hence, $v(ij) = 12 + 12 - 16 = 8$. The coalition $N = \{1, 2, 3\}$, on the other hand, can do no better than building two houses in one location and one house in another location, for a total surplus $v(N) = 12 + 12 + 12 - 16 - 10 = 10$. Clearly the core is empty, since $v(12) + v(23) + v(13) > 2v(N)$.[13] Note that the example does not depend on the symmetry in utilities and costs: small changes in individual utilities and in the costs at various locations still yield an empty core.

A symmetrical example with multiple inputs proves the last statement of Theorem 5.1. We have three agents $1, 2, 3$ and three inputs $1, 2, 3$. Agent 1 holds one unit of input 2 and one of input 3. Giving up one unit of any input brings a disutility of 10 (measured in units of output) and giving up both units brings a disutility of 30. Similarly, agent 2 holds one unit of input 1 and 3, and so on. The production function is additively separable in the three inputs with $f_i(1) = 10$ and $f_i(2) = 25$. One computes $v(i) = 0, v(ij) = 5, v(N) = 5$. \hfill Q.E.D.

The point of Theorem 5.1 is that the stand alone core is a fragile concept for cooperative production problems involving several inputs and/or several outputs. As noted in Section 2.8, another Achilles' heel of this core is the possibility of direct trade of output between the agents: in Theorem 5.1, the (one-input, one output) stand alone core is nonempty under the assumption that the output share y_i is nonnegative for all agent i. This must be true if our agents have no initial endowment of output, or if output is not tradable (e.g., the power company

[13] An allocation $(\alpha_1, \alpha_2, \alpha_3)$ of the 10K of total surplus would have to satisfy $\alpha_i + \alpha_j \geq 8$ for all i, j to be in the core. Summing up these three inequalities yields $2v(N) \geq \Sigma v(ij)$ See the general discussion of core existence in transferable utility games in Section 7.5.

prohibits its customers from reselling electric power). When this assumption is removed, the stand alone core can easily be empty even with convex individual preferences. Scarf's celebrated example makes precisely this point; see Exercise 5.12.

In the equal private property regime, an empty stand alone core leads to destructive competition and calls for the normative interference of the regulator. One extreme form of interference (the only one we discuss in this volume) is the common property regime: the technology c is viewed as the indivisible property of all its users, and the regulating authority embodies the normative concerns of this community for fairness and efficiency. These are the ethical premises of much of the literature on the regulation of "natural monopolies," namely, of industries where marginal costs are decreasing (see, e.g., Zajac [1985] or Sharkey [1982b]).

In the common property regime, the stand alone idea still plays a central role, this time as a normative test of equity. Start by the one-person version of this idea, namely, the stand alone test.

Stand Alone Test: agent i's final utility should not be lower than his stand alone utility $\mathrm{sa}(u_i) = \max_{y \geq 0} u_i(y; c(y))$.

When the cost function is subadditive (even if several outputs are produced together) the stand alone lower bound is always feasible. To see this, pick, for all i, a solution y_i of the above program; then $c(\Sigma_i y_i) \leq \Sigma_i c(y_i)$ implies that the allocation $z = ((y_i, c(y_i)); i = 1, \ldots, n)$ is feasible.

Recall that the stand alone utility is an *upper* bound on individual utilities in the case of increasing marginal costs. Under decreasing marginal costs, the stand alone test is a very appealing lower bound on individual utilities.[14] In particular, the stand alone test on utilities implies the (stand alone) upper bound on cost shares $x_i \leq c(y_i)$. (Indeed, $u_i(y_i, x_i) \geq \mathrm{sa}(u_i)$ implies $u_i(y_i, x_i) \geq u_i(y_i, c(y_i))$, hence the announced inequality, since u_i decreases in x_i.)

The stand alone core plays the same central role in the normative discussion of regulated natural monopolies as in the positive analysis of stable agreements under private ownership. Both facets (one positive, one normative) of the concept are often taken as mutually reinforcing one another (see, e.g., Baumol [1986]); this is also the position taken in this volume.

---

[14] In fact, it is the highest feasible individual rationality constraint when voluntary participation is guaranteed. The proof of this fact is similar to that of Exercise 4.17.

An important consequence of the stand alone core property is the following bounds on the cost borne by various coalitions of agents. Given a cooperative production problem $(u_1, \ldots, u_n; c)$ (satisfying the assumptions of Definition 5.2), and an allocation $z = ((y_i, x_i); i = 1, \ldots, n)$ in the stand alone core, we have for all coalition S, $S \subseteq N$,

$$c\left(\sum_N y_i\right) - c\left(\sum_{N \setminus S} y_i\right) \leq \sum_S x_i \leq c\left(\sum_S y_i\right). \qquad (7)$$

The right-hand inequality follows at once from Definition 5.2. Applying this to the coalition $N \setminus S$ (and taking $c(\sum_N y_i) = \sum_N x_i$ into account) yields the left-hand inequality.

Inequalities (7) are easy to check and to interpret: a coalition of agents should not pay more than its stand alone cost or less than its incremental cost (the additional cost of serving it once all other demands are met). Failure of one such inequality opens the door to a subsidization argument. The incremental cost of serving the coalition S is the most favorable computation of its cost share, because it allows S to take maximal advantage of the benefits afforded by decreasing returns. Similarly, the stand alone cost of serving S is its least favorable cost share because it gives S no share of the positive externality caused by other users. Thus (7) says that every coalition should get a non-negative share of the production externalities.

Requiring inequalities (7) for all S, is naturally, much less demanding than the stand alone core property. In a problem where the stand alone core is empty, these inequalities offer a reasonable alternative. For instance, in the house construction story (proof of Theorem 5.1), the stand alone core is empty but there are efficient outcomes satisfying the inequalities (7). Consider the efficient outcome where agents 1 and 2 build a house each at location A and agent 3 builds at B. Inequalities (7) imply the following bounds on the cost shares x_i:

$$x_i \leq 10, \qquad i = 1, 2, 3, \qquad x_1 + x_2 \leq 16.$$

Total cost is 26; therefore, the cost share $(8, 8, 10)$ meet (7).

When we take a normative view of the stand alone core, we are no more worried by its emptiness than we are with any other desirable equity property that is sometime too demanding. However, we must be concerned if the core is very large, because this means that the property has little discriminating power, and should by supplemented by additional requirements.[15]

---

[15] By contrast, in the positive interpretation of the core as a set of stable agreements, emptiness is a serious problem but a larger core is not.

Example 5.6.   Example 5.5 Continued

Suppose each agent is endowed with quasi-linear utilities of the form $u_i(y_i) - x_i$, with the normalization $u_i(0) = 0$. Define

$$\sigma_i = \max_{y \geq 0} \{u_i(y) - \beta \cdot y\}$$

to be the surplus enjoyed by agent i if the "price " of output is β (we assume that all above programs have bounded solutions). Therefore, agent i's stand alone surplus sa(u_i) is

$$\text{sa}(u_i) = (\sigma_i - b)_+$$

(agent i chooses $y = 0$ if $\sigma_i < b$). Similarly, the joint surplus available to any coalition S is

$$\text{sa}(S) = \left(\sum_{i \in S} \sigma_i - b \right)_+ .$$

Suppose first that σ_i exceeds b for all i: every agent wants to use the machine even if no one else can share the fixed cost. Then the stand alone core consists of all allocations where agent i pays an arbitrary (nonnegative) share of the fixed cost b and receives σ_i:

$$\pi_i = \sigma_i - \lambda_i, \quad \text{where } \lambda_i \geq 0 \text{ for all } i \quad \text{and} \quad \sum_{i=1}^{n} \lambda_i = b.$$

In other words, the stand alone core puts no restriction whatsoever on the division of the fixed cost provided everyone receives his "competitive" surplus at price β. In general (when some or all of the σ_i are below b), the stand alone core is isomorphic to that in the provision of a binary public good with cost b and benefit σ_i to agent i (see Section 5.7). Assuming that there are some cooperative benefits from using the technology c (i.e., assuming $\sum_{i=1}^{n} \sigma_i > b$), the surplus allocation $(\pi_1, \dots, \pi_n)$ is in the stand alone core if and only if

$$\pi_i = \sigma_i - \lambda_i. \quad \text{where } 0 \leq \lambda_i \leq \sigma_i \text{ for all } i, \quad \text{and} \quad \sum_{i=1}^{n} \lambda_i = b.$$

We proceed to show now that the stand alone core in a general one-input, one-output model is always quite large. To show this, we use a certain algorithm, known as the *greedy algorithm*, to construct a core allocation. We state the property of the greedy algorithm without proof, and refer the reader to the proof of Theorem 5.2, where a similar property is established for the provision of a public good problem. Theorem 7.1 provides a general explanation of the link between the greedy algorithm and the convexity property of cooperative games (see, in particular, Example 7.5).

To define the greedy algorithm, pick an arbitrary ordering of the agents, say $1, 2, \ldots, n$. First, give his stand alone utility u_1^* to agent 1. Next, give to agent 2 the highest possible utility u_2^* such that (u_1^*, u_2^*) is feasible for coalition $\{1, 2\}$ standing alone. Next, give to agent 3 the highest possible utility u_3^* such that (u_1^*, u_2^*, u_3^*) is feasible for coalition $\{1, 2, 3\}$ standing alone; and so on. The utility profile constructed in this way corresponds to an allocation in the stand alone core.

When utilities are conveniently quasi-linear, namely, $u_i(y_i, x_i) = v_i(y_i) - x_i$ (with x_i of an arbitrary sign) we write sa(S) for the stand alone surplus of coalition S, namely, the maximal joint utility $\sum_S v_i(y_i) - c(\sum_S y_i)$ over all nonnegative choices of y_i, $i \in S$. The greedy algorithm generates the vector of marginal contributions corresponding to our (arbitrary) ordering

$$u_1^* = \text{sa}(1),$$

$$u_2^* = \text{sa}(12) - \text{sa}(1),$$

$$u_3^* = \text{sa}(123) - \text{sa}(12), \ldots, \tag{8}$$

$$u_i^* = \text{sa}(1, \ldots, i) - \text{sa}(1, \ldots, i - 1), \ldots,$$

$$u_N^* = \text{sa}(N) - \text{sa}(N \setminus \{n\}).$$

The utility vectors discovered by the greedy algorithm are the extreme points of the stand alone utility core (every utility vector in the core is a convex combination of several marginal contribution vectors attached to different orderings of N; see Theorem 7.1).

From the marginal contribution formula (8), it follows that agent 1 gets his stand alone utility sa(1) if he is ranked first, and his incremental stand alone utility sa(N) − sa($N \setminus 1$) if he is ranked last. Now every core utility vector must yield a utility between those two bounds (as one sees by applying the core property to coalitions $\{1\}$ and $\{N \setminus 1\}$; this argument was used repeatedly in Chapters 2 and 3; see also inequality (7)). Therefore agent 1's utility in the stand alone core varies from sa(1) to sa(N) − sa($N \setminus 1$), including both bounds. By the same token, the utility of any coalition S varies from sa(S) to sa(N) − sa($N \setminus S$) in the core, including both bounds.

Here is a numerical example with indivisible outputs and binary demand:

y	1	2	3	4	5	6	7	8	9	10
∂u	20	18	16	14	12	10	8	6	4	2
∂c	10	9	8	7	6	5	4	3	2	1

Recall from Chapter 2 the notation ∂u for the aggregate marginal utility. We have 10 agents ranked by decreasing marginal utility. Agent 1 is willing to pay \$20 (input is money), and so on. Efficiency requires providing one unit to every agent, for a total surplus $sa(N) = 55$. Consider agent 1: we have $sa(1) = 10$ and $sa(N \setminus \{1\}) = 36$; therefore, her stand alone core utility u_1 varies from 10 to 19. Similar computations yield the range of every agent's utility in the core:

agent	1	2	3	4	5	6	7	8	9	10
highest	19	17	15	13	11	9	7	5	3	1
lowest	10	8	6	4	2	0	0	0	0	0

If we give his lowest core utility to everyone, we distribute only 55% of total surplus; if we give his highest core utility, we spend 182% of the surplus!

5.5. DECREASING MARGINAL COSTS: DETERMINISTIC SOLUTIONS

The simplest idea is the ω-egalitarian-equivalent solution. The construction is based, as in Section 5.3, on the λ-equivalent utilities:

$$\phi_\lambda(u_i) = \max_{x \geq 0} u_i\left(\lambda \cdot f\left(\frac{x}{\lambda}\right), x\right) = \max_{y \geq 0} u_i\left(y, \lambda c\left(\frac{y}{\lambda}\right)\right),$$

only this time, the utility $\phi_\lambda(u_i)$ decreases in λ, as $\lambda \cdot c(y/\lambda)$ increases in λ if c is concave, or equivalently, $\lambda \cdot f(x/\lambda)$ decreases in λ if f is convex and $f(0) = 0$. (Exercise: prove this claim.)

Lemma 5.2. Suppose the cost function c is concave (or equivalently, the production function is convex and $f(0) = 0$). Suppose utility functions are continuous. Then there exists a number λ^*, $1/n \leq \lambda^* \leq 1$, such that the utility vector $(\phi_\lambda^*(u_1), \ldots, \phi_\lambda^*(u_n))$ is Pareto optimal. The corresponding allocation(s) is called the ω-egalitarian-equivalent allocation. It gives to each agent at least his stand alone utility and at most his unanimity utility.

The proof rests on the observation that the vector of unanimity utilities $(\phi_{1/n}(u_1), \ldots, \phi_{1/n}(u_n))$ is either not feasible or Pareto optimal, whereas the vector of stand alone utilities is feasible. Therefore, the (nonincreasing) utility path $\lambda \to (\phi_\lambda(u_1), \ldots, \phi_\lambda(u_n))$ must pierce the Pareto frontier somewhere between $1/n$ and 1. We omit the details.

Example 5.7. Indivisible Output and Binary Demands

Each agent wants at most one car, and the utilities are ordered as $u_1 \geq u_2 \geq \cdots \geq u_n$. Marginal costs $\partial c(y) = c(y) - c(y - 1)$ decrease in the quantity y.

We use the linear interpolation to define c for noninteger values of y; consequently, the function $\lambda c(1/\lambda)$ is increasing (nondecreasing) in λ, and we have

$$\phi_\lambda(u_i) = \left(u_i - \lambda \cdot c\left(\frac{1}{\lambda}\right) \right)_+ .$$

Changing variables to $\mu = \lambda \cdot c(1/\lambda)$, we find the value of μ for which the vector $(\phi_\lambda(u_i), i = 1, \ldots, n)$ is efficient, by solving

$$\sum_{i=1}^{n} (u_i - \mu)_+ = v(N) = \left(\sum_{i=1}^{y^*} u_i \right) - c(y^*), \qquad (9)$$

where y^* is the efficient production.

Consider a numerical example with seven agents:

y	1	2	3	4	5	6	7
∂u	14	12	10	8	6	4	2
∂c	10	9	8	7	6	5	4

Efficient production is $y^* = 4$ or 5, for a total surplus $v(N) = 10$. Equation (9) is solved at $\mu = 8.67$, and brings the surplus shares

$$\alpha_1 = 5\tfrac{1}{3}, \qquad \alpha_2 = 3\tfrac{1}{3}, \qquad \alpha_3 = 1\tfrac{1}{3}, \qquad \alpha_i = 0 \quad \text{for } i \geq 4.$$

Therefore the active agent 4 does not get any share of surplus: in the ω-EE allocation, he must buy a car at price 8. Agents 1, 2, and 3, on the other hand, pay 8.67 for their car, whence they envy agent 4 (envy bears on consumption vectors, not on net surplus gains!). The situation in this numerical example is typical of the binary demand model: a subset of active agents with the lowest willingness to pay must each pay their reservation price for a car, and end up being envied by agents with higher reservation prices.

Next, we compute the ω-EE solution in Example 5.6. We have

$$\phi_\lambda(u_i) = \max_{y \geq 0}\left\{u_i(y) - \lambda\left(\beta\frac{y}{\lambda} + b\right), 0\right\} = (\sigma_i - \lambda \cdot b)_+ \ .$$

Therefore, the critical value λ^* solves

$$\sum_{i=1}^{n} (\sigma_i - \lambda \cdot b)_+ = \left(\sum_{i=1}^{n} \sigma_i - b\right)_+ \ .$$

This is formally the same equation as in Example 5.7: every agent pays the share $\lambda \cdot b$ of the fixed cost or σ_i, whichever is less. If each surplus σ_i is no less than b/n, then the fixed cost b is shared equally among all. Otherwise a subset of agents with the lowest surplus σ_i end up with no net gain whatsoever while the other agents pay an equal cost-share.

Exercises 5.8, 5.9, and 5.15 give further examples of the ω-EE solution in economies with divisible or indivisible outputs.

If the ω-EE solution passes the stand alone test, it may not, however, fall within the stand alone core (although it does in the two above examples).

This point is made by our next example, where preferences are represented by the Leontief utility functions.

Example 5.8. The Case of Leontief Utilities

A Leontief utility function takes the form

$$u_i(y, x) = \min\left\{\frac{y}{a_i}, \omega_i - x\right\} \quad \text{for } y \geq 0, 0 \leq x \leq \omega_i.$$

Here ω_i represents agent i's initial input endowment. His indifference contours are straight angles with the kink along the line Δ_i with equation $y = a_i(\omega_i - x)$. Leontief utilities represent preferences where the two goods (input and output) are strongly complementary: they must be consumed in the exact ratio a_i, otherwise the excess output (or excess input) is simply wasted. If input represents leisure time (so agent i works for x_i hours and is free for $(\omega_i - x_i)$ hours) and output is money, one possible story is that the agent uses all his leisure time flying. One hour of flight costs a_i and he derives no utility from leisure

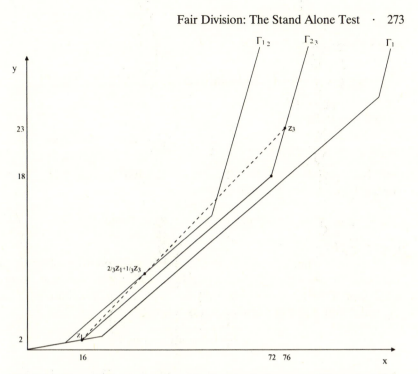

Fig. 5.4. The ω-EE solution is not in the stand alone core (Example 5.8).

if he cannot afford to fly (he is indifferent between working and staying idle in that case).

Fix λ, $1/n \leq \lambda \leq 1$, and draw the cost curve Γ_λ with equation $x = \lambda \cdot c(y/\lambda)$. The utility level $\phi_\lambda(u_i)$ obtains at the intersection $z_i(\lambda)$ of Γ_λ and Δ_i. Therefore the critical value λ^* defining the ω-EE solution obtains when the sum of the points $z_i(\lambda^*)$ belongs to Γ_1, or equivalently, when their average belongs to $\Gamma_{1/n}$ (this is often easier to visualize). See Figure 5.4.

To give a three-agent example where the ω-EE solution is not in the stand alone core, we construct a configuration where agents 1 and 2 have identical Leontief utilities and the critical value λ^* is larger than $1/2$. The the points $z_1(\lambda^*) = z_2(\lambda^*)$ will be below the curve $\Gamma_{1/2}$ and both agents prefer $z_1(1/2) = z_2(1/2)$ to $z_1(\lambda^*)$; but the coalition $\{1,2\}$ standing alone can achieve $(z_1(1/2), z_2(1/2))$. To summarize, we want a cost function c, a parameter $\lambda^* > 1/2$, and two points z_1, z_3 on Γ_{λ^*} such that the point $2z_1 + z_3$ belongs to Γ_1. Here is a numerical example with $\lambda^* = 2/3$, depicted in Figure 5.4:

$$c(y) = \min\{8y, (3.5)y + 13.5, (0.8)y + 86.4\}. \qquad (10)$$

The curve $\Gamma_{2/3}$ is

$$\Gamma_{2/3}: x = \frac{2}{3}c\left(\frac{3y}{2}\right) = \min\{8y, (3.5)y + 9, (0.8)y + 57.6\}.$$

Check that the points

$$z_i(2, 16), \qquad z_3 = (23, 76)$$

both belong to $\Gamma_{2/3}$ and that $2z_1 + z_3 = (27, 108)$ belongs to Γ_1. Therefore, a Leontief utility profile such as

$$u_1 = u_2 = \min\{y, 18 - x\}, \qquad u_3 = \min\{y, 99 - x\} \qquad (11)$$

does the job.

There is a simple variant of the ω-EE solution that chooses an outcome in the stand alone core. It is called the *constant returns equivalent* (CRE) solution. Fix a profile $(u_1, \ldots, u_n)$, a technology f (or c), and define the functions $\psi_\mu(u_i)$ as follows:

$$\psi_\mu(u_i) = \max_{x \geq 0} u_i(\mu x, x).$$

Agent i can reach the utility level $\psi_\mu(u_i)$ if he can use a linear technology with marginal product μ (i.e., he can buy any amount of output at price $1/\mu$). Note that $\psi_\mu(u_i)$ increases in μ, that the utility vector $(\psi_0(u_1), \ldots, \psi_0(u_n))$ is feasible (zero production is feasible), and that for μ large enough, the vector $(\psi_\mu(u_1), \ldots, \psi_\mu(u_n))$ is not feasible (assuming we cannot reach arbitrarily high levels of utility with the real technology; e.g., inputs endowments are finite and marginal cost is bounded away from zero). Therefore, by the same continuity argument as in Lemmas 5.1 and 5.2, there exists a critical value μ^* such that the vector $(\psi_{\mu^*}(u_1), \ldots, \psi_{\mu^*}(u_n))$ is precisely on the Pareto frontier. The corresponding allocations define the CRE solution.

In one way, the CRE solution is more robust that the ω-EE solution: it is defined for any technology c whether the marginal cost decreases, increases, or both (see Exercise 5.16 for the difficulty in defining ω-EE when returns are neither increasing nor decreasing). On the other hand, the CRE solution is defined only when a single output is produced from a single input, contrary to ω-EE. The striking property of the CRE solution is to always pick an allocation in the stand alone core.

Lemma 5.3 (Mas-Colell [1980a]). *Assume preferences are continuous, monotonic, but not necessarily convex. Assume marginal costs decrease. Then a constant returns equivalent allocation is in the stand alone core.*

The (omitted) proof is essentially the same as that of Lemma 2.5, except for the fact that preferences are not represented by quasi-linear utilities anymore.

We give some examples. With indivisible output and binary demands (Example 5.7), the CRE and ω-EE solutions coincide. Not so in the Example 5.6 with a fixed cost and constant marginal cost (as shown in Exercise 5.8) or in Example 5.8 with Leontief utilities. Consider the numerical example (10), (11). For a given technology with constant returns μ, agents 1 and 2 achieve the utility $\psi_\mu(u_1)$ at the allocation $z_1(\mu) = z_2(\mu) = 18/(\mu + 1)(\mu, 1)$ (or at any allocation requiring less input contribution and offering a greater share of output). Similarly, agent 3 achieves the utility $\psi_\mu(u_3)$ at $z_3(\mu) = 99/(\mu + 1)(\mu, 1)$. The highest μ such that $(\psi_\mu(u_1), \psi_\mu(u_2), \psi_\mu(u_3))$ is feasible when the true technology is given by (10) is μ^*, where $z(\mu^*) = (z_1(\mu^*) + z_2(\mu^*) + z_3(\mu^*))$ belongs to Γ_1, namely,

$$z(\mu) = \frac{135}{\mu + 1}(\mu, 1) \quad \text{and} \quad \frac{135}{\mu + 1} = c\left(\frac{135\mu}{\mu + 1}\right).$$

This equation is solved at $\mu^* = 1/4$ with the corresponding allocation

$$z_1(\mu^*) = z_2(\mu^*) = (3.6, 14.4), \qquad z_3(\mu^*) = (19.8, 79.2),$$

depicted in Figure 5.4.

Note that the aggregate production plan $z(\mu^*) = (27, 108)$ is the same as for the ω-EE solution, but the CRE allocation gives a bigger share of surplus (almost double) to the first two agents.

In general, the CRE solution (unlike the ω-EE solution) does not meet the unanimity bound, a lower bound on individual utilities if marginal costs increase, an upper bound if marginal costs decrease.[16] A striking illustration of this is the economy of Example 5.3, with linear utilities and quadratic cost. There we have

$$\psi_\mu(u_H) = 0 \qquad \text{if } \mu \leq 1,$$

$$= (\mu - 1) \cdot \omega_H \quad \text{if } \mu > 1,$$

$$\psi_\mu(u_L) = 0 \qquad \text{if } \mu \leq 3,$$

$$= (\mu - 3) \cdot \omega_L \quad \text{if } \mu > 3,$$

---

[16] Note that under decreasing marginal costs, it is possible to pick a stand alone core allocation satisfying the unanimity upper bound; see Theorem 7.2 and the discussion thereafter. However, no simple solution meeting these two objectives is known.

(where ω_i is agent i's endowment of input). When ω_H is large enough (namely, $\omega_H \geq 18$), the CRE solution gives all 18 units of surplus to agent H (because $\psi_\mu(u_H) + \psi_\mu(u_L) = 18$ yields $\mu = 1 + (18)/\omega_H$), $1 \leq \mu \leq 2$; hence, $\psi_\mu(u_L) = 0$. Note that agent L's unanimity utility is 3, and yet the CRE solution denies him any surplus. Thus the CRE solution picks an unfair allocation in this particular example. Note that the same configuration arises in every economy where marginal costs increase and all agents have (distinct) linear utilities: the agent with the largest marginal utility for output relative to input pockets the entire surplus. For instance, Figure 5.2 shows the CRE allocation of Example 5.4: $z_H = (36, 18)$, $z_M = z_L = (0, 0)$.

On the other hand, the CRE solution (unlike the ω-EE solution) meets two very natural monotonicity requirements. The first one is population monotonicity: under decreasing marginal costs, adding new agents to share the technology never results in a net loss for any preexisting agent.[17] Similarly, under increasing marginal costs, adding new agents never results in a net benefit for a preexisting agent.

Next, we have cost monotonicity: whenever the technology improves (we switch from c_1 to c_2, where $c_1(y) \geq c_2(y)$ for all y), none of the agents suffers a loss of utility. This is the exact analog of resource monotonicity for fair division.[18]

In the cooperative production of a private good, it turns out that the combination of cost monotonicity and the stand alone test is enough to characterize the CRE solution among all efficient solutions. This result holds both for a technology with always decreasing marginal costs (where the stand alone utility sets a lower bound on the actual utility) and for a technology with always increasing marginal costs (where the stand alone utility sets an upper bound on the actual utility); see Moulin [1987].

To summarize the discussion of Sections 5.3 to 5.5: The choice of a deterministic and efficient solution in the cooperative production of private goods (under common property) is controversial at the normative level. Even in the particularly simple case of a technology with one input and one output, and with always decreasing or always increasing marginal costs, several solutions compete for the normative prominence. After discussing three of them (the CEEI, ω-EE, and CRE solutions), few unambiguous conclusions emerge. The CEEI solution

---

[17] Note that population monotonicity is but a strengthening of the Stand Alone core property.

[18] And a statement analogous to Lemma 4.4 holds: an efficient and cost-monotonic solution cannot meet the unanimity upper (resp. lower) bound at all profiles, nor can it avoid the domination property.

has no serious claim under decreasing marginal costs (it violates voluntary participation), just as the CRE solution is not convincing under increasing marginal costs (because it may deny any share of the surplus to an agent who would be willing to work); and both the ω-EE and the CRE solutions must be eliminated if we cannot stand domination (namely, the property that agent i gets more output than agent j for less input).[19]

If we allow multiple outputs or multiple inputs, the CRE solution is no longer defined, and under decreasing marginal costs the stand alone core may be empty. In that case, the ω-EE solution appears to make a reasonable recommendation, but certainly not the only reasonable recommendation. Exercise 7.9, for instance, suggests another possible approach.

5.6. PUBLIC GOODS: THE STAND ALONE CORE

In this section, we introduce public goods and show first that equal cost shares for all agents (as implied by the no envy test) is normatively untenable (the argument is similar to that of Section 5.4). Then we discuss the stand alone core.

A public good is a desirable commodity from the consumption of which no agent can be excluded. A radio broadcast, street lamps, and the law are typical examples. The equal beneficiaries of the public good may contribute different amounts to its provision: the teammates may not work equally hard to win the game but they will equally share the rewards of victory; all workers benefit from the compensation package negotiated by the union, even the nonunion members; all firms in an industry benefit from the tariff obtained by the lobbying effort of a few firms. (Olson's book provides many more examples.) The methodology formulated in Chapter 1 yields three central questions of the public good provision problem. As individual preferences about the public good differ, can we expect a direct agreement stable in the sense of the core? In the justice mode, how much inequality in the private contributions is fair? In the decentralized mode, how can we decentralize the contribution process and provide the efficient level of public good (can we avoid the free-rider problem)? We address the first question in this section, and the second one in the next two sections. We postpone until Sections 6.2 and 6.3 the discussion of the third one.

---

[19] To give an idea of how little is known at this time about our normative flexibility, consider the following open question: under decreasing marginal costs, can we select an allocation in the stand alone core without domination between any two agents?

In one kind of public good model, there is a single technology to produce the public good from private good contributions (a single legislative body passes bills as a result of lobbying, a single antenna broadcasts a program, and so on). The technology is the common property of all agents. The situation is formally equivalent to the common property of a technology producing private output, under the constraint that all the output must be shared equally among all agents irrespective of their input contribution. The second kind of model allows each agent or coalition of agents free access to the (same) technology to produce the public good. In other words, each agent i owns a machine (a copy of the technology) transforming x_i units of input into $y_i = f(x_i)$ units of output. When different agents produce the levels $y_1, \ldots, y_n$ of output, everyone enjoys $\bar{y} = \max_i(y_i)$ units of public good.[20] Examples include street lights and unpatented scientific discoveries.

In both interpretations (common ownership or free access), the set of feasible allocation consists of the vectors $(y; x_1, \ldots, x_n)$, such that

$$x_i \geq 0 \quad \text{all } i, \qquad y = f\left(\sum_{i=1}^{n} x_i \right). \tag{12}$$

Here, x_i is agent i's (nonnegative) contribution to the public good, y is the public good level consumed by every agent, and f is the increasing production function (we always assume $f(0) = 0$).

The discussion of the core will follow closely that of the private good production problem under decreasing marginal costs (Section 5.4). Yet, we do not assume that the marginal cost of the public good is always increasing or always decreasing (we only rule out negative marginal costs).

Consider first the no envy property. Since all agents already consume the same level of public good, no envy simply says that they must contribute the same amount of input: $x_1 = \cdots = x_n = c(y)/n$, where c is, as usual, the cost function. This is undoubtedly equitable (inasmuch as final consumptions of all agents coincide), and compatible with efficiency as well, despite the diversity of individual preferences (and unlike the fair-division problem of Chapter 4, where equal split, in general, is not efficient). An easy proof of this claim uses the simplifying assumption that individual preferences are represented by quasi-linear utilities of the form $u_i(y, x_i) = u_i(y) - x_i$. An allocation is then Pareto optimal if (and only if) it maximizes the joint surplus $(\sum_i u_i(y)) - c(y)$.

[20] Other forms of externalities are discussed in Section 5.9.

Any division of the costs, in particular equal division, yields an efficient allocation. A general argument (valid even when preferences are not quasi-linear) is given by Diamantaras [1992].

The main objection against equal cost shares (hence against no envy) in the provision of a public good was already spelled out in Example 1.2. Sharing equally the cost of the efficient level of public good may force an agent to pay more than the benefit he derives from consuming the public good. The simple argument is transparent in the case of a *binary* public good, namely, an indivisible public good (y takes only the value 0 or 1) with cost c, $c > 0$, and for which agent i is willing to pay u_i, $u_i > 0$ (his willingness to pay is measured in the input good). An interesting example is the common property technology with a fixed cost and constant marginal cost thereafter (Examples 5.5 and 5.6). Once the fixed cost is paid for, the access to the constant marginal cost technology is a public good (because usage of that technology entails no externalities in production).

In a binary public good problem, efficiency commands producing the public good if $\sum_{i=1}^{n} u_i > c$ (and not producing it if $\sum_{i=1}^{n} u_i < c$). With uniform cost shares c/n, an agent such that $u_i < c/n$ ends up with a net loss.

The voluntary participation axiom says that when a number of agents own in common certain resources (in this case, the technology to produce a public good), the cooperative use of these resources should not result in a net loss for any individual agent (that is, comparing his final utility to his utility before the resources were available). If this property is not satisfied, the agent suffering a net loss is coerced by the rest of society, and common ownership means nothing more than the obligation to submit to the general will (in the spirit of the authoritarian interpretation of the Social Contract as articulated by Rousseau and Bentham).

I take instead the "liberal" viewpoint (going back to Wicksell [1896] in the case of the public good provision problem) insisting that common property should mean at least the individual right to veto net utility losses; formally, this is the voluntary participation axiom.

Because there may not exist an efficient and envy-free outcome meeting voluntary participation (see above), we must dismiss no envy. (In this chapter we only discuss efficient solutions; however, in the next chapter, we will find interesting inefficient allocations meeting no envy and voluntary participation; see Section 6.2.)

In the free access regime, the argument against no envy turns into a positive statement and does not rely on the efficiency postulate. Just as in cooperative production with decreasing marginal costs, the stand alone logic forces a tighter lower bound on individual utilities than voluntary participation. Define agent i's stand alone utility

$$\text{sa}(u_i) = \max_{y \geq 0} u_i(y, c(y)) = \max_{x \geq 0} u_i(f(x), x).$$

Agent i will not agree to an allocation if she does not receive at least her stand alone utility. Now an easy but important observation: there may exist no envy-free allocation (whether efficient of not) giving her stand alone utility to each agent.

To prove the claim, we need only to consider a binary public good with cost c. Suppose agent 1 is ready to pay the full cost of the public good, because $u_1 > c$. On the other hand, suppose agent 2 is not even willing to pay the average cost, as $u_2 < c/n$ (as usual, n is the number of agents). We check that in any feasible outcome, either some agent does not get his stand alone utility, or some agent envies another agent. If the public good is not produced, then agent 1 can reach his stand alone utility $\text{sa}(u_1) = u_1 - c$, $\text{sa}(u_1) > 0$, only by receiving some transfer of input: $x_1 < -\text{sa}(u_1)$.[21] In that case, the budget balance condition $\Sigma_i x_i = 0$ implies that at least one x_i is strictly positive, and agent i envies agent 1. If the public good is produced, agent 2 reaches his stand alone utility $\text{sa}(u_2) = 0$ by paying at most u_2, whence $x_2 < c/n$. By feasibility, at least one agent must pay more than c/n, and that agent envies agent 2. A similar configuration with a divisible public good is Example 6.5 in the next chapter.

The stand alone core is defined in much the same way as in the case of private goods and decreasing marginal costs (but recall that the marginal costs of the public good technology vary in arbitrary fashion). Throughout the discussion of public goods, we assume that agents have a finite endowment of private good, to rule out the possibility of producing an unbounded amount of public good.

Definition 5.3. Given is a public good provision problem $(u_1, \ldots, u_n; c)$. If $(y; x_1, \ldots, x_n)$ is a feasible allocation ((12)), we say that coalition S has a stand alone objection against it if there exists an S-allocation $(y'; x_j', j \in S)$, such that
(i) $\Sigma_S x_j' = c(y')$, $x_j' \geq 0$, all $j \in S$,
(ii) $u_j(y', x_j') \geq u_j(y, x_j)$ for all $j \in S$, with at least one strict inequality.
The stand alone core is the set of feasible allocations against which no stand alone objection exists.

---

[21] If negative contributions are ruled out, as we generally assume, the argument stops there.

As in Section 5.4, the stand alone core has a positive interpretation as the set of stable direct agreements if each agent has free access to the technology (each one owns a machine), and a normative interpretation if a single copy of the technology is the common property of all agents.

Definition 5.3 is equally valid for the case where several public goods are produced simultaneously: just think of y as a vector in $\mathbb{R}_+^K$ (where K is the number of public goods) instead of a vector in $\mathbb{R}_+$ (see, however, Remark 5.2).

Only in the case where a single public good is produced ($K = 1$) are we guaranteed that the stand alone core exists.

Theorem 5.2

(*a*) *In the case of a single public good, suppose that c is continuous, increasing, and $c(0) = 0$. Suppose that each utility $u_i(y, x_i)$ is nondecreasing in y, nonincreasing in x_i, and continuous in (y, x_i). Then the stand alone core of the public good provision problem $(u_1, \ldots, u_n; c)$ is nonempty. Moreover, the following "greedy algorithm" always picks (a utility vector corresponding to) a core allocation. Fix an arbitrary ordering of the agents, denoted $1, \ldots, n$ for simplicity.*

- u_1^* *is agent 1's stand alone utility* $[u_1^* = \max_{y \geq 0} u_1(y, c(y))]$.
- u_2^* *is agent 2's best utility such that* (u_1^*, u_2^*) *is feasible for coalition* $\{1, 2\}$ *standing alone.*

  $\vdots$

- $\dot{u}_n^*$ *is agent n's best utility such that* $(u_1^*, \ldots, u_n^*)$ *is feasible for the grand coalition.*

(*b*) *When several public goods are jointly produced, the stand alone core may be empty, even if individual preferences are convex and the marginal cost is constant.*

Proof of Statement (a). For simplicity, assume that utilities are quasi-linear: $u_i(y, x_i) = u_i(y) - x_i$ for all i. Given quasi-linear utilities, coalition S has no objection against the allocation $(y, x_1, \ldots, x_n)$ if and only if

$$\left(\sum_{i \in S} u_i(y) \right) - c(y) \geq \mathrm{sa}(S) = \max_{y' \geq 0} \left(\sum_S u_i(y') \right) - c(y'), \quad (13)$$

where we abuse notation by denoting as $\mathrm{sa}(S)$ (instead of $\mathrm{sa}(\Sigma_{i \in S} u_i)$) the net surplus of coalition S standing alone.

Now we pick an efficient level of public good y^*, namely, a solution of the program $\max\{\Sigma_i u_i(y)\} - c(y)$. The greedy algorithm with ordering $1, \ldots, n$ picks the following allocation $(x_1^*, \ldots, x_n^*)$ of the cost $c(y^*)$:

$$u_1(y^*) - x_1^* = \text{sa}(\{1\}),$$

$$\sum_{i=1}^{2} (u_i(y^*) - x_i^*) = \text{sa}(\{1, 2\}), \tag{14}$$

$$\sum_{i=1}^{m} (u_i(y^*) - x_i^*) = \text{sa}(\{1, 2, \ldots, m\}), \quad \text{all } m = 1, \ldots, n.$$

Check that the outcome $(y^*, x_1^*, \ldots, x_n^*)$ is in the stand alone core. We prove by induction on m the property P_m: no coalition contained in $\{1, \ldots, m\}$ has a stand alone objection against $(y^*, x_1^*, \ldots, x_n^*)$. Property P_1 is obvious, so we assume P_m and prove P_{m+1}. Pick a coalition $S = S_0 \cup \{m + 1\}$, where $S_0 \subseteq \{1, \ldots, m\}$. Then

$$\sum_{S_0} (u_i(y^*) - x_i^*) \geq \text{sa}(S_0),$$

$$u_{m+1}(y^*) - x_{m+1}^* = \text{sa}(\{1, \ldots, m + 1\}) - \text{sa}(\{1, \ldots, m\}).$$

Therefore inequality (13) for S follows from

$$\text{sa}(S_0) + \text{sa}(\{1, \ldots, m + 1\}) - \text{sa}(\{1, \ldots, m\}) \geq \text{sa}(S_o \cup \{m + 1\}). \tag{15}$$

The above inequality is a particular case of the property known as the submodularity of the function sa, namely,

for all $T, \quad T' \subseteq N, \quad$ all $i \in N \quad \{T \subseteq T' \text{ and } i \notin T'\}$

$$\Rightarrow \text{sa}(T \cup \{i\}) - \text{sa}(T) \leq \text{sa}(T' \cup \{i\}) - \text{sa}(T'). \tag{16}$$

This says that the marginal contribution of an agent to a coalition T (namely, the difference $\text{sa}(T \cup \{i\}) - \text{sa}(T)$) is nondecreasing in T (see Section 7.4). Property (15) is the particular case where $T = S_0$ and $T' = \{1, \ldots, m\}$. To check property (16), fix T, T', and i as in the premises of (16) and denote by y, y', respectively, the stand alone production level of coalition $T \cup \{i\}$ and T', respectively:

$$\text{sa}(T \cup \{i\}) = \sum_{T \cup \{i\}} u_j(y) - c(y),$$

$$\text{sa}(T') = \sum_{T'} u_{j'}(y') - c(y').$$

If $y \leq y'$, we have

$$sa(T \cup \{i\}) + sa(T') \leq \sum_T u_j(y) - c(y) + u_i(y')$$

$$+ \sum_{T'} u_{j'}(y') - c(y')$$

$$\leq sa(T) + sa(T' \cup \{i\});$$

if $y' \leq y$, we have

$$sa(T \cup \{i\}) + sa(T') \leq \sum_{T \cup \{i\}} u_j(y) - c(y)$$

$$+ \sum_{T' \setminus T} u_{j'}(y) + \sum_T u_{j'}(y') - c(y')$$

$$\leq sa(T' \cup \{i\}) + sa(T),$$

as was to be proved. The proof of statement (b) is given at the end of this section.   Q.E.D.

The striking feature of the existence result (statement (a)) is the lack of any convexity assumption either on preferences or on the cost function itself. Compare with the existence of a stand alone core in the decreasing marginal cost case (Theorem 5.1).

The above proof will be generalized in Section 7.4 to show that the greedy algorithm picks a core outcome in every convex cooperative game.

Just as in Section 5.4, a consequence of statement (a) is that the stand alone core is "large." Since the systems (8) and (14) are equivalent, the argument given after (8) shows that the surplus share of any coalition S varies from $sa(S)$ to $sa(N) - sa(N \setminus S)$.

Consider, for instance, a binary public good problem. If efficiency commands producing the public good, namely, $\sum_{i \in N} u_i > c$, an allocation $(x_1, \ldots, x_n)$ is in the stand alone core if and only if

$$\sum_{i=1}^n x_i = c \quad \text{and for all } i = 1, \ldots, n, \qquad 0 \leq x_i \leq u_i. \qquad (17)$$

In other words, only the stand alone objections by single-agent coalitions and by coalitions with $|N| - 1$ agents matter.[22] The constraints

---

[22] Indeed, $\{i\}$ does not object implies $u_i - x_i \geq sa(\{i\})$, whence $x_i \leq u_i$; $N \setminus \{i\}$ does not object implies $\sum_{N \setminus i}(u_i - x_i) \geq sa(N \setminus i) \geq \sum_{N \setminus i} u_i - c$, therefore $\sum_{N \setminus i} x_i \leq c$, or equivalently, $x_i \geq 0$. Conversely, inequalities (17) imply, for all S, the core property (13).

(17) are fairly loose. Within the core, agent i's cost share varies by as much as u_i, c, or sa(N), whichever is less.[23]

Example 5.9. A Public Good Variant of Example 5.4

The three agents share a machine that produces y units of public good at cost $c(y) = y^2/8$. This corresponds to the technology $f(x) = 6\sqrt{2x}$, as in Examples 5.3 and 5.4, when the output must be divided in three equal shares; in order to give y units of output to each agent, total cost x is $c(3y)$.

The correct quasi-linear representation of utilities is

$$\tilde{u}_3(y, x_3) = y - x_3, \qquad \tilde{u}_2(y, x_2) = \tfrac{3}{4}y - x_2, \qquad \tilde{u}_1(y, x_1) = \tfrac{1}{6}y - x_1.$$

The numeraire must be the input good because the output is not transferable anymore. Note the change of notation: the three players are now 1, 2, and 3.

The efficient level of public good solves

$$\max_{y \geq 0} (1 + \tfrac{3}{4} + \tfrac{1}{6})y - \frac{y^2}{8}.$$

Hence, $y^* = 7.67$. The technology f produces 23 units of outputs divided three ways; by contrast, in Examples 5.3 and 5.4, 36 units of output were produced. Total surplus is sa($\{123\}$) = 7.35. Therefore, agent 3's surplus share in the core varies from sa($\{3\}$) = 2 to sa(N) − sa($\{12\}$) = 5.67, whereas agent 2's share varies from 1.125 to 4.63 and agent 1's share varies from 0.05 to 1.225. Interestingly, individual cost shares vary in precisely the same interval.[24] This represents a variation of 280% for agent 1, and of more than 2400% for agent 3!

An example borrowed from Sharkey [1993] proving statement (b) in the theorem concludes this section.

Example 5.10   Street Lights

Three agents are located at the vertices of a triangle of which the edges are three streets. A street light can be installed in the middle of any one of these streets at a cost of $40 per light. Call α_i the location opposite

---

[23] Exercise: prove this claim.

[24] Exercise 5.20 gives the general formula for surplus share ranges, and explains the equality of cost and surplus share ranges.

to agent i's vertex. Agent i's benefit from a subset of lights is as follows:

- zero if there is no light anywhere or only in α_i,
- 30 if there is one nearby light,
- 45 if both nearby lights are installed.

The three lights are the three public goods, and the set A is $\{0, 1\}^3$ (note that convexity of preferences if meaningless—automatically true —on this set).

We check that the stand alone core of this economy is empty. Any coalition of two agents can guarantee a net surplus of 20 by paying for a light in the street connecting them $(30 + 30 - 40 = 20)$. Now the grand coalition generates the largest surplus by installing two lights, say in α_1 and α_2, generating a net surplus of $30 + 30 + 45 - 2 \times 40 = 25$. Note that a third light only brings an additional benefit of $15 + 15 = 30$, not enough to cover its cost. Any allocation of the surplus 25 among the agents, say, $25 = a_1 + a_2 + a_3$, must raise an objection of at least one two-person coalition as

$$a_i + a_j \geq 20 \quad \text{all } i, j \Rightarrow a_1 + a_2 + a_3 \geq 30.$$

This is the desired contradiction.

Exercise 5.18 gives a variant of the street lights example with divisible public goods.

Remark 5.2

When each agent owns a copy of a technology producing several public goods at once, it may not even be efficient to run a single copy of the technology if the following inequality holds:

$$\text{for some } y, y' \in \mathbb{R}_+^K : \quad c(y) + c(y') < c(y \vee y')$$

(where K is the number of public goods and $\vee$ represents the coordi- natewise supremum). Recall that when y and y' are independ- ently produced, each agent consumes $\max(y_k, y'_k)$ units of good k. When this inequality holds, the stand alone core of Definition 5.2 is trivially empty (even the stand alone test for individual agents is no longer feasible). Indeed, when the n agents, each one owning a machine c, coordinate the production of public goods, they achieve the following cost function $\tilde{c}$:

$$\tilde{c}(y) = \inf\left\{ \sum_{i=1}^{n} c(z_i) \right\},$$

where the infimum runs over all $(z_1, \ldots, z_n)$ such that $(z_1 \vee z_2 \vee \cdots \vee z_n) = y$. So Definition 5.2 applied to multiple public goods is interesting only if c satisfies

$$c(y \vee y') \le c(y) + c(y') \quad \text{for all } y, y' \text{ in } \mathbb{R}_+^K. \tag{18}$$

This holds true in particular if $K = 1$ or if the cost function is linear (as in Example 5.10).

To guarantee a nonempty stand alone core, we need very strong assumptions on both utilities and costs. An example of sufficient conditions is the subject of Exercise 5.19.

5.7. PUBLIC GOODS: THE RATIO EQUILIBRIUM

As in the case of private goods produced under decreasing marginal costs, the stand alone core sets loose bounds on the division of surplus among the beneficiaries of the public good. In this section and the next one, we look for deterministic selections from the core. As in the private goods case, a version of egalitarian-equivalence provides a simple and convincing selection. Unlike in the private goods case, we also define another selection with a competitive flavor, called the ratio equilibrium. The comparison of these two solutions raises important normative questions.

Throughout this section and most of the next one, we have a single public good. This is not just for convenience, since neither the ratio equilibrium nor the public good equivalent solution of the next section is easily generalizable to multiple public goods; see the discussion at the end of Section 5.8.

In the case of a binary public good, the principle "pay according to your own benefit" (an idea going back to J. S. Mill) suggests sharing the cost c in proportion to individual benefits from the public project, or

$$x_i = \frac{u_i}{\sum_N u_j} \cdot c \quad \text{for all } i = 1, \ldots, n.$$

When efficiency commands producing the public good, this selects a "fair" allocation in the stand alone core ((17)).

When the public good can be produced at different levels, sharing costs in proportion to individual benefits (a well-defined solution at least when utilities are quasi-linear so that we measure benefits in input units) no longer selects an allocation in the stand alone core. See Example 5.12 or Exercise 5.21. The ratio equilibrium suggests instead paying in proportion to *marginal* benefits, which, in turn, selects an allocation in the stand alone core.

Definition 5.4. Given is a public good provision problem $(u_1, \ldots, u_n; c)$. A set of ratios is a vector $(r_1, \ldots, r_n)$ where $r_i \geq 0$ for all i and $\sum_{i=1}^{n} r_i = 1$. We say that the outcome $(y^*; x_1^*, \ldots, x_n^*)$ is a ratio equilibrium with ratios $(r_1, \ldots, r_n)$ if we have, for all i

$$x_i^* = r_i \cdot c(y^*) \quad \text{and} \quad u_i(y^*, x_i^*) = \max_{y \geq 0} u_i(y, r_i \cdot c(y)). \quad (19)$$

Lemma 5.4. Assume that every utility u_i is decreasing in x_i. A ratio equilibrium is in the stand alone core. In particular, it is Pareto optimal.

Proof. Say that coalition S objects against the ratio equilibrium $(y', x_1^*, \ldots, x_n^*)$. There exists $(y, x_i, i \in S)$, feasible for coalition S standing alone, and such that $u_i(y^*, x_i^*) \leq u_i(y, x_i)$ for all $i \in S$, with at least one strict inequality. By definition of the ratio equilibrium, we have $u_i(y^*, x_i^*) \geq u_i(y, r_i c(y))$. Combining this with the above inequalities and the fact that u_i decreases in x_i, we deduce

$$x_i \leq r_i c(y) \quad \text{for all } i \in S \text{ with at least one strict inequality.}$$

Thus $(y, x_i, i \in S)$ is not feasible for coalition S, a contradiction.
Q.E.D.

Taking the first-order conditions of the program (19) (as implied by the ratio equilibrium property if utilities are differentiable and $y^* > 0$), we get

$$\frac{u_{iy}}{|u_{ix}|}(y^*, x_i^*) = r_i \cdot c'(y^*),$$

whence ratios (as well as cost shares) are proportional to individual marginal rates of substitution between the private and public good (in particular, to marginal benefits $u_i'(y^*)$ if utilities take the quasi-linear form $u_i(y) - x_i$).

Figure 5.5 illustrates the ratio equilibrium for a two-person economy with $c(y) = y$ and utilities quasi-linear in input.

The ratio "equilibrium" is often compared to the competitive equilibrium of an exchange economy (as suggested by the equilibrium terminology): agent i receives a personalized "price" signal (you will bear the fraction r_i of total costs); facing this "price," he demands the same amount of public good as every other agent.[25] However, one cannot

---

[25] Note that this price is nonlinear in the quantity of public good, unless c itself is linear. An alternative concept using a truly linear price—even if c is not linear—is the Lindahl equilibrium which, however, is not always a core selection; see Exercise 5.23.

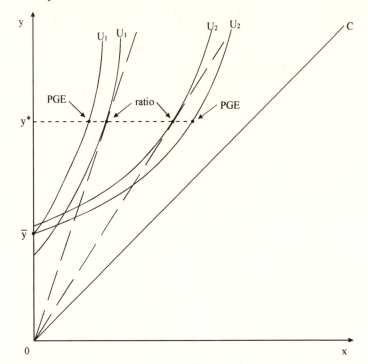

Fig. 5.5. The ratio equilibrium and PGE solution in a two-person public good problem.

extend the decentralization interpretation of the competitive price for exchange economies (Chapter 2) to the ratio equilibrium: these ratios are not anonymous and the "Lindahlian" auctioneer must monitor the demands of each and every beneficiary of the public good, a far more complex endeavor than the recording of aggregate net demands performed by the Walrasian auctioneer.

The concept of cost share equilibrium allows an interpretation of all the stand alone core outcomes similar to that of the ratio equilibrium; see Exercise 5.26.

When does the ratio equilibrium define a single-valued selection of the stand alone core? The answer depends upon the divisible or indivisible character of the public good.

When the public good is produced in indivisible units, we must expect multiple ratio equilibria. For instance, if the public good is binary ($y = 0$ or 1), every allocation in the stand alone core (i.e., any cost-sharing of c such that $0 \le x_i \le u_i$) is a ratio equilibrium (with ratio

$r_i = x_i/c$.[26] Our next example, where up to three units of public good can be produced, illustrates a typical configuration where the set of ratio equilibria does not reduce to a single point, yet cuts a subset of the stand alone core. The example shows also that we may have no ratio equilibrium whatsoever when the preferences and/or the cost function are not convex.

Example 5.11.   Indivisible Units of Public Good

Assume two agents with (quasi-linear) utilities:

y	cost	u_1	u_2
0	0	0	0
1	1	4	2
2	3	5	4
3	6	5	6

The efficient level is $y^* = 2$, with corresponding net surplus sa($\{12\}$) = 6. The stand alone utilities are

$$sa(1) = 3, \qquad sa(2) = 1.$$

A cost-sharing (x_1, x_2) is in the core if

$$0 \le x_1 \le 2, \qquad x_i + x_2 = 3.$$

A ratio profile (r_1, r_2) yields a ratio equilibrium if both agents demand two units of public good, or

$$5 - 3r_1 \ge 4 - r_1, 5 - 6r_1 \quad \text{and} \quad 4 - 3r_2 \ge 2 - r_2, 6 - 6r_2,$$

which reduces to $r_2 \ge 2/3$. The corresponding cost shares are $0 \le x_1 \le$ 1, or "half" of the core.

Change now agent 2's preferences as follows: $u_2(1)$ raises to 3 (instead of 2), everything else unchanged. Now both levels $y = 1$ or $y = 2$ are efficient and sa(1) = 3, sa(2) = 2. The core is the set of efficient allocations where agent 1 gets at least 3 and at most 4 of the 6 units of surplus. Yet there is no longer any ratio equilibrium in this problem, as no ratio r_2 can ever induce agent 2 to demand $y = 2$ (the inequalities $4 - 3r_2 \ge 3 - r_2$ and $4 - 3r_2 \ge 6 - 6r_2$ being incompati-

---

[26] By contrast, the ω-EE solution picks (uniquely) the cost shares proportional to benefits; see Section 5.8.

ble), and it will induce agent 2 to demand $y = 1$ only if $r_2 \geq 3/5$ (as we need $3 - r_2 \geq 6 - 6r_2$). On the other hand, we need $r_1 \geq 1/2$ to induce the demand $y = 1$ from agent 1 (i.e., $4 - r_1 \geq 5 - 3r_1$). Exercise 5.24 provides a numerical example in the same format where (i) preferences are convex, (ii) marginal costs *de*crease, and (iii) there is no ratio equilibrium.

When the public good is divisible, the existence of a ratio equilibrium is also threatened when preferences are not convex and/or marginal cost decreases. See Exercise 5.24 for an example with divisible goods where no ratio equilibrium exists due to nonconvex preferences.

Lemma 5.5

(*a*) *Indivisible units of public good with quasi-linear utilities. If marginal cost is nondecreasing and marginal utility is nonincreasing, then at least one ratio equilibrium exists.*

(*b*) *Divisible public good with arbitrary utilities. If marginal cost is nondecreasing and preferences are convex and continuous, then at least one ratio equilibrium exists.*

(*c*) *Divisible public good with quasi-linear utilities. If marginal cost is nondecreasing and marginal utilities for the public good are decreasing (or if marginal cost is increasing and marginal utilities are nonincreasing), then we have a unique ratio equilibrium.*

The easy proof of statements (a) and (b) is omitted. Exercise 5.25 generalizes statement (c) and outlines its proof. A very simple illustration of statement (c) is Example 5.9, or in general, any economy where the utilities for the public good take the form

$$u_i(y, x_i) = \lambda_i \cdot v(y) - x_i \quad \text{for some } \lambda_i \geq 0, \tag{20}$$

where the concave function v is the same for every agent. At any level y, absolute benefits from the public good $(\lambda_i \cdot v(y))$ and marginal benefits $(\lambda_i \cdot v'(y))$ remain proportional to the fixed coefficients λ_i; consequently, the unique ratio equilibrium divides surplus and cost shares in proportion to the coefficients λ_i. For instance, in Example 5.9, the surplus shares at the ratio equilibrium are as follows:

$$\pi_1 = 0.64, \quad \pi_2 = 2.88, \quad \pi_3 = 3.83.$$

Our next example illustrates a situation where, on the contrary, the ratios of marginal benefits across individuals vary widely at different levels of the public good. An agent who becomes satiated for the public good before the efficient level of production ends up paying nothing at all in the ratio equilibrium, even though her absolute benefit from the public good may be large (perhaps the largest).

Example 5.12. Where the Ratio Equilibrium Gives an Unfair Free Ride to Some Agents

The technology produces one unit of public good for one of private good ($c(y) = y$). Each one of the three agents is satiated once he or she reaches one unit of benefit from the public good; his or her utility is otherwise linear. Specifically, we have

$$u_i(y) = \min\{a_i \cdot y, 1\} \text{ for all } y \geq 0, \text{ with } a_1 = 3, \; a_2 = 1, \; a_3 = 0.5.$$

Efficiency commands producing $y^* = 1$. Indeed, an increase dy beyond y^* benefits only agent 3 (whose willingness to pay for more public good does not match the cost) and a decrease dy below y^* hurts both agents 2 and 3 (whose joint marginal utility exceeds marginal cost).

The unique ratio equilibrium has

$$r_1 = 0, \qquad r_2 = r_3 = .5.$$

To check this, consider agent 1: if the public good is free, she demands any y such that $y \geq 1/3$. Note that if she were charged a positive price, she could not possibly demand more that $1/3$, in contradiction of the ratio equilibrium property. Agent 3 faces the ratio .5, equal to his marginal utility for the public good. Hence he demands any y such that $0 \leq y \leq 2$. If he were charged any other price, he would demand either $y = 0$ or $y \geq 2$. Finally, agent 2 faces a price lower than his marginal utility, hence he demands precisely $y^* = 1$.

The ratio equilibrium is quite unfair in this example. The first agent pays nothing and gets a fully free ride at the expense of the third agent, who gets no share of the cooperative surplus ($u_3(y^*) - r_3 y^* = 0!$). Because she is quickly satiated by the public good, she can exploit the rest of the users. Exercise 5.21 generalizes this example and compares the ratio equilibrium to the two egalitarian-equivalent solutions defined in the next section.

The example also demonstrates the property stated just before Definition 5.4, namely, that allocating cost burdens in proportion to absolute benefits may not yield an allocation in the stand alone core.

Cost shares proportional to benefits $u_i(y^*)$ yield $x_1 = .4$, hence a net gain $u_1(y^*) - x_1 = .6$. On the other hand, agent 1's stand alone utility is

$$\text{sa}(1) = u_1(\tfrac{1}{3}) - \tfrac{1}{3} = .66.$$

Our next example displays another troubling property of the ratio equilibrium: the welfare of an agent may go up as a result of an overall increase of costs (a worsening of the technology).

Example 5.13.   Where the Ratio Equilibrium is not Cost-Monotonic

The two utility functions are shaped as in the previous example: linear in the public good level up to a satiation point. In this example, the two utilities share the same linear part (same marginal utility for the good near the origin), but have different satiation points:

$$u_1(y) = \min\{2y, 10\}, \qquad u_2(y) = \min\{2y, 9\}.$$

Consider first the cost function $c_1(y) = y$. With this technology, the efficient level of the public good is $y_1^* = 5$. Since agent 2 is satiated at this level, the ratio equilibrium imputes the full cost to agent 1, so that net benefits are

$$u_1(y_1^*) - c_1(y_1^*) = 5, \qquad u_2(y_1^*) - 0 = 9.$$

Now assume the marginal cost rises to 5 after four units of public good are produced, whence the new, higher cost function

$$c_2(y) = \max\{y, 5y - 16\}.$$

Efficiency commands producing $y_2^* = 4$ units of the good, and we have a continuum of ratio equilibria: any pair r_1, r_2 such that $r_i \geq 2/5$, $i = 1, 2$ (and $r_1 + r_2 = 1$) can be used. In any event, agent 1's net benefit is at least

$$u_1(y_2^*) - \tfrac{3}{5}c_2(y_2^*) = \tfrac{28}{5},$$

definitely more than his ratio equilibrium benefit under c_1.[27]

To summarize: the ratio equilibrium solution is guaranteed to exist only if we have convex preferences and increasing marginal costs.

---

[27] Neither the satiation of utilities for the public good nor the multiplicity of ratio equilibria under c_2 is essential to the example. As an exercise, the reader may construct a similar example with unique ratio equilibrium and strictly monotonic utilities.

Moreover, it may give a complete free ride to some agents ($x_i = 0$) and/or give no surplus whatsoever to some agent ($u_i(y, x_i) = u_i(0, 0)$), and finally, it violates cost monotonicity. In the next section, we present a solution avoiding all these problems.

5.8. PUBLIC GOODS: TWO EGALITARIAN-EQUIVALENT SOLUTIONS

In a public good provision problem where the technology of production is the common indivisible property of the group, the property of cost monotonicity is as natural as the resource monotonicity property in the fair-division problem (Section 4.7).

Cost Monotonicity: The welfare of no agent should go up as a result of an increase in the cost function.

When the cost function goes up (namely, $c_1(y) \leq c_2(y)$ for all $y \geq 0$), the common property technology becomes objectively less valuable; therefore, the community as a whole suffers. The axiom requires that every individual agent takes a share of this collective loss.

Of course, when each agent and each coalition of agents has access to the technology (namely, in the context where the stand alone core can be interpreted as a positive statement), the cost monotonicity axiom is less natural. As the stand alone utility of each agent and of each coalition deteriorates, it is conceivable that the stand alone utility of agent i worsens considerably more than that of agent j, resulting in a better bargaining position for i and, perhaps, a better overall share of surplus. Remarkably, within the single public good provision model, it is always possible to select a stand alone core outcome satisfying cost monotonicity; this selection is in fact unique. Contrast this result with the very unappealing consequences of resource monotonicity in the fair-division problem (see Section 4.7).

Definition 5.5. The Public Good Equivalent Solution. Assume the public good is divisible. The outcome (y^*; $x_1, \ldots, x_n$) is public good equivalent if it is feasible and there is a public good level $\bar{y}$, $\bar{y} \geq 0$, such that

$$u_i(y^*, x_i) = u_i(\bar{y}, 0) \quad \text{for all } i. \tag{21}$$

The public good equivalent (PGE) solution picks an efficient and public good equivalent outcome.

This definition can be extended to the case of public goods produced in indivisible units. See below after Lemma 5.6.

Taking the public good for numeraire, the PGE solution equalizes individual benefits measured as units of free public good. An alternative choice of the numeraire (in the model where x_i can be negative) is the (private) input good. This does not, however, lead to a selection of the stand alone core. (See Exercise 5.20, question (d).)

Lemma 5.6. A public good equivalent solution exists if utilities are continuous, nondecreasing in y, and decreasing in x_i. All corresponding outcomes yield the same utility profile; in other words, the utility level $u_i(\bar{y}, 0)$ in (21) is unique (and the public good level $\bar{y}$ is unique, too, if the joint utility $\Sigma_i u_i$ is strictly increasing). Moreover, a public good equivalent outcome is in the stand alone core, and the public good equivalent solution meets the cost monotonicity axiom.

Proof. For simplicity, we assume quasi-linear utilities of the form $u_i(y) - x_i$. Let y^* be an efficient level of the public good. When y varies from 0 to y^*, the joint utility $u_N = \Sigma_i u_i$ increases from 0 to $u_N(y^*)$. For simplicity, we rule out the special case $y^* = 0$ (where the proof is easily adapted), so we assume $y^* > 0$, implying $u_N(y^*) > \text{sa}(N)$ (remember the cost function c is increasing). Because u_N is continuous, there exists a level $\bar{y}$, $\bar{y} < y^*$, such that $u_N(\bar{y}) = \text{sa}(N)$. Note that $\bar{y}$ is unique if u_N is strictly increasing, but even when $\bar{y}$ is not unique, the utility level $u_i(\bar{y})$ is unique. Equation (21) then defines a feasible profile of cost shares x_i, because $c(y^*) = u_N(y^*) - \text{sa}(N) = u_N(y^*) - u_N(\bar{y})$. This completes the construction of an efficient and public good equivalent outcome. To show that it is in the stand alone core (a remark due to Mas-Colell [1980a]), we prove a stronger property.

Population Monotonicity: Consider two problems, one with the agents $1, 2, \ldots, n$, and one with only the agents $1, 2, \ldots, n - 1$. The utility levels that a certain agent i, $1 \le i \le n - 1$, enjoys at the public good equivalent solution of the first problem is never lower that her utility in the second problem.

To prove this, denote $N = \{1, 2, \ldots, n\}$, $N_0 = \{1, 2, \ldots, n - 1\}$, and y, y_0 the public good equivalent levels:

$$u_N(y) = \text{sa}(N), \qquad u_{N_0}(y_0) = \text{sa}(N_0).$$

We will show $u_N(y_0) \le u_N(y)$, implying $u_i(y_0) \le u_i(y)$, as desired. Let y_0^* be an efficient level for society N_0. In the problem within society N, it is feasible to give $\text{sa}(N_0)$ units of joint surplus to N_0 while agent n gets $u_N(y_0^*)$ (simply pick an efficient outcome for N_0 and set n's cost share to zero). Therefore

$$sa(N_0) + u_n(y_0^*) \leq sa(N).$$

Because $y_0 \leq y_0^*$ (see above), we deduce

$$u_{N_0}(y_0) + u_n(y_0) = sa(N_0) + u_n(y_0) \leq sa(N) = u_N(y).$$

This completes the proof of population monotonicity.

To show the stand alone core property, we apply repeatedly population monotonicity (PM). Consider a coalition $S = \{1, 2, \ldots, s\}$, and for $i \leq s$, let $\pi_i(s)$ denote agent i's net public good equivalent benefit within society S. The PM property implies

$$\pi_i(\{1, \ldots, s\}) \leq \pi_i(\{1, \ldots, s+1\}) \leq \cdots \leq \pi_i(N).$$

By efficiency of the egalitarian-equivalent solution, $\sum_1^s \pi_i(S) = sa(S)$; hence,

$$\sum_{i \in S} \pi_i(N) \geq sa(S),$$

which is the stand alone core property for coalition S.

It remains only to check the cost monotonicity property. Fix a utility profile u_i, $i = 1, \ldots, n$, and two cost functions c_1, c_2, with $c_1 \leq c_2$ ($c_1(y) \leq c_2(y)$ for all y). The corresponding levels of public good $\bar{y}_1, \bar{y}_2$ (from (21)) satisfy

$$u_N(\bar{y}_2) = sa(N, c_2) \leq sa(N, c_1) = u_N(\bar{y}_1).$$

If u_N is strictly increasing, this implies $\bar{y}_2 \leq \bar{y}_1$, hence $u_i(\bar{y}_2) \leq u_i(\bar{y}_1)$ for all i, as was to be proved. The case where u_N is not strictly increasing requires adjusting the above argument.   Q.E.D.

Remark 5.3

The public good equivalent solution can actually be characterized by the combination of cost monotonicity and the stand alone core property (Moulin [1987]). Thus there is no other cost-monotonic selection of the stand alone core. By contrast, the ratio equilibrium is neither cost-monotonic (Example 5.13) nor population-monotonic (Exercise 5.25).

A few examples illustrate the differences between the PGE and ratio solutions. The two solutions coincide in the quasi-linear context if the different willingness to pay (for the public good) remain in fixed proportions at every level of the public good; that is, utilities take the form (20) (e.g., Example 5.9). Equation (21) then reads

$$\lambda_i \cdot v(y^*) - x_i = \lambda_i \cdot v(\bar{y}) \quad \text{for all } i,$$

implying that cost shares are proportional to $\lambda_1, \ldots, \lambda_n$.

In the economy of Figure 5.5, the PGE solution is more favorable to agent 1 (and less to agent 2) that the ratio equilibrium. Next, consider Example 5.13 with cost c_1. Efficiency forces $y^* = 5$. Equation (21) yields

$$10 - x_1 = \min\{2\bar{y}, 10\}, \qquad 9 - x_2 = \min\{2\bar{y}, 9\}. \tag{22}$$

Summing up gives $14 = \min\{2\bar{y}, 10\} + \min\{2\bar{y}, 9\}$; hence, $\bar{y} = 3.5$ and $x_1 = 3$, $x_2 = 2$. In contrast with the ratio equilibrium, agent 2 does not get a free ride. Both agents share equally the net surplus measured, as usual, in input. (Recall that equal surplus sharing does not, in general, pick a core outcome.) Note that in the case of the cost function c_2, similar computations give $\bar{y} = 3$, $x_1 = x_2 = 2$. In contrast with the ratio equilibrium, the PGE solution picks a unique allocation (Lemma 5.6 tells us that this property is perfectly general).

As with other egalitarian-equivalent solutions, a very desirable feature of the PGE solution is that its existence and virtual uniqueness are guaranteed without making any convexity assumption on individual preferences and/or the technology. Contrast with Lemma 5.5 for the ratio equilibrium. This feature is preserved in the case of a public good produced in indivisible units, provided we adapt the definition by means of a simple linear interpolation trick.

Suppose, for simplicity, that utilities are quasi-linear and that y takes integer values. Extend the utility function u_i into a piecewise linear utility $\tilde{u}_i$ (and the cost function c into $\tilde{c}$) defined for all nonnegative numbers y, by "connecting the dots" (see the numerical examples below). If the indivisible good problem $(u_1, \ldots, u_n, c)$ has a unique efficient level of provision y, then y is still uniquely efficient in the extended divisible good problem $(\tilde{u}_1, \ldots, \tilde{u}_n, \tilde{c})$. If the indivisible good problem has several efficient levels, these levels are still efficient in the extended divisible good problem. Therefore we can solve (21) for the divisible good problem while insisting that y^* be an integer. Of course, the reference level $\bar{y}$ may not be an integer; this does not matter because no actual production takes place at that level.

Consider the example of a binary public good, when it is efficient to produce: $\sum_{i=1}^n u_i > c$. The extended utility is $u_i(y) = u_i \cdot y$ for all $y \geq 0$;

hence (21) reads

$$u_i - x_i = u_i \cdot \bar{y} \quad \text{all } i, \qquad \sum_{i=1}^{n} x_i = c.$$

Thus $\bar{y} = 1 - (c/\Sigma_i u_i)$ and costs are divided in proportion to benefits: $x_i = (u_i/\Sigma_j u_j) \cdot c$.

For a slightly less simple computation, consider Example 5.11. The net surplus is $sa(\{1,2\}) = 6$, and 1 unit of free public good gives 6 units of surplus as well. Thus $\bar{y} = 1$ in (21) and the PGE allocation is $x_1 = 1$, $x_2 = 2$ (it is an extreme point of the core interval). Next, we change agent 2's preferences to $u_2(1) = 3$, $u_2(2) = 4$, $u_2(3) = 6$. We now have $u_1(1) + u_2(1) > sa(\{1,2\})$, hence $\bar{y} < 1$. Equation (21) yields $\bar{y} = 6/7$, and the 6 units of surplus are divided as 3.43 for agent 1 and 2.57 for agent 2 ($x_1 = 1.57$, $x_2 = 1.43$).

The result of Lemma 5.6 still holds in the case of an indivisible public good: the PGE solution is an essentially single-valued solution selection of the stand alone core.

We turn to another important normative advantage of the PGE solution over the ratio equilibrium. The latter may give a free ride to some agents ($x_i = 0$) and may deny any share of the surplus to some other agents ($x_i = u_i(y^*)$, or $u_i(y^*, x_i) = u_i(0, 0)$). Example 5.12 shows that both difficulties can occur in the same public good problem. The PGE solution, on the other hand, avoids both difficulties. Suppose $u_i(y, x_i)$ increases in y and decreases in x_i, and c is increasing. Then if the zero production outcome ($y = x_1 = \cdots = x_n = 0$) is not efficient (that is, if there are cooperative opportunities to begin with), a PGE allocation ($y^*, x_1^*, \ldots, x_n^*$) has $0 < \bar{y} < y^*$ (where $\bar{y}$ is defined by (21)). Therefore,

$$u_i(0,0) < u_i(y^*, x_i^*) = u_i(\bar{y}, 0) < u_i(y^*, 0).$$

Thus $x_i^* > 0$ (no one gets a free ride) and everyone gets a positive share of surplus (i.e., in the quasi-linear case, $x_i < u_i(y^*)$).[28]

Finally, we define and discuss briefly the egalitarian-equivalent solution using the technology itself as a numeraire. The construction is analogous to that of Section 5.3 (see Lemma 5.1) and Section 5.5 (Lemma 5.2). As before, we call it the ω-EE solution.

Say that n agents share a technology to produce a public good with cost function c. For every λ, $1/n \leq \lambda \leq 1$, we define the λ-equivalent

[28] Note that if the function u_i is constant in y, for $\bar{y} \leq y \leq y^*$, agent i does get a free ride.

utility level

$$\psi_\lambda(u_i) = \max_{y \geq 0} u_i(y, \lambda \cdot c(y)).$$

The function $\psi_\lambda(u_i)$ is nonincreasing in λ, and coincides with the stand alone utility for $\lambda = 1$: $\psi_1(u_i) = \text{sa}(u_i)$. Therefore the utility vector $(\psi_1(u_1), \ldots, \psi_1(u_n))$ is always feasible. On the other hand, for $\lambda = 1/n$, the vector $(\psi_{1/n}(u_1), \ldots, \psi_{1/n}(u_n))$ is either unfeasible or Pareto optimal.

Lemma 5.7. *In the public good provision problem* $(u_1, \ldots, u_n; c)$, *agent i's unanimity utility is defined as*

$$\text{una}(u_i) = \max_{y \geq 0} u_i\left(y, \frac{c(y)}{n}\right).$$

The vector $(\text{una}(u_1), \ldots, \text{una}(u_n))$ *is either unfeasible or Pareto optimal. Suppose the functions c and* u_i *(for all i) are continuous and that* u_i *decreases in* x_i. *Then there exists a number* λ^*, $1/n \leq \lambda^* \leq 1$, *such that the utility vector* $(\psi_{\lambda^*}(u_1), \ldots, \psi_{\lambda^*}(u_n))$ *is Pareto optimal. The corresponding allocation(s) is called the* ω-*egalitarian-equivalent allocation. It gives to each agent at least her stand alone utility and at most her unanimity utility.*

Proof. Suppose a feasible allocation $(y; x_1, \ldots, x_n)$ gives to each agent his unanimity utility or more. For all i, we have

$$u_i(y, x_i) \geq \text{una}(u_i) \geq u_i\left(y, \frac{c(y)}{n}\right). \tag{23}$$

As u_i decreases in x_i, this implies $x_i \leq c(y)/n$ for all i. Taking the budget balance into account ($\sum_i x_i = c(y)$), we get $x_i = c(y)/n$ for all i, whence all inequalities in (23) are equalities.   Q.E.D.

Note that the definition of the ω-EE solution is equally valid if the good is divisible or indivisible, and if more than one public good is produced by the common property technology. In the latter case, we simply need to assume the inequality in Remark 5.2 to ensure that the vector of stand alone utilities is feasible.

Consider, once again, a binary public good with $\sum_{i=1}^n u_i > c$. The functions ψ_λ are $\psi_\lambda(u_i) = (u_i - c \cdot \lambda)_+$. The critical value λ^* is determined by the equation

$$\sum_{i=1}^{n} (u_i - c \cdot \lambda^*)_+ = \sum_{i=1}^{n} u_i - c.$$

The ω-EE solution makes every agent pay either $c \cdot \lambda^*$ or u_i, whichever is less. For instance, if every benefit u_i is at least c/n, everyone will pay the same cost share c/n, but if agent i's benefit is smaller than c/n, then agent i must pay exactly u_i (as $u_i \leq c/n$ implies $\psi_\lambda(u_i) = 0$ for all $\lambda > 1/n$) and is denied any share of the joint surplus. Thus the ω-EE solution simply gives equal cost shares under the voluntary participation constraint. It gives the entire surplus to those agents whose benefit exceeds the average cost.

Figure 5.6 depicts the ω-EE allocation in Example 5.9 (splitting total surplus as $\pi_1 = 0.128$, $\pi_2 = 2.60$, $\pi_3 = 4.62$) and compares it with the ratio equilibrium allocation (also recommended by the PGE solution).

Our next example illustrates the fact that the ω-EE solution may not be in the stand alone core.

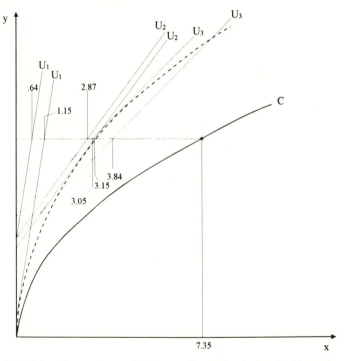

Fig. 5.6. The ratio equilibrium and ω-EE solution in Example 5.9.

Example 5.14. A Variant of Example 5.9

The cost function is $c(y) = y^2/8$ and the quasi-linear utilities of the three agents are

$$u_1(y, x_1) = 5y - x_1, \qquad u_i(y, x_i) = y - x_i, \qquad i = 2, 3.$$

Total surplus is

$$\text{sa}(N) = \max_{y \geq 0} \left\{ 7y - \frac{y^2}{8} \right\} = 98,$$

and the efficient level of public good is $y^* = 28$. Both ratio and PGE solutions coincide; they share surplus and costs in proportion to marginal utilities for the public good; hence,

$$x_1 = \pi_1 = 70, \qquad x_i = \pi_i = 14, \qquad i = 2, 3$$

(where π_i represent agent i's net surplus).

We compute the ω-EE solution:

$$\psi_\lambda(u_1) = \frac{50}{\lambda}, \qquad \psi_\lambda(u_2) = \psi_\lambda(u_3) = \frac{2}{\lambda}.$$

Therefore, agent 1's surplus share is 25 times larger than agent 2's share (compare with the $5:1$ ratio in the PGE solution). The critical value of λ is $\lambda^* = 27/49$, with corresponding surplus and cost shares:

$$\pi_1 = 90.74, \quad x_1 = 49.26, \qquad \pi_i = 3.63, \quad x_i = 24.37, \quad i = 2, 3.$$

Agent 1 benefited greatly in the shift from the PGE solution to the ω-EE solution, a general feature given linear utilities and quadratic costs. (For instance, in Example 5.9 the solution recommends $\pi_1 = 0.13$, $\pi_2 = 2.60$, $\pi_3 = 4.62$. See also Exercise 5.20.) In fact, agents 2 and 3 are hurt so badly by the ω-EE solution that they have a stand alone objection:

$$\text{sa}(\{23\}) = \max_{y \geq 0} \left\{ 2y - \frac{y^2}{8} \right\} = 8 > 7.26 = \pi_2 + \pi_3.$$

Note that we may easily pick an allocation in the stand alone core meeting the unanimity upper bound on individual utilities:

$$\text{una}(u_1) = \max_y \left\{ 5y - \frac{1}{3} \cdot \frac{y^2}{8} \right\} = 150,$$

$$\text{una}(u_i) = \max_y \left\{ y - \frac{1}{3} \cdot \frac{y^2}{8} \right\} = 6, \qquad i = 2, 3.$$

An example is $\pi_1 = 86$, $\pi_2 = \pi_3 = 6$ ($x_1 = 54$, $x_2 = x_3 = 44$).[29] The systematic analysis of public good problems with linear utilities and quadratic costs is the subject of Exercise 5.20.

The ω-EE solution denies any share of the surplus to any agent whose unanimity is simply $u_i(0, 0)$. It may also happen that an agent gets a "free ride"; Exercise 5.22 gives an example.

Although it may lie outside the stand alone core (hence violates population monotonicity) and is not cost-monotonic, the ω-EE solution deserves our attention because it accommodates easily the joint production of several public goods. The PGE solution does not generalize at all to multiple public goods. The same is true of the ratio equilibrium solution (see, however, Mas-Colell and Silvestre [1989] and Diamantaras and Wilkie [1994]).

5.9. PUBLIC BADS AND OTHER FORMS OF EXTERNALITIES

In general, a public bad provision problem has a process d (a nonde-creasing mapping such that $d(0) = 0$) transforming a desirable good Y into a bad X consumed by all agents in identical amounts. The quantity $x = d(\sum_{i=1}^n y_i)$ of public bad is produced if each agent i consumes y_i units of good Y. The resulting utility to agent i is $u_i(y_i, x)$, where, as usual, u_i increases in y_i and decreases in x.

As already discussed in the Introduction, one way to derive a public bad model starts from the cooperative production of a (single) private output from a (single) private input and imposes the exogenous constraint that input contributions must be equal. The feasible allocations then take the form $(y_i, c(\sum_{j=1}^n y_j)/n$, $i = 1, \ldots, n$ (similarly, the exogenous constraint that output shares must be equal yields the public good provision problem)

Examples include a party sharing a restaurant bill equally (here y is

---

[29] In any public good provision problem with quasi-linear utilities, we can pick a stand alone core allocation meeting the unanimity upper bound; see the discussion after Theorem 7.2 (Section 7.4). We can use the descending algorithm described in the Appendix to Chapter 7 and define yet another selection from the stand alone core.

the food and $d(y)$ is the per capita bill) and, more importantly, pollution models: y_i is the amount of polluting activity (e.g., how much waste is dumped in the lake) and x is the resulting pollution level (e.g., pollution level in the water supply).

We discuss only the case of one single process d that is the common property of all agents (the case where each agent owns a process d is practically less relevant and theoretically not different). We compare the impact of the same normative arguments that were used in the discussion of public goods. The no envy test suggests equalizing the shares of good Y: the feasible, envy-free allocation make all agents consume the same vector (y, x), where $x = d(ny)$. This is formally equivalent to a feasible and envy-free allocation in a public good provision problem, and, not surprisingly, this leads to the same difficulties as in Section 5.6.

We always normalize the goods X and Y in such a way that $d(0) = 0$: if none consume any amount of good Y, the "zero" level of public bad ensues. If this zero outcome is the "status quo ante," before any amount of public bad was produced, then *voluntary participation* says that no one should end up with a net loss from the status quo. This is a basic tenet of our interpretation of common property, discussed at length in Sections 5.4 and 5.6.

In even the simplest example of the public bad problem, we observe that no envy and voluntary participation together are incompatible with efficiency, exactly as in the public good case (Section 5.6) and in the case of decreasing marginal costs to produce a private good (Section 5.4). Say that if the aggregate consumption of good Y is positive but does not exceed 1, the level of public bad brings a disutility v_i to agent i (measured in good Y); if any more good Y is consumed, the level of public bad becomes unbearably high. Efficiency commands splitting 1 unit of good Y among the n agents if $1 > \sum_{i=1}^{n} v_i$. Voluntary participation requires that agent i consumes at least v_i units of good y, whereas no envy gives $1/n$ units to each.

In the public bad problem, the stand alone test is a vacuous requirement.[30] On the other hand, the unanimity utility is an upper bound on utilities with much bite. We define, as above,

$$\text{una}(u_i) = \max_{y \geq 0} u_i(y, d(ny)).$$

---

[30] At a feasible allocation $(y_1, \ldots, y_n; x)$, we have

$$x = d\left(\sum_j y_j\right) \geq d(y_i) \Rightarrow u_i(y_i, x) \leq u_i(y_i, d(y_i)) \leq \max_{y \geq 0} u_i(y, d(y)).$$

The interpretation is by now familiar: if all agents share agent i's preferences, they would implement the egalitarian outcome where every agent enjoys his unanimity utility. The vector of unanimity utilities is unfeasible or Pareto optimal (the proof of Lemma 5.7 applies without change). A natural solution of the public bad problem uses, once again, the ω-egalitarian-equivalence. Define for any λ, $0 \le \lambda \le 1/n$, the utility level $\phi_\lambda(u_i)$:

$$\phi_\lambda(u_i) = \max_{y \ge 0} \left\{ u_i\left(y, d\left(\frac{y}{\lambda}\right)\right)\right\}.$$

The level $\phi_\lambda(u_i)$ decreases with λ, and for a critical value λ^*, the utility vector $(\phi_{\lambda^*}(u_1), \ldots, \phi_{\lambda^*}(u_n))$ is Pareto optimal.

We turn now to other forms of public good externalities. Throughout the discussion of public goods, we used a specific model of externalities whereby if each agent produces y_i units of public good on his own machine (from x_i units of inputs, $y_i = f(x_i)$), the overall level $\bar{y} = \max_i(y_i)$ results. Moreover, we implicitly assumed that input was transferable across different machines, so that the efficient cooperation of n agents, each owning a machine, requires them to apply the total input effort $\sum_i x_i$ to a single machine, resulting in the level $f(\sum_i x_i)$ of public good for all to consume. Let us call this model the *standard* model of public good.

A more general model combines the public goods independently produced at levels $y_1, \ldots, y_n$ into an overall level $\bar{y} = \phi(y_1, \ldots, y_n)$, where ϕ is a symmetrical function (giving to each agent an equal opportunity to influence the public good level). As long as we maintain the assumption that input is freely transferable across machines, the general model is virtually identical to the standard one. Indeed, the optimal cooperative use of $x = \sum_i x_i$ units of input consists of solving the following program:

$$\max \phi(f(z_1), f(z_2), \ldots, f(z_n)) \quad \text{such that} \quad \sum_{i=1}^{n} z_i = \sum_{i=1}^{n} x_i,$$

of which the optimal value defines a function $g(\sum_i x_i)$. We are back to the standard model, with the only difference that agent i standing alone has access to the production function $y_i = \phi(0, \ldots, 0, f(x_i), 0, \ldots, 0)$ instead of the (superior) function $y_i = g(x_i)$. This will enlarge the stand alone core in the positive interpretation (because each agent has inferior opportunities when standing alone), but leaves intact the normative justification of the stand alone core associated with the production function g.

Now suppose that inputs are not transferable across different machines. For instance, several neighbors contribute to the beauty of the street by providing flowers in front of their houses. The aggregate flower output $\sum_i y_i$ is enjoyed by all (we neglect the fact that neighbor i enjoys his own flowers more than other flowers). The overall production function of the public good is $y = \sum_{i=1}^{n} f(x_i)$. The resulting model is not much different from the standard model (it coincides with the standard model if we measure input contributions with the variable $z_i = f(x_i)$).

A more interesting aggregation function is

$$\phi(y_1, \ldots, y_n) = \min_i \{y_i\}.$$

Think of a lakeside occupied by n consecutive towns. Each town builds a dike against floods on its portion of the lakeside. The level of protection afforded by a collection of dikes of various heights is just that of the lowest dike (assume that all towns will be equally flooded if any one dike breaks up). Another example is the macroeconomic games of coordination by Bryant [1983] and Cooper and John [1988]. This model is markedly different from the standard model. Efficiency here requires equal contribution of input by every agent, so that $y_i = f(x_i)$ is independent of i. Therefore, upon restricting attention to efficient allocations, we have the fully egalitarian model of cooperative production discussed in Section 6.2 (all agents consume exactly the same (output, input) pair). In particular, an efficient and fair outcome can be implemented by a decentralized mechanism; see Section 6.2.

Finally, consider the aggregation function

$$\phi(y_1, \ldots, y_n) = \max_i \{y_i\}.$$

When input is nontransferable, this model epitomizes the free-rider problem. The technology to produce the public good is a one-man job: slaying the dragon that terrorizes the town, opening the ball, or hosting the club's annual party are good examples (see Bliss and Nalebuff [1986]). Here efficiency and fairness conflict with one another, as efficiency requires that $(n - 1)$ agents get a free ride. On the other hand, the voluntary contribution mechanism discussed in Section 6.3 easily implements an efficient outcome. See Remark 6.4.

EXERCISES ON CHAPTER 5

Exercise 5.1.   A Variant of Example 5.1 (private good, increasing marginal costs (imc))

The output comes in indivisible units. Four agents have the following quasi-linear utilities:

y	u_1	u_2	u_3	u_4
1	3	4	5	6
2	6	8	10	12
3	8	11	14	17
4	10	14	18	22
5	11	16	21	26
6	12	18	24	30
7	12	19	26	33

Each agent owns the same (increasing marginal costs) machine γ:

y	1	2	3	4	5	6	$\cdots$
$\partial\gamma$	1	2	3	4	5	6	$\cdots$

.

(a) Show that the efficient production brings a total surplus $v(N) = 32$. Show that the vector $(\pi_1, \pi_2, \pi_3, \pi_4)$ of surplus shares is (corresponds to an allocation) in the core if and only if

$$\pi_1 + \pi_4 = 18, \quad \pi_2 + \pi_3 = 14, \quad 8 \leq \pi_3 \leq 9, \quad \pi_3 + 4 \leq \pi_4 \leq \pi_3 + 6.$$

(b) Show that the price p of the CEEI solution can be anywhere between 3 and 4 and that the corresponding profit distributions cover a one-dimensional interval from $(6, 6, 8, 12)$ to $(3, 5, 9, 15)$.

(c) Show that the set of surplus distributions corresponding to envy-free allocations contains all the core distributions and more (the dimension of this set is higher than that of the core).

Exercise 5.2.   A Property of Binary Demand Problems (private good, imc)

Suppose the agents are ranked by decreasing willingness to pay for one unit of the good: $u_1 \geq u_2 \geq \cdots \geq u_n$. The common property technology has increasing marginal costs $\partial c(1) \leq \partial c(2) \leq \cdots \leq \partial c(n)$.

(a) Recall why, in the CEEI solution, inactive agents receive more than their stand alone utilities.

(b) Give the necessary and sufficient system of inequalities guaranteeing the existence of (at least) one envy-free and efficient allocation such that no agent i receives more than his stand alone utility.

Exercise 5.3. Linear Utilities and Quadratic Costs (private goods)

The cost of producing the private good is $c(y) = y^2/2$. The technology is the common property of all agents. Agent i has a linear utility with constant marginal rate of substitution λ_i between the output and the input:

$$u_i(y_i, x_i) = \lambda_i \cdot y_i - x_i \quad \text{for } i = 1, \dots, n.$$

We assume that the input is transferable across agents (and that they have unbounded reserves of input). We order the agents in such a way that $\lambda_1 \geq \lambda_2 \geq \cdots \geq \lambda_n$.

(a) Show that efficiency requires producing λ_1 units of output and giving them to those agents with the highest parameter λ_i. Of course other agents can receive a compensation in input instead of output.

(b) Compute the CEEI allocation. (*Hint:* The competitive price is $p = \lambda_1$.) Show that this allocation is the worst among the envy-free and efficient ones from the point of view of agent 1.

(c) Give an example (some numerical values of λ_i) where every envy-free and efficient allocation fails the stand alone (upper bound) test.

(d) Compute the ω-EE solution (for arbitrary values of parameters λ_i).

In the last two questions, we suppose that the input is not transferable across agents.

(e) Show that efficiency commands producing λ_1 units of output and letting only those agents with the highest parameter λ_i contribute any input (other agents thus get some output for free).

(f) Answer the same question as in (b), (c), (d) above. Show that the CEEI (resp. ω-EE) solution under transferable input is Pareto superior (or indifferent) to the CEEI (resp. ω-EE) solution under nontransferable input.

Exercise 5.4. An Empty Core from Equal Split (private good, imc)

The three agents have identical (nonconvex) preferences represented by the quasi-linear utility

$$u_i(y_i, x_i) = w(y_i) - x_i$$

$$
\begin{aligned}
\text{where} \quad w(y) &= 7y && \text{for } 0 \leq y \leq 5, \\
&= \tfrac{2}{5}y + 33 && \text{for } 5 \leq y \leq 10, \\
&= 6y - 23 && \text{for } 10 \leq y \leq 11, \\
&= 43 && \text{for } 11 \leq y.
\end{aligned}
$$

Each agent owns the following technology:

$$f(x) = \min\{2x, 8\}.$$

To show that the core of this private ownership economy is empty, compute the (joint) surplus available to a coalition of any two agents:

$$v(12) \geq w(5) + w(11) - 2x_4 = 70.$$

Then show that the overall surplus-maximizing allocation has $y_1 = 11$, $y_2 = y_3 = 5$, for a total surplus $v(N) = 102.5$.

Exercise 5.5. Where the Core and the Stand Alone Upper Bound Conflict (private good, imc)

Each agent owns a machine γ producing output at nondecreasing marginal costs:

$$\gamma(y) = \max\{y, 2y - 1\}.$$

There are four agents, with quasi-linear utilities $u_i(y_i) - x_i$, where

$$u_1(y_1) = (y_1) = \min\{y_1, 1\}, \qquad u_i(y_i) = \min\{2y_i, 3\} \quad \text{for } i = 2, 3, 4.$$

Check that the stand alone utility of agent 1 is zero. Check that a coalition $\{1i\}$ (for any $i = 2, 3$, or 4) generates 1.5 units of surplus by coordinating the use of two machines. Check that the total surplus available in the economy is 4 (efficiency requires producing 4 units of output split between agents 2, 3, 4). Conclude that no allocation in the core of this economy passes the stand alone test.

Exercise 5.6. About the Uniqueness of the CEEI Solution (private good, imc)

(a) In the plane $z = (y, x)$, consider the four points

$$z_1 = (6, 1), \qquad z_1' = (3, 2), \qquad z_2 = (10, 15), \qquad z_2' = (11, 10).$$

Choose the preferences of agent i in such a way that the indifference contour through z_i has slope 3 ($dx/dy = 3$) and the indifference contour through z_i' has slope 1 (see Figure 5.7). Note that this is compatible for agent 1 with a quasi-linear utility of the form $u_1 = y_1 - v_1(x_1)$ with v_1 convex (but not with a form $u_1 = w_1(y_1) - x_1$ with w concave). Similarly, this is compatible with a utility u_2 of the form $u_2 = w_2(y_2) -$

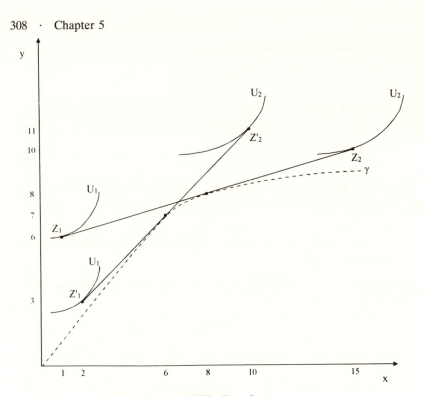

Fig. 5.7. An example with two CEEI allocations.

x_2 with w_2 concave (but not with a form $u_2 = y_2 - v_2(x_2)$ with v convex). Choose a convex cost function γ such that

$$\gamma(0) = 0, \quad \gamma(7) = 6, \text{ with } \frac{d\gamma}{dy}(7) = 1, \quad \gamma(8) = 8 \text{ with } \frac{d\gamma}{dy}(8) = 3.$$

(See Figure 5.7.) Consider the production problem where two agents have preferences u_1, u_2 and each owns a copy of γ (or where the machine c, $c(y) = 2\gamma(y/2)$, is the common property of both agents). Check that by construction there are (at least) two CEEI allocations in this problem, namely, (z_1, z_2) with price $p = 3$ and $r = 20$, and (z'_1, z'_2) with price $p = 1$ and $r = 1$.

(b) Show that if c is convex and all n agents have quasi-linear utilities $u_i = w_i(y_i) - x_i$ with w_i strictly concave, the CEEI allocation is unique (whether inputs are transferable or not). Show the same property if all utilities take the form $u_i = y_i - v_i(x_i)$ with v_i strictly convex.

Exercise 5.7.   Example 5.2 Continued (private good, imc)

The (common property) technology is $c(y) = (y - 1)_+$, a particular case of the family of cost functions discussed in Example 5.2. Recall that when the input is not transferable among agents, the vector of unanimity utilities is Pareto optimal and equals the CEEI utility vector as well. Throughout the exercise, it will be useful to sustain one's intuition with figures similar to Figure 5.1.

(a) Suppose the input is transferable among agents. Check that the unanimity utilities are not affected, that the CEEI solution may be Pareto superior to the vector of unanimity utilities. Specifically, consider n agents with utilities

$$u_i(y_i, x_i) = \alpha_i \cdot \sqrt{y_i} - x_i, \quad \text{all } y_i \geq 0, \text{ all } x_i.$$

Show that if $\alpha_i^2 \geq 4/n$ for all i, the CEEI solution does not involve any transfers and yields precisely the unanimity utilities, as before. Next, assume $\sum_j \alpha_j^2 \geq 4$, but $\alpha_i^2 < 4/n$ for some i, and show that the CEEI solution gives his unanimity utility to every agent i such that $\alpha_i^2 \geq 4/n$ and strictly more than his unanimity utility to every agent i such that $\alpha_i^2 < 4/n$. (*Hint:* The competitive budget constraint is $y \leq x + 1/n$, as in the previous case.) Finally, assume $\sum_j \alpha_j^2 < 4$ and set $\mu = 1/2(\sum_j \alpha_j^2)^{1/2}$. Check that in the (unique) CEEI allocation, every agent gets more than his unanimity utility. (*Hint:* The competitive budget set is $y \leq x/\mu + 1/n$.)

(b) Give an example where (input is transferable and) there is no efficient and envy-free allocation meeting the stand alone (upper bound) test.

(c) Compute the ω-EE solution. Recall that it coincides with the CEEI solution if input is not transferable. Suppose that input is transferable. Show that this equality (CEEI = ω-EE) holds true whenever the CEEI utility vector equals the unanimity utility vector, but only then. Show that the utility $\phi_\lambda(u_i)$ defined in (6) is as follows:

$$\phi_\lambda(u_i) = \lambda + \frac{\alpha_i^2}{4} \quad \text{if } \lambda \leq \frac{\alpha_i^2}{4}$$

$$= \alpha_i\sqrt{\lambda} \quad \text{if } \frac{\alpha_i^2}{4} \leq \lambda.$$

Deduce the ω-EE utility vector for the two following specifications:

$$n = 4 \quad \alpha_1 = \tfrac{1}{2}, \quad \alpha_2 = \alpha_3 = 1, \quad \alpha_4 = 2,$$

$$n = 4 \quad \alpha_1 = \alpha_2 = \tfrac{1}{2}, \quad \alpha_3 = 1, \quad \alpha_4 = \tfrac{3}{2}.$$

(d) Compute the CRE solution, assuming input is transferable. Check that the utility $\psi_\mu(u_i)$ is worth:

$$\psi_\mu(u_i) = \frac{\alpha_i^2}{4} \, \mu \quad \text{for all } \mu > 0.$$

Assuming $\sum_i \alpha_i^2 \geq 4$, deduce that the critical value of μ is

$$\mu^* = \frac{4 + \sum_i \alpha_i^2}{\sum_i \alpha_i^2},$$

and that the CRE allocation is

$$y_i = \frac{\alpha_i^2}{4}, \qquad x_i = \frac{\alpha_i^2}{4}(2 - \mu^*), \quad \text{for all } i.$$

Check that this allocation involves no transfer of input ($x_i \geq 0$ for all i). Hence the CRE solution is quite different from the CEEI and ω-EE solutions even when input is not transferable.

Exercise 5.8.   Examples 5.5 and 5.6 Continued (private good, dmc)

We fix the production function $f(x) = (x - 1)_+$, and the utility functions as in Exercise 5.7. In the current exercise, it does not matter whether the input is transferable or not.

(a) Check that efficiency forbids producing any output if $\sum_i \alpha_i^2 < 4$. If $\sum_i \alpha_i^2 > 4$, efficiency commands producing more than one unit of output.

(b) Compute the ω-EE allocation. (*Hint:* $\phi_\lambda(u_i) = (\alpha_i^2/4 - \lambda_+)$.) Interpret this solution as follows: let σ_i be the net surplus to agent i from the opportunity to buy (any amount of) output at price 1. The fixed cost is then shared equally among all agents, under the constraint that no one should pay more than σ_i. Show that the ω-EE allocation is in the stand alone core.

(c) Compute the CRE allocation and compare its utility vector to that of the ω-EE allocation.

Exercise 5.9.   The Case of Leontief Utilities (private good)

Three agents have the following utility functions:

$$u_i(y_i, x_i) = \min\{y_i, \beta_i - x_i\} \qquad i = 1, 2, 3,$$

where β_i is fixed and positive.

(a) Consider the (imc) cost function

$$c(y) = \max\{y, 2y - 30\}.$$

Compute the CEEI, ω-EE, and CRE solutions for the three following specifications of the parameters β_i:

(i) $\beta_1 = 4,$ $\beta_2 = 10,$ $\beta_3 = 22.$
(ii) $\beta_1 = 18,$ $\beta_2 = 26,$ $\beta_3 = 28.$
(iii) $\beta_1 = 14,$ $\beta_2 = 22,$ $\beta_3 = 34.$

(*Hint:* Draw a figure.)

(b) Consider the (dmc) cost function

$$c(y) = \min\left\{y, \frac{y}{2} + 15\right\}.$$

Answer the same questions as in (a).

(c) Fix an arbitrary number n of agents, and arbitrary Leontief utility functions of the form $u_i = \min\{y_i, \beta_i - \alpha_i x_i\}$, with $\beta_i, \alpha_i > 0$. Show that if c is any convex (imc) cost function, there is a unique CEEI allocation. Discuss the uniqueness of the CEEI if c is any concave (dmc) cost function.

Exercise 5.10.   *Where No Envy and the Stand Alone Test Conflict (private good, dmc)*

The cost function is

$$c(y) = \min\left\{10y, \frac{y}{10} + 9.9\right\}.$$

Two agents have the following utilities:

$$u_1 = \min\{y_1, 1\} - \frac{x_1}{10},$$

$$u_2 = \min\{y_2, 2\} - x_2.$$

Compute an efficient allocation where the stand alone test is met. Check that agent 1 receives $y_1 = 1$ and pays x_1, $8.2 \le x_1 \le 10$, whereas agent 2 gets $y_2 = 2$ and pays $0.2 \le x_2 \le 2$. Then check that there is no efficient and envy-free allocation where the stand alone test is met.

Exercise 5.11. An Empty Stand Alone Core With Multiple Outputs and Binary Demands (private good)

We have four agents with four outputs. The input (money) is transferable. Agent i is willing to pay u_i for one unit of good i and is not interested in good j, $j \neq i$ (think of a cable network and of good i as a line connecting agent i's house to the network). For any coalition S, $S \subseteq N = \{1, 2, 3, 4\}$, we write $c(S)$ for the cost of producing 1 unit of each good in S and no good j, for all $j \notin S$. Consider the following numerical specification:

$$c(\{i\}) = 4 \quad \text{for all } i, \qquad c(S) = 8 \quad \text{if } |S| = 2 \quad \text{and} \quad 1 \in S$$
$$= 4.5 \quad \text{if } |S| = 2 \quad \text{and } 1 \notin S,$$
$$c(S) = 8 \quad \text{if } |S| = 3, \qquad c(N) = 10,$$
$$u_1 = 2.5, \qquad u_i = 5, \qquad i = 2, 3, 4.$$

Check that the cost function is subadditive and show that the stand alone core is empty.

Exercise 5.12. An Empty Stand Alone Core With Initial Endowment of Output (Scarf [1986]) (private good, dmc)

The technology has a fixed cost and constant marginal costs thereafter (as in Example 5.5):

$$f(x) = 2(x - 1)_+, \quad \text{all } x \geq 0.$$

The three agents have identical Leontief utilities:

$$u_i = \min(y_i, \omega_i x_i), \qquad i = 1, 2, 3.$$

Initial endowments: agents 1 and 2 each have 2 units of input and no output; agent 3 has 2 units of output and no input.

Check $sa(u_i) = 2/3$ for $i = 1, 2$, and that any coalition of two agents can achieve the utility vector $(\lambda, 2 - \lambda)$ for any λ, $0 \leq \lambda \leq 2$. Deduce that if (z_1, z_2, z_3) is an allocation in the stand alone core, we must have $z_i + z_j \geq (2, 2)$ for all pairs i, j and deduce a contradiction.

Exercise 5.13. Where No Envy and Efficiency Conflict Under DMC (Vohra [1992]) (private good, dmc)

The technology has a fixed cost 1.6 and constant marginal costs thereafter:

$$c(y) = \frac{y}{3} + 1.6 \quad \text{if } y > 0, \qquad c(0) = 0.$$

Utilities of the two agents are quasi-linear and input is transferable (initially the agents have no output):

$$u_1 = \min\{y_1, 2\} - x_1, \qquad u_2 = y_2 - \max\{2x_2, 4x_2 - 2\}.$$

Check that the Pareto set is made of two disconnected pieces. On the first piece, no production takes place ($y_1 = y_2 = 0$) and $x_2 \geq 0.2$, $x_1 = -x_2$. On the second piece, production takes place as follows:

$$y_1 = 2, \quad x_1 \geq 1.8, \qquad y_2 = 3x_1 - \tfrac{19}{5}, \quad x_2 = 1.$$

See Figure 5.8. Deduce that every efficient allocation generates envy. Note that this example is similar to Varian's example for fair division (Exercise 4.15). The only difference is that here, both agents have convex preferences.

Exercise 5.14. The Proportional Solution (Roemer and Silvestre [1993])
    (private good)

The proportional solution selects an efficient allocation such that the ratios y_i/x_i are equal for all $i = 1, \ldots, n$. In other words, each agent is "paying" average cost for her output share and enjoying average product for her input contribution:

$$x_i = y_i \cdot \frac{c(\Sigma_j\, y_j)}{\Sigma_j\, y_j} \Leftrightarrow y_i = x_i \cdot \frac{f(\Sigma_j\, x_j)}{\Sigma_j\, x_j}.$$

We are not guaranteed that such an allocation exists unless we make further assumptions on preferences.

(a) Suppose that preferences are quasi-linear of the form $u_i = w_i(y_i) - x_i$. Then a proportional allocation exists: simply take any solution $y_1^*, \ldots, y_n^*$ of the program

$$\max_{y_1, \ldots, y_n \geq 0} \sum_1^n w_i(y_i) - c\left(\sum_1^n y_i\right),$$

and divide costs accordingly. Check that in Examples 5.3 and 5.4, the proportional solution gives all the cooperative surplus to the most productive agent. Check that under decreasing marginal costs and binary demands, the proportional solution violates voluntary participation.

(b) Show that under increasing marginal costs and convex, continuous preferences, at least one proportional allocation exists.

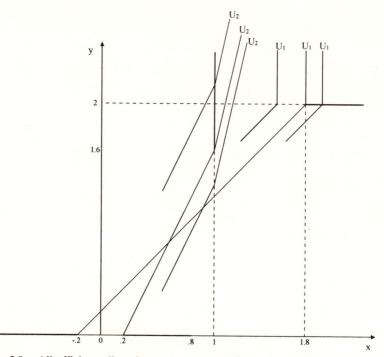

Fig. 5.8. All efficient allocations generate envy (Exercise 5.13).

(c) In the economies discussed in Exercise 5.9, what does the proportional solution recommend?

Exercise 5.15. Computing the ω-EE and CRE Solutions with Indivisible Output (private good, dmc)

Consider the numerical example at the end of Section 2.6, with seven agents willing to consume at most two units of indivisible good and a decreasing marginal cost technology. Efficiency requires producing two units for agent 1 and for agent 2, and one unit for agent 3 and for agent 4, for a total surplus of 41.

(a) Show that the CRE solution distributes the surplus as follows:

$$\pi_1 = 23.25, \qquad \pi_2 = 13.25, \qquad \pi_3 = 3.25, \qquad \pi_4 = 1.25.$$

(*Hint:* Compute $\psi_\lambda(u_1) = \max\{60 - \lambda, 90 - 2\lambda\}$.)

(b) Show that the ω-EE solution proposes the following distribution:

$$\pi_1 = 26.18, \qquad \pi_2 = 14.18, \qquad \pi_3 = 0.64, \qquad \pi_4 = 0.$$

(*Hint:* The functions $\lambda \cdot c(1/\lambda)$ and $\lambda c(2/\lambda)$ are computed by means of the linear interpolation of c. Compute them explicitly, and the intervals in which they take a simple linear form. Use trial and error to determine the correct interval where the quantity $\Sigma \phi_\lambda(u_i)$ takes the value 41.)

Exercise 5.16. Mixed Returns (private good)

When the marginal costs are neither increasing nor decreasing, the vector of stand alone utilities $\mathrm{sa}(u_i)$ and the vector of unanimity utilities $\mathrm{una}(u_i)$ may or may not be feasible.

(a) In this example, the vector $(\mathrm{sa}(u_1), \mathrm{sa}(u_2))$ is feasible, the vector $(\mathrm{una}(u_1), \mathrm{una}(u_2))$ in unfeasible, and there is no utility vector (a_1, a_2) such that $\mathrm{sa}(u_i) \le a_i \le \mathrm{una}(u_i)$ for $i = 1, 2$. The output is indivisible and the cost and utilities are

q	1	2	3	4
$c(q)$	10	14	23	30

,

q	u_1	u_2
1	15	25
2	19	40

.

Compute $\mathrm{sa}(u_1) = 5$, $\mathrm{sa}(u_2) = 26$, and $\mathrm{una}(u_1) = 8$, $\mathrm{una}(u_2) = 25$, and check the claim.

(b) In the next example, both vectors $(\mathrm{sa}(u_1), \mathrm{sa}(u_2))$ and $(\mathrm{una}(u_1), \mathrm{una}(u_2))$ are unfeasible, and there is no efficient utility vector (a_1, a_2) such that $a_i \le \mathrm{sa}(u_i)$ and a $a_i \le \mathrm{una}(u_i)$ for $i = 1, 2$. Utilities are as in (a) and the cost function is

q	1	2	3	4
$c(q)$	10	15	28.5	40

.

Check the claim.

(c) In the third example, both the vector of unanimity utilities and that of stand alone utilities are feasible. There is, however, no feasible (a_1, a_2) such that $a_i \ge \mathrm{sa}(u_i)$ and $a_i \ge \mathrm{una}(u_i)$ for $i = 1, 2$.
The divisible output is produced at the following cost:

$$c(y) = y \qquad \text{if } 0 \le y \le 2,$$
$$= y + 2 \quad \text{if } 2 < y.$$

(A fixed cost must be paid when we reach $y = 2$ to produce any more output. Think of buying a new machine, once the first machine reaches its capacity 2.) The two agents have Leontief utilities:

$$u_1 = \min(y_1, 4 - x_1), \qquad u_2 = \min(y_2, 9 - x_2).$$

Compute $sa(u_1) = 2$, $sa(u_2) = 3.5$, $una(u_1) = 1.5$, $una(u_2) = 4$, and check the claim.

(d) Construct a two-person problem where the unanimity vector is feasible, the stand alone vector is not feasible, but there is no (a_1, a_2) such that $una(u_i) \leq a_i \leq sa(u_i)$ for $i = 1, 2$.

Exercise 5.17. Samuelson's Efficiency Equation for Public Goods

Fix a public good provision problem $(u_1, \ldots, u_n; c)$ with a single public good. Assume the cost function and all utility functions are differentiable. Assume also that $\partial u_i / \partial x_i$ is never zero ($\partial u_i / \partial x_i < 0$). Consider an interior allocation $(y; x_1, \ldots, x_n)$, namely, such that $x_i > 0$ for all i, and $y = f(\sum_i x_i) > 0$ (f is the production function).

(a) Show that if the allocation is Pareto optimal, Samuelson's equation holds:

$$\sum_{i=1}^{n} \frac{\partial_y(u_i)}{-\partial_{x_i}(u_i)}(y, x_i) = \frac{dc}{dy}(y). \tag{24}$$

(*Hint:* Consider small variations $dx_1, \ldots, dx_n$ of individual contributions around a given interior allocation. They affect the public good level as $dy = (\sum_i dx_i)/c'$ (where $c' = dc/dy$).) Write the system of inequalities expressing that each agent (strictly) benefits from the move, and show this system has a solution if and only if (24) does not hold.

(b) What is the analog of Samuelson's equation for public bads?

Exercise 5.18. An Empty Stand Alone Core with Multiple Public Goods

This is a divisible goods example analog to Example 5.10. We have three public goods produced at constant marginal cost $c(y, z, t) = 3(y + z + t)$, where y, z, t are the respective quantities of the three goods. Three agents share the public good technology. Their utility functions are concave and quasi-linear:

$$u_1(y, z, t; x_1) = \min\{2(y + z), \tfrac{1}{5}(y + z) + \tfrac{9}{5}\} - x_1,$$

$$u_2(y, z, t; x_2) = \min\{2(z + t), \tfrac{1}{5}(z + t) + \tfrac{9}{5}\} - x_2,$$

$$u_3(y, z, t; x_3) = \min\{2(y + t), \tfrac{1}{5}(y + t) + \tfrac{9}{5}\} - x_3.$$

Show that any two-person coalition generates one unit of surplus by standing alone (*hint:* Agents 1 and 2 produce one unit of good Z), and that the grand coalition cannot generate any more surplus. Conclude that the stand alone core is empty.

Exercise 5.19. A Nonempty Stand Alone Core with Multiple Public Goods

We have k public goods and one input (transferable or not). Denote by y a vector in $\mathbb{R}^k_+$ representing the level of production of the different public goods. We denote by $\vee$ and $\wedge$ respectively the coordinatewise maximum and coordinatewise minimum in $\mathbb{R}^k$. We say that a real-valued function h defined on $\mathbb{R}^k_+$ is submodular (resp. supermodular) if the following inequality (resp. its opposite) holds:

$$h(y) + h(y') \geq h(y \vee y') + h(y \wedge y') \quad \text{for all } y, y' \text{ in } \mathbb{R}^k_+ .$$

Consider a k public goods provision problem $(u_1, \ldots, u_n; c)$ where (a) the cost function c is submodular, (b) each utility takes the form $u_i = w_i(y) - x_i$, with a supermodular function w_i. Then show that the stand alone cooperative game is convex, implying that its core is nonempty (see Theorem 7.1 for the definition of a convex cooperative game).

Exercise 5.20. Public Good with Linear Utilities and Quadratic Costs

The cost function is $c(y) = y^2/2\alpha$, where α is a positive parameter. Agent i's utility is $u_i = \mu_i \cdot y - x_i$, where we assume $\mu_1 \geq \mu_2 \geq \cdots \geq \mu_n > 0$. As usual, we use this quasi-linearity with respect to input (money) to measure surplus in monetary terms.

(a) Show that in the stand alone core, the surplus share of the coalition S varies between $(\alpha/2)\mu_S^2$ and $(\alpha\mu_S/2) \cdot (2\mu_N - \mu_S)$, where we write $\mu_S = \Sigma_{i \in S} \mu_i$. Thus, if S supplies half of the total marginal willingness to pay, its surplus share varies by a factor of 3 in the core. Show that the cost share of coalition S varies in precisely the same interval as its surplus share.

(b) Check that the ratio equilibrium and the PGE solution coincide and urge us to split total surplus in proportion to the marginal willingness to pay for the public good (μ_i).

(c) Show that the ω-EE solution recommends splitting total surplus in proportion to the square of individual marginal willingness to pay. Therefore, agent 1 prefers ω-EE to PGE, and agent n has the opposite preferences. Show that coalition S has a stand alone objection against the ω-EE solution if and only if

$$\frac{(\mu_N)^2}{\Sigma_N \mu_i^2} < \frac{(\mu_S)^2}{\Sigma_S \mu_i^2}. \tag{25}$$

(Example 5.14 is a particular case.)

(d) Consider the solution dividing equally the surplus among all agents. Find a formula similar to (25) expressing that coalition S has a stand alone objection.

(e) Suppose $n = 2$ and denote by (π_1, π_2) a distribution of the efficient surplus sa(12). Consider the interval

$$I = \{(\pi_1, \pi_2)/\pi_1 + \pi_2 = sa(12); \quad sa(i) \leq \pi_i \leq una(u_i)\}.$$

Show that the variation of π_1 in I is at most $1/9$ of available surplus sa(12).

(f) Suppose $n = 3$, and $\mu_1 = \frac{1}{2}$, $\mu_2 = \frac{1}{3}$, $\mu_3 = \frac{1}{6}$. Draw in the simplex $\pi_1 + \pi_2 + \pi_3 = sa(123)$ the area delineated by the stand alone core *and* the unanimity upper bound. Show that this area is "small" vis-à-vis the available surplus. Compare this area with the PGE and the ω-EE solutions. Answer the same questions for $\mu_1 = .7$, $\mu_2 = .2$, $\mu_3 = .1$.

Exercise 5.21. Generalization of Example 5.12

The cost is $c(y) = y$ and the n quasi-linear utilities are $u_i(y) = \min\{a_i y, 1\}$. The fixed coefficients a_i satisfy $a_1 \geq a_2 \geq \cdots \geq a_n > 0$, and $\Sigma a_i > 1$. (Note that if $\Sigma a_i \leq 1$, the stand alone core reduces to the zero outcome.)

(a) Define the index i^* by the inequalities

$$\sum_{i^*}^{n} a_i > 1 + \sum_{i^*+1}^{n} a_i \quad \text{with the convention} \quad \sum_{n+1}^{n} a_i = 0.$$

Show that the efficient level of the public good is $y^* = 1/\alpha_{i^*}$. Show that the ratio equilibrium is unique and the ratio profile is

$$r_i = 0 \text{ if } 1 \leq i \leq i^* - 1, \quad r_{i^*} = 1 - \sum_{i+1}^{n} a_i, \quad r_i = a_i \text{ if } i^* + 1 \leq i \leq n.$$

(b) Compare the ratio equilibrium, the PGE, and the ω-EE solutions in the economy of Example 5.12.

(c) In this question, we consider the four-agent example $a_1 = 1.5$, $a_2 = 1$, $a_3 = 0.8$, $a_4 = 0.5$. Check that total cooperative surplus (measured in the input numeraire) is 2.375 and that the ratio equilibrium distributes it as follows:

$$\pi_1 = \pi_2 = 1, \qquad \pi_3 = 0.375, \qquad \pi_4 = 0.$$

In particular, check that agents 1 and 2 get more than their unanimity utility.

Show that the PGE solution is much more favorable to agents 3 and 4. Specifically, it distributes surplus as follows:

$$\pi_1 = 0.9375, \qquad \pi_2 = 0.625, \qquad \pi_3 = 0.5, \qquad \pi_4 = 0.3125.$$

Show that the ω-EE solution is even more favorable to agents 3 and 4, at the expense of agent 1 only. It yields

$$\pi_1 = 0.78, \qquad \pi_2 = 0.67, \qquad \pi_3 = 0.575, \qquad \pi_4 = 0.34.$$

(d) Check that the ω-EE solution is in the stand alone core for the two above numerical examples. Is this a general property in the economies of Example 5.12?

Exercise 5.22. Free Riders in the ω-EE Solution (public good)

(a) The cost of the public good is $c(y) = y$, and the three agents have the following utilities:

$$u_i = \min\{y_i, 4\} - x_i \quad \text{for } i = 1, 2, \qquad u_3 = \min\{5y_1, 40\} - 6x_1.$$

Show that the ω-EE solution and the ratio equilibrium coincide and give a free ride to agent 3. Contrast this outcome with that of the PGE solution.

(b) If the economy has only two agents with strictly increasing utility for the public good, show that the ω-EE solution never has an agent free-riding (unlike the ratio and PGE solutions).

Exercise 5.23. The Lindahl Equilibrium (public good)

This solution concept is the ancestor of the ratio equilibrium. Given a public good provision problem $(u_1, \ldots, u_n; c)$, we consider a price vector $(p_1, \ldots, p_n)$ (with $p_i \geq 0$ for all i) and a level y^* of public good such that:

(i) y^* maximizes the utility $u_i(y, p_i. y)$ over all $y \geq 0$.

(ii) y^* maximizes the profit $(\Sigma_i \, p_i) \cdot y - c(y)$ over all $y \geq 0$.

A vector $(p_1, \ldots, p_n; y^*)$ is a *Lindahl equilibrium* if it satisfies these two properties. Note that if we charge $p_i \cdot y^*$ to agent i, the overall revenue exceeds cost. In order to define a full-fledged solution to the cooperative production problem, we divide this surplus equally among all agents. Thus agent i's net cost share is

$$x_i = p_i \cdot y^* - \frac{1}{n}\left\{\left(\sum_j p_j\right) \cdot y^* - c(y^*)\right\}.$$

(a) Check that this solution is Pareto optimal if utilities are quasi-linear, but not in general.

(b) Compute the Lindahl equilibrium solution in the following numerical example (with imc):

$$c(y) = 5y^2, \qquad u_1 = 9y - x_1, \qquad u_2 = y - x_2.$$

Compare it to the ratio equilibrium solution. Check that it fails the stand alone test.

(c) Same as question (b) for the following example with dmc:

$$c(y) = 10\sqrt{y}, \qquad u_1 = \min\{9y, 36\} - x_1, \qquad u_2 = \min\{y, 4\} - x_2.$$

Note that we have many Lindahl equilibria in this example.

Exercise 5.24.   Where No Ratio Equilibrium Exists (public good)

In each example, show that there is no ratio equilibrium. Then compute the PGE and ω-EE solutions.

(a) Our first example has a decreasing marginal cost technology and two agents with convex preferences represented by quasi-linear utilities:

$$c(y) = \tfrac{3}{2}\sqrt{y}, \qquad u_1 = \min\{y, 1\} - x_1, \qquad u_2 = \min\{y, 2\} - y.$$

(*Hint:* The efficient level of public good is $y^* = 2$.)

(b) In the second example, the public good comes in indivisible units, marginal utilities decrease, and so do marginal costs:

q	c	u_1	u_2
1	3	4.5	4
2	5	5.5	5
3	5.5	6.5	5

(c) In out third example, marginal costs are constant, one agent has nonconvex preferences, the other has convex preferences:

$$c(y) = 4y, \qquad u_1 = \sqrt{2y} - x_1,$$

$$u_2 = \max\{\min(4y, y + 3), \min(4y - 6, y + 6)\} - x_2.$$

(*Hint:* Compute the demand curves $d_i(r_i)$, namely, the quantity (or quantities) of public good demanded by agent i if he pays $r_i \cdot c(y)$.)

Exercise 5.25. Existence and Uniqueness of the Ratio Equilibrium (pubic good)

Fix a convex, increasing cost function c, and n agents with utility functions u_i, $i = 1, \ldots, n$. Assume that u_i has nonzero partials ($\partial_{x_i}(u_i)$ $< 0 < \partial_y(u_i)$) and represent convex preferences. We define the *normality assumption*: the input and the output are *normal* goods, namely, the slope supporting the indifference contour of u_i through (y, x_i) is nondecreasing with respect to x_i and nondecreasing with respect to y (where the slope is defined as $dy/dx = -\partial_{x_i}(u_i)/\partial_y(u_i)$). Note that a quasi-linear utility $w_i(y) - x_i$ or $y - v_i(x_i)$ satisfies the normality assumption if w_i is concave or v_i is convex.

(a) In this question, the normality assumption holds true. For a given y, $y \geq 0$, define $r_i(y)$ to be this number r_i, $r_i \geq 0$, such that the slope supporting the indifference contour of u_i through $(y, r_i \cdot c(y))$ equals $1/r_i \cdot c'(y)$. Show that $r_i(y)$ is uniquely defined, and that $r_i(y)$ is continuous and nonincreasing in y. Show that r_i is decreasing in y if c is strictly convex and/or the upper contours of u_i are strictly convex.

Deduce that a unique ratio equilibrium exists under the additional assumption H: c is strictly convex and/or at least one agent has strictly convex preferences. If H does not hold, show that all ratio equilibria have the same utility vector.

(b) We still assume normality, as well as assumption H. Show that the ratio equilibrium solution is population-monotonic: deleting one agent cannot make any other agent better off.

(c) We do not assume normality (or H) anymore. Find an economy (c, u_1, u_2) satisfying the properties spelled out at the beginning of the exercise, where several ratio equilibria (all different utility wise) coexist. Give a three-agent example showing that the ratio equilibrium is not population-monotonic.

Exercise 5.26. Cost-Share Equilibria (Mas-Colell and Silvestre [1989])
(public good)

Fix a public good provision problem $(u_1, \ldots, u_n; c)$ with an arbitrary number of public goods (or bads), and one input (money). A feasible allocation is a list $(y; x_1, \ldots, x_n)$, where y is the vector of public goods (or bads) and x_i is agent i's net input contribution (positive if i gives up some money, negative is i gets a net transfer). Feasibility requires $c(y) = \sum_i x_i$. Note that we do not need any monotonicity or continuity assumption on c (hence y could represent goods, bads, or abstract "public decisions").

All utility functions $u_i(y, x_i)$ are decreasing in x_i. Moreover, any change in the public good vector can be compensated by a monetary transfer: for every (y, x_i) and every y', there exists x_i' such that $u_i(y, x_i) = u_i(y', x_i')$. These assumptions hold in particular if u_i is quasi-linear of the form $u_i(y) - x_i$, with "large" reserves of cash.

A system of cost shares is a list $(t_1, \ldots, t_n)$, where each t_i associates a monetary transfer $t_i(y)$ to every public good level y, and where the allocation $(y; t_1(y), \ldots, t_n(y))$ is feasible for all y. Finally, a *cost-share equilibrium* consists of an allocation $(y^*; x_1^*, \ldots, x_n^*)$ and a system of cost shares $(t_1, \ldots, t_n)$ such that
  (i) $t_i(y^*) = x_i^*$ for all i.
  (ii) (y^*, x_i^*) maximizes u_i over all pairs $(y, t_i(y))$.

A particular case is the ratio equilibrium where the transfers take the form $t_i(y) = r_i \cdot c(y)$. The interpretation of a cost-share equilibrium is in a similar competitive vein: given the nonlinear price t_i, agent i demands the public good level y^*; all individual demands are compatible.

(a) Show that every cost-share equilibrium is a Pareto optimal allocation. Conversely, show that every Pareto optimal allocation is a cost-share equilibrium for an appropriate system of cost shares. (*Hint:* If $(y^*, x_1^*, \ldots, x_n^*)$ is Pareto optimal, and y is arbitrary, define x_i by $u_i(y^*, x_i^*) = u_i(y, x_i)$ and show $\sum_i x_i \leq c(y)$.)

(b) Show that every cost-share equilibrium with a nonnegative system of transfers ($t_i(y) \geq 0$ for all i and all y) is in the stand alone core. (*Hint:* Let $(y^*, x_1^*, \ldots, x_n^*)$ be such an allocation and let $(y, x_i, i \in S)$ be a stand alone objection by coalition S. Show $t_i(y) \geq x_i$ for all $i \in S$, and deduce from the nonnegativity of all t_j that all inequalities must be equalities.

(c) Conversely, show that every allocation in the stand alone core is a cost-share equilibrium with a nonnegative system of transfers. (*Hint:* Let $(y^*, x_1^*, \ldots, x_n^*)$ be a core allocation. Pick any y and define x_i by $u_i(y^*, x_i^*) = u_i(y, x_i)$. Denote by S_0 the set of agents such that $x_i \geq 0$ and show $\sum_{S_0} x_i \leq c(y)$. Then choose $(t_1(y), \ldots, t_n(y))$ nonnegative and bounded below by $(x_1, \ldots, x_n)$.)

CHAPTER 6

Production Externality Games

6.1. INTRODUCTION

We continue the discussion of cooperative production initiated in Chapter 5, looking now at the decentralized mode of cooperation (Section 1.6) in the same production problems, namely, the provision of a public good (or bad), and the production of a private good under increasing or decreasing marginal costs.

We study half a dozen simple mechanisms illustrating several fundamental patterns of strategic externalities.[1] These are (i) voting games to select the level of provision of a public good, or bad (Section 6.2), (ii) the provision of a public good by voluntary contributions (Section 6.3), (iii) the exploitation of commons by equal return to individual contributions of input (Section 6.4) or by equal unit cost to individual demands of output (Section 6.5), and (iv) the exploitation of commons by the serial mechanism (Sections 6.6 and 6.7).

Three salient conclusions from this chapter are now presented, and then illustrated in a brief overview of the different sections.

The first common feature to these various mechanisms is the lack of efficiency of the equilibrium outcomes. This important and simple observation (of which the underprovision of public goods by voluntary contribution and the tragedy of the commons are the two most familiar instances) reveals a fundamental limitation of the decentralized mode of cooperation. It can be overcome only by using more complex mechanisms where the nature of individual message forces the players to coordinate their strategic choices. Such mechanisms are beyond the scope of this volume.[2]

The second lesson of this chapter is that the strategic properties of the various mechanisms are quite different in the case of convex preferences (decreasing marginal utilities) and increasing marginal costs, and in the case of nonconvex preferences and/or decreasing marginal

---

[1] Indeed, the most famous archetypal games in strategic normal form, such as the Prisoner's Dilemma, the Battle of the Sexes, or the oligopolistic competition à la Cournot are all part of our discussion.

[2] In these mechanisms, an individual message involves a (summary of) other agents' preferences, and the mechanism typically penalizes divergent reports. General references on mechanism design are Maskin [1985] and Groves and Ledyard [1987].

costs. In the former case, our mechanisms have good or very good strategic features (leading to a unique Nash equilibrium and/or a dominant strategy equilibrium in some cases), whereas in the latter case, even the existence of a Nash equilibrium is in jeopardy.[3] Accordingly, we concentrate our discussion on the former case.

The third lesson is that the tests of equity central to the discussion of efficient (first-best) solutions in Chapter 5 (most notably the stand alone and no envy tests) continue to play a key role in the discussion of the (second-best) equilibrium outcomes of our mechanisms. Thus this chapter illustrates the interplay of the two modes of justice and or decentralized behavior (Sections 1.5 and 1.6) in the absence of efficiency.

In Section 6.2, a public good is produced by a common property technology and its cost is shared equally among the participants (as required by no envy). Agents vote to select the quantity of public good enjoyed by all. Under convexity of individual preferences and nondecreasing marginal costs, several simple voting rules achieve full decentralization of the decision (in the sense of dominant strategy behavior; see Section 1.6). We study the two most natural rules, namely, majority voting (where the preferred choice of the median voter prevails) and unanimity voting (where the lowest preferred level of public good is produced). The former is generally more efficient than the latter; on the other hand, the latter respects the voluntary participation constraint (no agent may end up paying more than her willingness to pay for the good actually produced), whereas the former does not.

In Section 6.3, the agents voluntarily contribute some amount of input (money) toward the provision of the public good, and the entire input so collected is used to produce the public good. In equilibrium, individual contributions normally differ (sometimes to the point where an agent enjoys a complete "free ride"), generating some envy ex post. On the other hand, the stand alone test is met in equilibrium, unlike in the voting equilibria discussed in Section 6.2. Of course, the equilibrium level of public good is inefficiently low, but it is not systematically lower than the level achieved under unanimity or majority voting rule.

In Sections 6.4 and 6.5, a technology to produce a private good is the common property of all participants. Two natural mechanisms equalizing the returns across all agents are successively discussed. In the first one, agents send their input contribution (as in the voluntary contribution mechanism of Section 6.3) and receives a share of output proportional to their own input. We call it the *average-return* mechanism (Section 6.4). In the second mechanism, each agent demands some

---

[3] This observation corroborates the role of convexity assumptions in pure exchange economies and fair division; see Chapters 2 to 4.

amount of output, and must cover total costs in proportion to his own demand. This is the *average-cost* mechanism (Section 6.5). The exploitation of commons (such as forests, fisheries, or underground water) is often described as an average-return mechanism; we give several more examples ranging from rent-seeking games to oligopolistic competition à la Cournot. Average-cost mechanisms are no less common (see Section 6.5 for examples).

Both the average-return and the average-cost mechanisms yield (a different form of) the tragedy of the commons: when marginal costs increase (resp. decrease), the Nash equilibrium behavior leads to overproduction (resp. to underproduction) of the private good, because each agent does not internalize the negative (resp. positive) externality she causes to other agents by raising her activity level. In the terminology of Example 1.4, the pasture is overgrazed if its marginal returns decrease, and it is undergrazed if its marginal returns increase.

The average-return and the average-cost mechanisms also share identical strategic features: both possess at least one Nash equilibrium if preferences are convex and marginal costs are nondecreasing;[4] both may fail to have any Nash equilibrium otherwise. However, the average-cost mechanism yields typically[5] a *smaller* production level than the average-return one; hence, it is less (resp. more) inefficient if marginal costs increase (resp. decrease).

In Section 6.6, we drop the proportionality requirement (namely the proportionality of output shares to input contributions) that was the key feature of the two average mechanisms. We define the remarkable *serial* mechanism. In its input-sharing form, the mechanism elicits individual demands q_i (one for every agent i) and charges agent i as if all agents demanding no less than q_i were in fact demanding exactly q_i. Thus, for instance, the agent with the lowest demand pays the average cost of serving n demands identical to his. The serial mechanism in its output-sharing form is defined in a similar way. If marginal costs are nondecreasing, the serial mechanism (whether in its input- or output-sharing form) has a unique Nash equilibrium outcome provided individual preferences are convex; this equilibrium is an extremely compelling strategic prediction.[6] Moreover, both games (input- or output-sharing) have the same equilibrium outcome. If marginal costs decrease, the serial mechanism still displays much better strategic properties than the

---

[4] This equilibrium is actually unique if preferences are *binormal*; see Theorem 6.2.

[5] That is, if preferences are binormal.

[6] It obtains by successive elimination of strictly dominated strategies. It is robust to coalitional deviations. Moreover, it can be implemented by a strategy-proof mechanism. See Theorem 6.3.

average-return or average-cost mechanism. In particular, typically it has at least one Nash equilibrium (Remark 6.10).

The equilibrium of the serial mechanism is no less remarkable from the normative viewpoint. It guarantees a positive share of surplus to any agent who cares to use the technology when standing alone. This is not true for any of the other mechanisms just discussed. Consider the three equity tests central to the discussion of Chapter 5: no envy, the stand alone test (an upper bound on individual utilities if marginal costs increase, a lower bound if marginal costs decrease), and the unanimity test (a lower bound on individual utilities if marginal costs increase, an upper bound if they decrease). Unlike for any other mechanism discussed here, any Nash equilibrium of the serial mechanism passes all three tests, whether marginal costs increase or decrease.[7] By contrast, a typical equilibrium of the average-return or average-cost mechanism passes the stand alone test but fails the other two.

Section 6.7 adapts the serial mechanism to the public good provision problem when partial exclusion is possible, that is to say, when we can force some agents to consume less public good than is actually produced. The corresponding mechanism compares favorably to voluntary contribution, even though partial exclusion is unambiguously an inefficient move.

As already mentioned in Chapter 5 (Section 5.1), we can think of a public good as a private good output of which every individual agent must consume the same amount. In this fashion, we may think of the collection of production externality games in this chapter as so many different methods to exploit a common property technology transforming a (private good) input into a (private good) output). The voting mechanisms of Section 6.2 are constrained by the equality of all output shares and of all input shares (in brief, by the equality of individual consumption vectors). The mechanism of voluntary contribution to a public good (Section 6.3) is constrained by the equality of all output shares, and similarly, the equality of all input shares leads to the mechanism called free access to a public bad in Remark 6.5. The average-return and average-cost mechanisms are constrained by the requirement that output shares should be proportional to input contributions. No constraint limits the serial mechanism.

The advantage of the above viewpoint is to allow welfare comparison across mechanisms. Our intuition suggests that the fewer constraints we place on the individual allocations, the higher their equilibrium utility. An instance where this intuition is correct is the following observation:

---

[7] However, note that two of the tests are not well defined if marginal costs are sometimes increasing and sometimes decreasing.

the serial equilibrium outcome is always Pareto superior to that of any voting mechanism (be it by the unanimity or the majority rule; see Lemma 6.7).[8] On the other hand, if we ask whether the constraint "equal output shares" is more or less damaging (in equilibrium) than the constraint "output shares proportional to input contributions" (in other words, if we compare the voluntary contribution mechanism of Section 6.3 and the average-return mechanism of Section 6.4), the answer is ambiguous (see Section 6.4, in particular, Example 6.8).

6.2. VOTING OVER A PUBLIC GOOD; MAJORITY VERSUS UNANIMITY

Throughout this section, co-operative production takes a strictly egalitarian modus operandi. Each agent contributes the same amount of input x_i and consumes the same amount of output y_i. If both goods are private goods, the equal-consumption constraint is the easiest way to achieve equity (just as equal split of the resources is the simplest equitable solution of the fair-division problem), hence its appeal.

The most familiar example comes from the public good production problem (Sections 5.6 to 5.8) where an exogenous constraint forces equality of output shares, so that we drop the index i and speak of a quantity y of output consumed by all agents. In the public good problem, equality of input contributions (cost shares) is equivalent to no envy: since we consume the same amount of output, I will envy you if you pay less than I do. Most public goods provided by the state are in effect financed by a uniform tax among all citizens.[9] As we know from Section 5.6, some powerful arguments suggest accepting unequal cost shares in the provision of a public good,[10] but for now we impose equal cost shares and explore the corresponding decentralized mechanisms.

We denote by y the level of public good and by x the common level of input contribution. The production function writes $y = f(nx)$, where n is the number of agents. We assume throughout this chapter (as in Chapter 5) that agents' preferences are increasing in y and decreasing in x.

---

[8] A similar statement holds true when we compare serial cost-sharing of an excludable public good with unanimity voting; see Lemma 6.9, property (ii).

[9] This includes all goods financed out of the general tax revenue, because taxes are collected independently of the particular combination of public goods actually provided. Of course, progressive taxation means that the cost is not shared equally but according to some fixed weights. The point is that those weights are not influenced by individual preferences for the public goods.

[10] When each agent has free access to (owns a copy of) the technology, the stand alone test is often incompatible with no envy. See Section 5.6 as well as Example 6.5 in the next section.

The agents face a voting problem: they all have an opinion about which level y of public good should be chosen, and must find a compromise between their different opinions.

Example 6.1. Binary Public Good

The indivisible public good can be produced ($y = 1$) at cost c, $c > 0$, or not produced ($y = 0$) at no cost. Agent i is willing to pay u_i, $u_i \geq 0$, for the good. There are two possible decisions: $y = 0$ and no money changes hand, or $y = 1$ and everyone pays c/n. Therefore, majority voting is always a practical rule (Condorcet cycles cannot appear). Of course, we need to break possible ties (which can be done, for instance, by selecting always the same decision—say $y = 0 -$). As noted in Section 1.5, majority voting is the compelling decision rule if we wish to avoid any bias in favor of a particular voter or in favor of a particular decision.

Yet, recall for Example 1.2 that this rule may force an agent to pay more for the good than he is willing to. The numerical example was: $c = 2000$, $u_1 = 800$, $u_2 = 600$, $u_3 = 450$, $u_4 = 350$, $u_5 = 300$. The majority {123} favors provision of the good and so agent 4 suffers a loss $50, whereas agent 5's loss is $100.

The voluntary participation principle was formulated by early writers in public finance, such as Wicksell [1896]: everyone should approve the provision of the public good. Voluntary participation (VP) introduces a bias between the two outcomes $y = 0$ or $y = 1$. It views the outcome "$y = 0$, nobody pays" as a status quo ante, and the corresponding utilities as a lower bound for each individual agent. Therefore, it is not surprising that majority voting fails VP. Consider instead *unanimity* voting: the good is provided if and only if all agents are willing to pay c/n; otherwise, the outcome is $y = x = 0$. One checks easily that this unanimity voting rule is nonmanipulable (by single agents or by coalitions of agents). Actually, unanimity voting is the only strategy-proof voting rule that respects the voluntary participation constraint and treats all voters equally.[11]

Now let us compare majority voting and unanimity voting from the point of view of the surplus they generate. Suppose first that it is efficient to produce the good (because $\sum_{i=1}^{n} u_i > c$). If unanimity voting picks the efficient decision, so does majority voting, but the converse is not true (as shown by the above numerical example). Conversely, if efficiency commands not producing the good ($\sum_i u_i < c$), unanimity

---

[11] This follows easily from the analysis of strategy-proof voting rules in the single-peaked context; see Appendix 6.1.

voting always takes the correct decision, whereas majority voting may not do so. These two discrepancies (where unanimity fails to produce when it should, and majority does produce when it should not) are not equally likely. In fact, unanimity errs more often on the side of excessive restraint from producing, than majority errs on the side of excessive production. Our next example illustrates this statistical fact, under the assumption that individual willingnesses to pay are identically and independently distribution (iid). Note that we must be careful in interpreting this example with a statement such as "majority voting is, on average, *more efficient* than unanimity voting," for the "efficiency" in question refers to a situation where the cost of the public good can be split in unequal shares, so that by virtue of the transferable utility assumption, a higher surplus brings us closer to efficiency.[12]

Example 6.2. Where Majority Voting Raises an Average, More Surplus Than Unanimity Voting

Each one of the three agents draws his utility (for the public good) with uniform probability from the five values

$$0, \frac{c}{4}, \frac{c}{2}, 3\frac{c}{4}, c. \tag{1}$$

The surplus σ is given by

$$\alpha = \max\{b_1 + b_2 + b_3 - c, 0\}.$$

Here the probability that the decision $y = 1$ is the only efficient decision is 73% (it is an efficient decision with probability 85%) and the expected surplus is $E\sigma = (0.564)c$. Unanimity voting raises, on average, only $(0.27)c$, whereas majority raises $(0.54)c$, or exactly twice as much (and has an expected relative loss of only 4%). Exercise: explain these computations.

The distribution (1) is rather favorable to majority voting because $y = 1$ is likely to be the efficient decision. Consider next the uniform

---

[12] Within the constraints of the voting model, where the only feasible outcomes involve equal cost shares, both voting rules pick efficient outcomes.

distribution over

$$0, \frac{c}{6}, \frac{c}{3}, \frac{c}{2}, \frac{2c}{3}, \tag{2}$$

where, assuming again that the u_i are iid, the probability that $y = 1$ is the unique efficient decision drops to 45%, with an expected surplus $E\sigma = (0.164)c$. Unanimity voting raises 66% of that surplus, whereas majority voting raises 88% of it. Unlike with the previous distribution, we now have cases where unanimity picks the efficient decision ($y = 0$), whereas majority does not (for instance, $b_1 = c/3$, $b_2 = c/2$, and $b_3 = 0$). However, on average, the cases where majority is right and unanimity is wrong are more numerous.[13]

We turn now to the general case where the collective decision involves different levels of production of the public good. We ask whether and how the two above voting rules can be used in this context. The answer to this question hinges upon the nonemptiness of a certain core. Majority voting gives full control over the decision to any coalition containing more than half of the voters, and the corresponding core outcome is called a Condorcet winner. This is an outcome y such that, when we compare it to any feasible outcome y', at most one-half of the agents strictly prefer y over y' (see Section 1.4). Unanimity voting requires the consent of all participants to raise the level of public good (hence any single agent can lower this level as much as he or she pleases). An outcome y is in the corresponding core if (a) everyone prefers y to any lower level of the public good, and (b) there is no higher level y', $y' > y$, that everyone prefers to the level y. Both of these core concepts typically contain a single outcome.[14] Consequently, if the core is nonempty, it defines unambiguously a voting rule; as shown in Theorem 6.1, this voting rule is a satisfactory decentralization device (with the same properties of strategy-proofness as in the binary example above). The key is to find a class of voting problems where the core is guaranteed to be nonempty. The next example shows that convexity assumptions, once again, play a critical role.

---

[13] We could also find a distribution where the two voting rules yield the same expected surplus, for instance, the uniform distribution over 0, $c/8$, $c/4$, $3c/8$, $c/2$, but notice that the probability of $y = 1$ being the only efficient decision drops to 15%. By pushing this example a little further, we may even get to the point where majority yields less surplus than unanimity (e.g., uniform distribution over 0, $c/8$, $c/4$, $3c/8$).

[14] We noted this for Condorcet winners in Section 1.4; for a unanimity winner, this follows at once from the definition.

Example 6.3. Where the Marginal Costs Increase or Decrease

The public good is produced in indivisible units and the three agents derive the following utilities from the good:

Public good level	u_1	u_2	u_3
0	0	0	0
1	10	11	9
2	13	21	17
3	16	22	25

(3)

Note that marginal utilities are nonincreasing. The first cost function has increasing marginal cost (imc):

$$c(0) = 0, \quad c(1) = 9, \quad c(2) = 21, \quad c(3) = 39. \quad (4)$$

Given equal cost shares, agent 1 would favor producing 1 unit of the good (with corresponding net benefit $10 - 3 = 7$ for herself), her second choice is to produce 2 units (net benefit 6), next to produce 3 units (net benefit 3), and finally her bottom choice is to produce nothing at all. Similarly, the preferences of agents 2 and 3 are determined by the inequalities

$$u_2(2) - \frac{c(2)}{3} > u_2(3) - \frac{c(3)}{3} > u_2(1) - \frac{c(1)}{3} > u_2(0) = 0,$$

$$u_2(3) - \frac{c(3)}{3} > u_3(2) - \frac{c(2)}{3} > u_3(1) - \frac{c(1)}{3} > u_3(0) = 0.$$

Majority voting picks the median vote, or $y = 2$: the allocation {2 units of public good at cost 7} is the Condorcet winner. Both agents 1 and 2 vote against an increase of the production to $y = 3$, and agents 2 and 3 oppose a decrease to $y = 1$ (all agents oppose stopping production altogether).

Next, consider a cost function with decreasing marginal cost (dmc):

$$c(0) = 0, \quad c(1) = 24, \quad c(2) = 39, \quad c(3) = 45. \quad (5)$$

Notice that agents 2 and 3 have the same preferences as before over the four possible levels of y. Agent 1's preferences are altered as follows:

$$u_1(1) - \frac{c(1)}{3} = 2 > u_1(3) - \frac{c(3)}{3} = 1 > u_1(2) - \frac{c(2)}{3} = 0 = u_1(0).$$

This seemingly minor modification destroys the majority winning property of the allocation ($y = 2, x = 13$), for both agents 1 and 3 prefer now to produce at the level $y = 3$ instead. In fact, the majority relation has a cycle:

		Majority
($y = 2, x = 13$) beaten by ($y = 3, x = 15$)		$\{1, 3\}$,
($y = 3, x = 15$) beaten by ($y = 1, x = 8$)		$\{1, 2\}$,
($y = 1, x = 8$) beaten by ($y = 2, x = 13$)		$\{2, 3\}$.

Thus for the utility profile (3), the core of majority voting is nonempty for the imc technology (4), but empty for the dmc technology (5). The same situation may occur for unanimity voting, too, but not at this particular utility profile.[15] Consider a profile identical to (3) for u_2 and u_3, but where agent 1's utility is

$$\bar{u}_1(0) = 0, \quad \bar{u}_1(1) = 10, \quad \bar{u}_1(2) = 13.8, \quad \bar{u}_1(3) = 17.5.$$

Under (4) (imc), the core of unanimity voting is $y = 1$. Under (5) (dmc), agent 1's preferences are

$$\bar{u}_1(3) - \frac{c(3)}{3} = 2.5 > \bar{u}_1(1) - \frac{c(1)}{3}$$

$$= 2 > \bar{u}_1(2) - \frac{c(2)}{2} = 0.8 > \bar{u}_1(0).$$

Now $y = 1$ is unanimously inferior to $y = 3$, and $y = 3$ is blocked by an objection of agent 2 (who prefers the lower level 2); in turn, $y = 2$ is blocked by an objection of agent 1, and the unanimity core is empty.

The crucial feature of Example 6.3 is that the preferences of agent 1 over the choice set $\{0, 1, 2, 3\}$ have, in the imc case, their "peak" at $y = 1$ and are decreasing on both sides of the peak, whereas in the dmc case, her preferences have two peaks, one at $y = 1$ and the other at $y = 3$, as illustrated by Figure 6.1, where a fictitious continuous utility function over the positive line connects the four "real" points of our choice set.

---

[15] Under the profile (3), the core of unanimity voting is $y = 1$ for any of the two cost functions (4), (5).

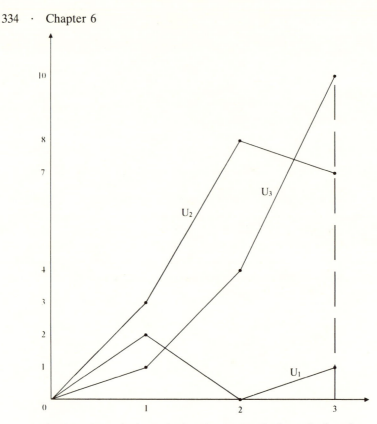

Fig. 6.1. Two single-peaked and one non-single-peaked preferences.

Definition 6.1. Consider a (finite or infinite) subset A of the real line. A preference (represented by a utility function) u is *single-peaked* if u has a best element a^* in A, called its peak, and moreover,

$$a < b < a^* \Rightarrow u(a) < u(b) < u(a^*)$$
$$a^* < b < a \Rightarrow u(a^*) > u(b) > u(a)$$

for all a, b in A.

We denote by SP(A) the set of single-peaked preferences on A.

For instance, the preference that coincides with the natural order of the real line is single-peaked (with its peak at the highest element of A) and so is the reverse preference (choosing a over b if a is smaller than b for the natural order of the real line).

Theorem 6.1. (Black [1948]). Consider a finite set of n agents and a subset A of the real line. Assume every agent is endowed with a single-peaked preference on A. Then the core of majority voting and the core of unanimity voting are both nonempty.

(*a*) *The unanimity core is the smallest peak; the corresponding voting rule is strategy-proof (even with respect to coalitional misreports).*

(*b*) *If n is odd, the majority core (Condorcet winner) is the median peak and defines a strategy-proof voting rule.*

(*c*) *If n is even, n = 2n′, the majority core is the interval from the n′-th peak to the (n′ + 1)-th peak. Each extreme point of the core defines a strategy-proof voting rule.*

Proof of Statement (a). Denote a_i for the peak of agent i's preferences and set $y = \min_i a_i$. Check that y is the only point in the unanimity core: for all y', $y' < y$, and all i, agent i prefers y over y' because her peak is not smaller than y; moreover, for all y', $y' > y$, an agent whose peak is y prefers y to y', hence y' is not Pareto superior to y (and y' is not in the core either).

Check now that the mechanism eliciting individual peaks and enforcing the lowest reported peak is strategy-proof (even in the coalitional sense). If $\bar{y}$ is the smallest of the true peaks, single-peakedness of preferences (Definition 6.1) implies that *everyone* prefers outcome $(\bar{y}, c(\bar{y})/n)$ to any allocation $(y, c(y)/n)$ such that $0 \leq y < \bar{y}$. Now let i be an agent such that $y_i = \bar{y}$: as agent i gets his best choice in A, a misreporting coalition T cannot contain i; hence, T can only lower the outcome below $\bar{y}$, a move from which no one in T can benefit.

Proof of Statement (b). The median of the peaks $a_1, \ldots, a_n$ is the number λ such that at least $(n + 1)/2$ of the peaks are bounded above by λ, and at least $(n + 1)/2$ peaks are bounded below by λ. For instance, the median of the set of peaks $\{3, 1, 4, 3, 2\}$ is 3. A voter i such that $a_i = \lambda$ is called a median voter; note that the median peak is a unique number but we have two median voters (agents 1 and 4).

Check that the median peak λ is the only Condorcet winner: opposing λ to any outcome a above λ, we find all the voters whose peak is bounded above by λ preferring λ over a, so a is defeated by λ in this majority contest. Similarly, λ wins the majority contest with an outcome b below λ because all the voters with their peak bounded below by λ prefer λ over b.

Finally, we check that the voting rule eliciting individual peaks and enforcing the median of the reported peaks is (coalitionally) strategy-proof. Call $(y^*, c(y^*)/n)$ the true Condorcet winner, and say that a coalition T misreports, after which the "apparent" Condorcet winner is $(y, c(y)/n)$. In the true profile, y^* defeats y majoritywise, but in the apparent profile, y defeats y^*. As the voters outside T send the same report in both profiles, there must exist at least one voter in T who in truth prefers y^* over y, yet misreports that he prefers y to y^*; that voter is strictly worse off after the misreport than before. The similar proof of statement (c) is omitted (see Remark 6.1 for a generalization).

Q.E.D.

Now we can explain the critical role of decreasing marginal costs in Example 6.3.

Lemma 6.1.   Consider a public good produced (in indivisible units or in divisible amounts) under nondecreasing marginal costs. Suppose an agent has strictly convex preferences (his marginal utility is strictly decreasing, or in general, the upper contour sets are strictly convex). Then the preferences of this agent over the allocation $\{(y, c(y)/n) \mid y \geq 0\}$ are single-peaked.

This result, a straightforward exercise in convex analysis (its proof is omitted), can be combined with Black's theorem. When a public good is produced under imc and individual preferences are convex, we conclude that majority and unanimity voting over the level of public good (the cost of which is shared equally) are impeccable decentralization devices. Figure 6.5 illustrates these two mechanisms in an economy with three agents and constant marginal costs.

We repeat the main differences between the two mechanisms. Unanimity voting respects the voluntary participation constraint, namely, the property that everyone benefits, or at least no one suffers, vis-à-vis the "status quo ante" where no public good is produced:

$$u_i\left(y, \frac{c(y)}{n}\right) \geq u_i(0, c(0)) \quad \text{for all } i. \text{[16]}$$

Majority voting, on the contrary, does not introduce any bias between the various levels of public good (no special role is given to the zero level of production). It may force some individual agents to pay for the public good more than they benefit from it. On the other hand, since no individual agent can veto the production of public good, majority voting raises, on average, more surplus above the status quo ante than unanimity voting.

When considered among all feasible allocations where the cost does not have to be shared equally, the equilibrium of unanimity voting never

---

[16] Note that this property holds true at every Nash equilibrium of the direct unanimity mechanism, whether or not preferences are single-peaked.

produces too much public good: either the equilibrium allocation is fully efficient (this is the case, for instance, if all individual preferences agree) or it produces too little public good. A proof of this important fact is given in the corollary to Lemma 6.4. By contrast, majority voting may yield too little, too much, or just the right amount of public good.

Example 6.4.   Linear Utilities and Quadratic Costs

This example is important because it is easily computable and allows a wide range of preference distributions:

$$c(y) = \frac{1}{2a}y^2, \quad u_i(y, x_i) = \lambda_i \cdot y - x_i \quad \text{all } i, \quad \text{all } y \geq 0, \quad \text{all } x_i \geq 0.$$

We assume n is odd, $n = 2n' - 1$ and, moreover, $\lambda_1 \leq \cdots \leq \lambda_n$. Agent i's peak when he pays c/n is $y_i = n \cdot \lambda_i a$. Therefore, unanimity voting results in $y_1 = n \cdot \lambda_1 a$, whereas majority voting yields $y_{n'} = n \cdot \lambda_{n'} a$. The surplus-maximizing public good level is $\bar{y} = \sum_{i=1}^{n} y_i$; therefore, unanimity voting always produces too little public good, whereas majority voting may produce too little or too much public good. Finally, we note that joint surplus is always larger (at least, not smaller) at the Condorcet winner level y_n, than at the smallest peak level y_1:

$$\sum_{i=1}^{n} u_i(y_1) - c(y_1) = n\lambda_1 \cdot \left(\lambda_N - \frac{n}{2}\lambda_1\right) \cdot a$$

$$\leq n\lambda_{n'} \cdot \left(\lambda_N - \frac{n}{2}\lambda_{n'}\right) \cdot a$$

$$= \sum_{i=1}^{n} u_i(y_{n'}) - c(y_{n'}),$$

where $\lambda_N = \sum_{i=1}^{n} \lambda_i$. For instance, consider the numerical example 5.9: $n = 3$, $a = 4$, $\lambda_1 = 1/6$, $\lambda_2 = 3/4$, $\lambda_3 = 1$. The unanimity equilibrium produces $y = 2$ units of public good (when efficiency—if unequal cost shares are allowed—commands producing $y^* = 7.35$) and the surplus shares:

$$\pi_1 = 0.17, \quad \pi_2 = 1.33, \pi_2 = 1.83. \tag{6}$$

The majority equilibrium (over)produces, $y = 9$, and the surplus shares are:

$$\pi_1 = -1.875, \qquad \pi_2 = 3.375, \qquad \pi_3 = 5.625. \qquad (7)$$

Thus total surplus at the unanimity equilibrium is less than half that at the majority equilibrium, and the latter surplus is 97% of the efficient surplus. On the other hand, the majority equilibrium violates voluntary participation.[17]

Example 5.14 ($n = 3$, $a = 4$, $\lambda_1 = \lambda_2 = 1$, $\lambda_3 = 5$) illustrates a different situation where the unanimity and majority equilibrium coincide and produce about $2/3$ of joint surplus.

Remark 6.1

When preferences are single-peaked over a subset of the real line, majority and unanimity voting are not the only mechanisms sharing these remarkable properties of nonmanipulability. Fix any integer k, $1 \le k \le n$, and define the k-rank mechanism: each agent i reports a peak z_i, reported peaks are rearranged in increasing order $z_1^* \le z_2^* \le \cdots \le z_n^*$, and the level z_k^* is enforced. For instance, $k = 1$ corresponds to unanimity voting, majority voting corresponds to $k = n'$ if $n = 2n' - 1$, and so on. For any k, the k-rank mechanism is strategy-proof (reporting one's true peak is a dominant strategy) and robust to coalitional deviations as well.

Appendix 6.1 and Exercise 6.22 describe the substantially richer class of strategy-proof direct mechanisms in the single-peaked context. Within this class, unanimity voting stands out as the only mechanism meeting the voluntary participation constraint.

However, if individual preferences are not restricted in any way, a famous theorem due to Gibbard and Satterthwaite tells us that the only strategy-proof voting rules are the dictatorial rules (where a certain agent always prevails) or the rules choosing between no more than two outcomes. See Appendix 6.2 for a precise statement.

Remark 6.2

When several public goods are jointly produced, the analog of the single-peaked preferences is not immediately obvious. Moreover, the existence of a Condorcet winner and the existence of equitable strategy-proof voting rules receive different answers. In a nutshell, it is

---

[17] Note that agents 2 and 3 even prefer the majority equilibrium to the ratio or ω-EE equilibrium (computed in Sections 5.7 and 5.8, respectively).

hard to find natural restrictions of individual preferences guaranteeing the existence of a Condorcet winner, yet we know of such restrictions that allow for strategy-proof voting rules; Appendix 6.3 gives the main example. A discussion of the nonexistence of Condorcet winners in the popular *spatial voting* model is also in Appendix 6.3.

Note that the absence of a core to majority voting, if it implies a definite instability of the political equilibrium, is not necessarily interpreted as "destructive competition" (as was the case in the production economies of Chapter 2). See the discussion of Section 1.4, in particular, notes 8 and 9.

6.3. VOLUNTARY CONTRIBUTION TO A PUBLIC GOOD

Wicksell's voluntary participation principle says that an agent should pay no more for the public good than his willingness to pay (measured from the initial position where no public good is produced). The stand alone lower bound is a stronger requirement than voluntary participation: each agent must be guaranteed the utility level she can reach in "autarchy" namely, by using alone the technology. Recall that this test is compelling in the free-access regime (where each agent owns a copy of the technology) and normatively appealing in the common property regime (see Section 5.6).

Unanimity voting fails the stand alone test (although it meets voluntary participation; indeed, we noted in Section 5.6 that the stand alone test is often incompatible with equal cost shares. This was shown in Section 5.6 in the case of a binary public good; we give now a continuous example.

Example 6.5.   Where No Envy and the Stand Alone Test are Incompatible

The marginal cost is constant, $c(y) = 2y$, and the quasi-linear utilities of the two agents are

$$u_1(y, x_1) = \min\left\{3y, \frac{y+5}{2}\right\} - x_1,$$

$$u_2(y, x_2) = \min\left\{3y, \frac{y+35}{2}\right\} - x_2.$$

The stand alone utilities are:

$$sa(u_1) = 1 \quad \text{by producing } y = 1, \qquad sa(u_2) = 7 \quad \text{by producing } y = 7.$$

Consider the allocation where y is produced and its cost equally shared. This allocation guarantees his stand alone utility to agent 1 if and only if

$$u_1(y) - \frac{c(y)}{2} \geq 1 \Leftrightarrow y \leq 3.$$

Similarly, the stand alone test for agent 2 gives

$$u_2(y) - \frac{c(y)}{2} \geq 7 \Leftrightarrow y \geq 3.5.$$

A simple allocation where each agent receives (at least) his stand alone utility is when $y = 7$, $x_1 = 0$, $x_2 = 14$: we say that agent 1 gets a free ride. This outcome has the following equilibrium property: agent 1 does not want to finance more public good (given that agent 2 is paying $14) and agent 2 does not want to decrease or increase his contribution (given that agent 1 is not paying anything).

Definition 6.2. In the voluntary contribution mechanism, each agent i chooses his level of contribution x_i, $x_i \geq 0$, and the public good is produced at the level $y = f(\Sigma_i x_i)$.

Lemma 6.2. Any Nash equilibrium of the voluntary contribution game passes the stand alone test.

Proof. Let $(y; x_1, \ldots, x_n)$ be a Nash equilibrium of the voluntary contribution mechanism and consider another contribution x_i' by agent i. Setting $y' = f(x_i' + \Sigma_{j \neq i} x_j)$, the equilibrium property, the monotonicity of f, and the monotonicity of preferences imply

$$u_i(y, x_i) \geq u_i(y', x_i') \geq u_i(f(x_i'), x_i').$$

The maximal value of $u_i(f(x_i'), x_i')$ is agent i's stand alone utility.
$$\text{Q.E.D.}$$

The voluntary contribution equilibrium passes the stand alone test but generates envy (of which free-riding is an extreme case), whereas the equilibria of both voting rules in the previous section pass no envy but fail stand alone.[18] We now compare these mechanisms from the

---

[18] In the public good provision problem, any strategy-proof mechanism must violate stand alone; see Saijo [1991].

point of view of their strategic properties and of efficiency. The first result shows the importance of the convexity assumptions.

Lemma 6.3. Suppose a divisible public good is produced under nondecreasing marginal costs, and that individual preferences are convex, continuous, and monotonic. Suppose also that no agent can or will contribute arbitrary large amounts of input. Then the voluntary contribution mechanism has at least one Nash equilibrium.

The proof is a straightforward consequence of Nash's theorem.[19] Exercise 6.17 shows that the voluntary contribution game may have no Nash equilibrium at all if either some individual preference is not convex, or marginal costs decrease at some point. In the case of a public good produced in indivisible units, the voluntary contribution game may have no Nash equilibrium even with convex preferences and constant marginal costs; see Exercise 6.17.

Note the similarity between the assumptions in Lemmas 6.3 and 6.1. However, Lemma 6.3, unlike Lemma 6.1, applies only to the case of a divisible public good. If the good comes instead in indivisible units, the above assumptions, surprisingly, are not sufficient to guarantee the existence of a Nash equilibrium. Exercise 6.17 gives an example.

The strategic properties of the voluntary contribution equilibrium are much poorer than those of the voting rules in Section 6.2. The simplest illustration of this fact comes in the case of an indivisible public good, where multiple Nash equilibria are normal.

Consider the case of a *binary public good* (Example 6.1). To fix ideas, suppose $n = 3$, $u_1 = 5$, $u_2 = 6$, $u_3 = 8$, and $c = 10$. Any vector of contributions (x_1, x_2, x_3) such that

$$x_1 + x_2 + x_3 = 10, \qquad x_1 \leq 5, \qquad x_2 \leq 6, \qquad x_3 \leq 8,$$

$$x_1 + x_2 < 10, \qquad x_2 + x_3 < 10, \qquad x_1 + x_3 < 10,$$

is a Nash equilibrium of the voluntary contribution game: these allocations cover precisely (the interior of) the stand alone core. Check, for instance, the equilibrium property for agent 3: he does not want to lower his contribution because this would imply that the good is not produced ($x_2 + x_3 < 10$) and he prefers the equilibrium outcome ($u_3 - x_3 \geq 0$). However, there is another equilibrium, namely, $x_i = 0$, $i = 1, 2, 3$; the corresponding outcome is inefficient.

---

[19] The pay-off function $u_i(f(\Sigma_j x_j), x_i)$ is quasi-concave in x_i by our assumptions.

The multiplicity of equilibria yields a typical "Battle of the Sexes" situation where each agent wants to commit himself to a low contribution in order to force the others to bear the largest share of the cost possible in equilibrium. For instance, if agent 1 is committed to contribute $x_1 = 2 + \epsilon$, he is sure that the public good will be provided by agents $\{2, 3\}$: because agent 3 alone is willing to pay the full cost, in every equilibrium of the restriction of the voluntary contribution game to agents 2 and 3, the good will be produced. A more daring agent 1 could commit himself to $x_1 = 0$ in the hope that agents 2 and 3 will avoid the only inefficient equilibrium of the restricted game, namely, $x_2 = x_3 = 0$. Similarly, agent 2 can safely commit herself to $x_2 = 2 + \epsilon$ and agent 3 to $x_3 = 4 + \epsilon$. Thus an agent capable of committing herself can guarantee a favorable equilibrium (and can even get a completely free ride if she is willing to take risks). The commitment tactics to force one's favorite equilibrium are a well-known factor of strategic indeterminacy (Schelling [1971]) and undermine the positive prediction of equilibrium behavior.

Multiple Nash equilibria and Battle of the Sexes are a normal configuration when the public good is produced in indivisible units. Exercise 6.3 gives an example. For a divisible public good, multiple equilibria in the voluntary contribution game are illustrated in Example 1.5. However, when both utilities and preferences are differentiable, we may expect a unique equilibrium outcome; see the discussion after Lemma 6.4.

We illustrate now another critical feature of the voluntary contribution mechanism, namely, its inefficiency. We take n identical agents and assume that each one holds one indivisible unit of input (e.g., labor) and may or may not contribute this input to the production of the public good. If t agents out of n contribute, the quantity $f(t)$ is produced. The utility of a given agent is $f(t)$ if t agents contribute and this agent does not, and $f(t) - v$ if he is among the contributors. We suppose that at all level t, $1 \leq t \leq n$, the following inequality holds:

$$\frac{v}{n} < \partial f(t) = f(t) - f(t - 1) < v.$$

The marginal benefit from one additional unit of input is lower than the cost of that input to an agent, but the joint benefit $(n \cdot \partial f(t))$ exceeds this cost. Therefore efficiency commands producing as much public good as possible (i.e., $f(n)$ units) but none is produced by voluntary contribution (because for each agent, contributing nothing is a dominant strategy). This is the well-known Prisoner's Dilemma configuration. See Exercise 6.4 for a nonsymmetrical version of this example.

Note that both unanimity and majority voting yield the efficient symmetrical outcome in this example, as they do for all problems with a unanimous preference profile (all individual preferences are equal). (Exercise: prove this claim.) In a general (nonunanimous) profile, the equilibrium of the unanimity and majority voting games may be more or less efficient than the voluntary contribution equilibrium (as the examples below demonstrate), but the former equilibria are always robust to coalitional deviations (because the mechanism forces equal cost shares no matter what). By contrast, every inefficient equilibrium of the voluntary contribution game is manipulable by the grand coalition (because every allocation is feasible in the voluntary contribution mechanism).

The above example with indivisible input contributions illustrates vividly the general proposition that the voluntary contribution mechanism results in the production of too little public good. This general fact is discussed in great detail in Olson [1965]. We give a formal statement in the case of a divisible public good.

Lemma 6.4. Assume u_i, c are differentiable and $u_{ix} < 0$ for all i. If the following inequality holds:

$$\sum_{i=1}^{n} \frac{u_{iy}}{|u_{ix}|}(y, x_i) > c'(y), \qquad (8)$$

and if it is possible to increase the level of the public good, then we can do so and (by sharing the cost increase appropriately) make every agent better off: too little public good is produced. If the opposite inequality holds true and it is feasible to decrease the level of public good, then we may do so and make every agent better off: too much public good is produced.

Corollary. Consider a Nash equilibrium of the voluntary contribution mechanism where at least one agent is active ($x_i > 0$ for at least one i), and where at least one other agent is not satiated of the public good ($u_{jy} > 0$ for at least one j, $j \neq i$). Then too little public good is produced; namely, inequality (8) holds.

In the (unique) Nash equilibrium of the unanimity voting game, too little public good is produced unless all agents agree on their preferred level of public good, in which case the equilibrium is efficient.

Proof of Lemma 6.4. Assume (8) and set $p_i = (u_{iy}/|u_{ix}|)(y, x_i)$. Consider the small variation $(dy; dx_1, \ldots, dx_n)$ given by $dy > 0$ and $dx_i = p_i/\sum_j p_j \cdot c'(y) \cdot dy$. This variation is feasible since $\sum_{i=1}^{n} dx_i =$

$c'(y) \cdot dy$. On the other hand, (8) implies $dx_i < p_i \cdot dy$; hence,

$$du_i = u_{iy} \cdot dy + u_{ix} \cdot dx_i = |u_{ix}| \cdot (p_i \cdot dy - dx_i) > 0,$$

as was to be proved. The proof of the second statement is similar.

Proof of the Corollary. Call x an equilibrium of voluntary contribution where x_i is positive. Because agent i does not want to lower his contribution, we must have

$$\text{if } dy < 0, \quad \text{then } u_{iy} \cdot dy + u_{ix} \cdot c'(y) \cdot dy \leq 0.$$

This is equivalent to

$$\frac{u_{iy}}{|u_{ix}|} \geq c'(y).$$

Hence (8) follows because at least one agent has $u_{jy} > 0$.

Next, we call y_i the demand of agent i when she must pay $(1/n)$th of the cost at all levels. Call y_1 the lowest such demand. By assumption, $0 < y_1 \leq y_i$ for all $i \geq 2$. The first-order optimality conditions for y_i are $(u_{iy}/|u_{ix}|)(y_i, c(y_i)/n) = c'(y_i)/n$. Under convexity of preferences, this implies

$$\frac{u_{iy}}{|u_{ix}|}\left(y_1, \frac{c(y_1)}{n}\right) \geq \frac{c'(y_1)}{n} \quad \text{for all } i.$$

Summing up yields $\Sigma_i(u_{iy}/|u_{ix}|) \geq c'(y_1)$ with equality if and only if y_1 is the common demand of all agents.   Q.E.D.

Consider the case of a linear cost function $c(y) = y$ and utilities quasi-linear in the private good: $u_i(y, x_i) = u_i(y) - x_i$. The efficient level of public good y^* solves the equation $\Sigma_{i=1}^n u_i'(y) = 1$. The demand y_i^u under equal cost shares is characterized (when u_i is concave and differentiable) by $u_i'(y_i^u) = 1/n$. Under unanimity voting, the lowest such demand is implemented.

The key to the equilibrium of the voluntary contribution game is the stand alone level y_i^{sa}, characterized by $u_i'(y_i^{sa}) = 1$. If the largest stand alone level is achieved by a single agent i, then the unique equilibrium of the voluntary contribution game has this agent paying the full cost of y_i^{sa} and every other agent free riding.[20] Of course, the smallest y_i^u may

[20] If several agents, say 1, 2, 3, have the highest stand alone level of public good, there is a continuum of equilibria where these three agents share the cost of y_1^{sa}, with a "Battle of the Sexes" pattern as in the binary public good above.

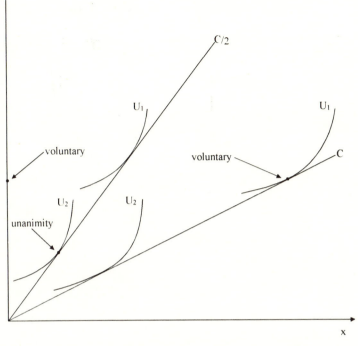

(a)

Fig. 6.2. Where more (Fig. 6.2a) or less (Fig. 6.2b) public good is produced under voluntary contribution than under unanimity voting.

be smaller or larger than the largest y_i^{sa}, so that we cannot tell in general which of unanimity voting or voluntary contribution brings about more surplus. Figure 6.2 shows two-person examples of both cases. Our next numerical example makes the same point.

Exercise 6.18 examines the case of utilities quasi-linear in the public good where the voluntary contribution game always has a unique equilibrium. It gives also more general conditions on preferences and costs guaranteeing uniqueness of the equilibrium to the voluntary contribution game. These conditions, known as binormality of preferences, play a crucial role in the discussion of the average-return and average-cost mechanisms. They are defined in the next section.

Example 6.6.   Example 6.4 Continued

We assume n odd, $n = 2n' - 1$, and $\lambda_1 \leq \lambda_2 \leq \cdots \leq \lambda_{n-1} < \lambda_n$. In the equilibrium of the voluntary contribution game, agent n pays for

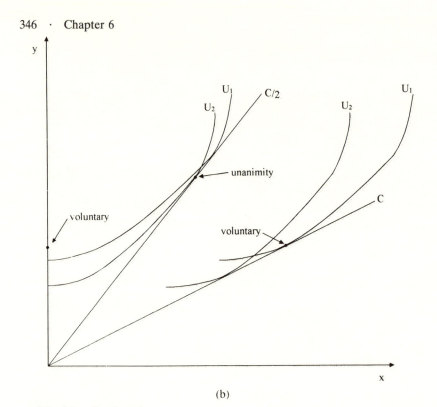

(b)

Fig. 6.2. Continued.

$a \cdot \lambda_n$ units of public good (λ_i is the stand alone level of agent i) and other agents free ride. In the unanimity (resp. majority) equilibrium, $na \cdot \lambda_1$ units (resp. $na \cdot \lambda_{n'}$ units) of public good are produced and paid for equally by all agents. The efficient level of public good is $a \cdot \lambda_N$ (using the notation $\lambda_N = \sum_{i=1}^{n} \lambda_i$). Compute the overall surplus in the three mechanisms:

$$\sigma^{vc} = a\lambda_n(\lambda_N - \tfrac{1}{2}\lambda_n),$$

$$\sigma^{una} = na\lambda_1(\lambda_N - \tfrac{1}{2}n\lambda_1),$$

$$\sigma^{maj} = na\lambda_{n'}(\lambda_N - \tfrac{1}{2}n\lambda_{n'}).$$

Recall from Example 6.4 that $\sigma^{una} \leq \sigma^{maj}$ for all $\lambda_1, \ldots, \lambda_n$. One checks that

$$\sigma^{vc} \leq \sigma^{una} \Leftrightarrow \lambda_n \leq n \cdot \lambda_1,$$

$$\sigma^{maj} \leq \sigma^{vc} \Leftrightarrow n \cdot \lambda_{n'} \leq \lambda_n.$$

In other words, the voluntary contribution mechanism gives the lowest surplus of the three (equivalently, it produces the least amount of public good) when the distribution of marginal utilities λ_i is not too spread (the ratio λ_i/λ_j never exceeds n); it gives the largest surplus of the three when the ratio $\lambda_n/\lambda_{n'}$ (largest λ_i to median λ_i) is at least n.

A similar conclusion holds when we look for those distributions of the 'types" λ_i such that *everyone* prefers the voluntary contribution equilibrium over both the unanimity and the majority equilibrium; such is the case if $2n\lambda_{n'} \leq \lambda_n$. Symmetrically, if $2\lambda_n \leq n\lambda_1$, everyone prefers the unanimity equilibrium over the voluntary contribution equilibrium. See Exercise 6.5 for details. For instance, in the numerical example 5.9, the voluntary contribution of agent 3 produces $y = 4$ (when efficiency requires producing $y^* = 7.35$) for the following surplus shares:

$$\pi_1 = 0.67, \qquad \pi_2 = 3, \qquad \pi_3 = 2. \qquad (9)$$

Thus, 77% of total available surplus is extracted, and this equilibrium is Pareto superior to that of unanimity voting (see formula (6) in the previous section). Agents 2 and 3 still prefer the outcome of majority voting, but the latter is unacceptable to agent 1. Here voluntary contribution strikes a reasonable compromise between the two voting rules. The situation is similar in the numerical example 5.14.[21]

Remark 6.3

Underprovision of the public good by the voluntary contribution mechanism as described in Lemma 6.4 is guaranteed only under differentiability of the utility functions. If the marginal utilities have "jumps" (the indifference contours are kinked), the voluntary contribution mechanism may have an efficient equilibrium. Example 6.5 is one instance: the efficient level $y = 7$ is also the highest stand alone level. Another instance is any economy with Leontief utility functions: there the voluntary contribution equilibrium is *always efficient* and always Pareto superior to the unanimity equilibrium. See Exercise 6.7.

Remark 6.4. Other Forms of Public Good Externalities

The alternative models of Section 5.9 lead to at least two interesting voluntary contribution games.

Suppose agent i produces the public good y_i and the resulting overall externality equals $y = \max_i y_i$. Call f the common production

---

[21] Where the surplus distribution at the voluntary contribution and at the majority (or unanimity) equilibrium are respectively $\pi = (20, 20, 50)$ and $\pi = (6, 6, 54)$. Exercise: check this claim.

function.[22] The voluntary contribution game has the following profit function:

$$u_i\!\left(f\!\left(\max_j x_j \right), x_i \right).$$

This is the epitome of the free-rider problem: in any efficient outcome, only one agent is contributing a positive amount of input. Let x_i be agent i's stand alone contribution [namely, the solution of $\max_{x_i} u_i(f(x_i), x_i)$], and denote by i^* an agent such that $x_{i^*} = \max_i x_i$. The corresponding outcome where all agents other than i take a free ride ($x_i = 0$ if $i \neq i^*$) is a Nash equilibrium of this game and is robust against coalitional deviations as well (it is a *strong equilibrium*); in particular, it is Pareto optimal. (Exercise: prove these claims.) Therefore the intense free riding has no social cost and voluntary contribution is a perfectly satisfactory decentralization device (see Bliss and Nalebuff [1986] for a discussion of this model involving mixed strategies).

Next, consider the aggregation pattern $y = \min_i y_i$. The resulting game has the profit function

$$u_i\!\left(f\!\left(\min_j x_j \right), x_i \right).$$

This game has been used by the literature on microcoordination failures (Bryant [1983]) and has interesting strategic properties. Call x_i (as above) agent i's stand alone contribution and denote by j^* an agent such that $x_{j^*} = \min_i x_i$. Assuming convex preferences and decreasing marginal costs, the game has a continuum of Nash equilibrium outcomes; in each equilibrium, all agents use the same strategy $x_i = \lambda$, all i, and λ can be any number between 0 and x_{j^*}. The larger λ, the better the equilibrium (from the point of view of the Pareto ordering), and for $\lambda = x_{j^*}$, the equilibrium is actually Pareto optimal. However, notice that a Pareto inferior equilibrium is also safer for every agent (when playing x_i, agent i may end up at the level $u_i(0, x_i)$ if someone fails to contribute).

Remark 6.5.   Public Bads

In the dual model of public bads (Section 5.9), individual agents consume different shares of output and contribute the same amount of input: the total cost of output is split equally between all participants

----

[22] The model where each public good has a different production function is very similar.

(recall the example of splitting the restaurant bill). The analog of the voluntary contribution mechanism is the *free-access* mechanism: every agent chooses his or her consumption of output (and the bill is paid equally by all participants). If the output is interpreted as a polluting activity, this means that individual agents are free to pollute at any level (and everyone suffers equally from the resulting pollution, irrespective of individual responsibilities).

The free-access mechanism results in the production of "too much" output in equilibrium, because an agent internalizes only a fraction of the real cost of increasing his demand of output. Even worse, the free-access mechanism badly fails the voluntary participation constraint: an agent who does not care for the output must bear his share of the cost anyway! Thus the mechanism combines the defects of majority voting (coercive participation) and of voluntary contribution (very inefficient level of production). The two interesting mechanisms in the public bad model are unanimity voting and majority voting. See several numerical examples in Exercise 6.9.

6.4. THE AVERAGE-RETURN MECHANISM: THE TRAGEDY OF THE COMMONS

The common property technology transforms x units of input (a private good) into $y = f(x)$ units of output (a private good). In this section and the next one, we impose the *equal-return* constraint: if agent i's net consumption is (y_i, x_i) (she gets y_i units of output and pays x_i units of input), then we have

$$\frac{y_i}{x_i} = \frac{y_j}{x_j} \quad \text{for all } i, j = 1, \ldots, n.$$

This may be the result of a physical constraint: if f represents a lake in which several fishermen operate, the catch will be proportional to individual effort. Similarly, in Example 1.4, the output (cow fattening) is distributed among herdsmen in proportion to the number of cows they send to the common pasture. In other models, the proportionality constraint is a normative principle; e.g., a production cooperative may decide to share its output in proportion to hours worked.[23]

In the *average-return* mechanism, every individual agent selects his contribution of input, and receives a share of total output proportional to his contribution. Thus agent i chooses her input contribution x_i, $x_i \geq 0$, and receives $y_i = (x_i/\Sigma_j x_j) \cdot f(\Sigma_j x_j)$.

---

[23] Proportionality is the oldest known equity principle: "Equals must be treated equally, and unequals unequally, in proportion to relevant similarities and difference." Aristotle, Ethics.

Together with the average-cost mechanism analyzed in the next section, this is the single most important game of production externalities. Examples include the exploitation of common property resources,[24] cooperatives of production,[25] rent-seeking games,[26] several models of armament race and advertising competition,[27] as well as oligopolistic competition.[28]

Throughout this section and the next one, it will be convenient to assume divisible inputs and divisible outputs. However, the average-return mechanism is well defined when the input comes in indivisible units (and the output is divisible); similarly, the average-cost mechanism (Section 6.5) is well defined when output is indivisible and input is divisible. See Exercise 6.10 for some examples.

The average-return mechanism is more flexible than any voting mechanism: the latter forces equal consumption vectors (y_i, x_i) to all agents, whereas the former allows personalized consumptions. This is clearly an advantage when individual preferences differ. Consider the case of constant marginal costs (e.g., $f(x) = 2x$ for all $x \geq 0$). In the average-return mechanisms, every agent can achieve his stand alone utility.[29] The corresponding outcome is fully efficient.[30] On the other hand, the majority voting and the unanimity voting equilibria are Pareto inferior to this outcome; for instance, with majority voting, the "median" agent obtains his stand alone utility and every other agent obtains *at most* her stand alone utility.

However, the rigidity of the voting mechanisms turns to the mutual advantage of the agents when their preferences are very similar. Consider a unanimity profile (i.e., all individual preferences are identical). The voting equilibrium (whether with majority rule, unanimity rule, or any other reasonable rule) yields the efficient equal-utility vector. By contrast, the average-return mechanism has an inefficient equilibrium as soon as marginal costs are not constant (this equilibrium is Pareto

---

[24] Such as fisheries, forests, or mineral riches: here input is the fishing, cutting, or digging effort, and output is the catch. See Ostrom [1991] for a survey.

[25] The input is labor, the output is the firm's profit. See Israelsen [1980], Sen [1966].

[26] The input is the lobbying effort and output is the favor allocated by the politician; see Tullock [1980].

[27] See Case [1979].

[28] Where the "input" is the quantity of good supplied by the oligopolist, and "output" is the revenue generated; see Example 6.7.

[29] Because an agent supplying x_i units of input gets $y_i = (x_i/\Sigma_j x_j) f(\Sigma_j x_j) = f(x_i)$ units of output.

[30] It is also the outcome chosen by the competitive equilibrium from equal income, by the constant returns equivalent, and by the ω-egalitarian-equivalent solution; see Sections 5.2, 5.3, and 5.5.

inferior to the voting equilibrium). Example 1.4 illustrates the tragedy of the commons at a unanimous profile: when marginal costs increase, the equilibrium of the average-return game entails inefficient *over*production of output (recall that if n agents use the common pasture, in the average-return equilibrium they only get $1/n$ of total available surplus). Similarly, when marginal costs decrease, this equilibrium entails inefficient *under*production of output. These properties are perfectly general.

Lemma 6.5. *Suppose both goods are divisible, and utilities are monotonic and differentiable. Moreover, assume $u_{iy} > 0$ for all i. Given a differentiable production function f (such that $f(0) = 0$), consider a Nash equilibrium x^* of the average-return game such that x^* is interior ($x_i^* > 0$ and x_i^* can be increased, for all i).*

(i) If marginal cost increases (i.e., marginal productivity decreases), then x^* entails overproduction: a (small) proportional reduction of x^* is a Pareto improvement.

(ii) If marginal cost decreases, then x^* entails underproduction: a (small) proportional increase of x^* is a Pareto improvement.

Proof. In an interior Nash equilibrium of the average-return game, the partial derivative of player i's utility with respect to his own strategy must be zero. Denoting $x_N^* = \sum_j x_j^*$, this means

$$\frac{\partial}{\partial x_i}\left(u_i\left(\frac{x_i}{x_N} \cdot f(x_N), x_i\right)\right) = 0$$

$$\Leftrightarrow u_{iy} \cdot \left(\frac{x_{N\setminus i}}{x_N^{*2}} f(x_N^*) + \frac{x_i^*}{x_N^*} f'(x_N^*)\right) + u_{ix} = 0. \quad (10)$$

Consider the homothetic expansion (or contraction) of x^* as $x = \lambda x^*$. We compute the following derivative at $\lambda = 1$:

$$\frac{d}{d\lambda}\left(u_i\left(\frac{x_i^*}{x_N^*} f(\lambda x_N^*), \lambda x_i^*\right)\right)\Bigg|_{\lambda=1} = u_{iy} \cdot x_i^* f'(x_N^*) + u_{ix} \cdot x_i^*. \quad (11)$$

Taking (10) into account, this is equal to

$$u_{iy} \cdot x_i^* \cdot \frac{x_{N\setminus i}^*}{x_N^*} \cdot \left(f'(x_N^*) - \frac{f(x_N^*)}{x_N^*}\right).$$

By assumption, u_{iy}, x_i^*, and $x_{N\setminus i}^*$ are all positive. If marginal productivity decreases, we have $f'(x) < f(x)/x$; hence, the derivative (11) is negative: a small homothetic contraction of x^* is a Pareto improve-ment. Similarly, if marginal productivity increases, $f'(x) > f(x)/x$ im-plies that a small homothetic expansion of x^* is Pareto improving.

Q.E.D.

Note that Lemma 6.5 applies unchanged to the average-cost mechanism discussed in the next section. The proof is similar, upon exchanging the role of x and y and of f and c.

Remark 6.6

The assumption of differentiability is critical to Lemma 6.5. For in-stance, when individual preferences are represented by Leontief utility functions, the equilibrium of the average-return mechanism is efficient, no matter how the marginal productivity varies. See Exercise 6.7.

Remark 6.7

Lemma 6.5 is a local statement: at a Nash equilibrium of the average-return mechanism, we can slightly improve everyone's welfare by a small proportional reduction (resp. expansion) of individual input con-tribution. When preferences are convex and binormal (as defined below, just before Theorem 6.2) and when marginal costs increase, we also have a global statement. Call x^* a Nash equilibrium of the average-return mechanism and $z_i^* = (y_i^*, x_i^*)$, $i = 1, \ldots, n$, the corresponding allocations. If $(z_1, \ldots, z_n)$ is both efficient and Pareto superior to $(z_1^*, \ldots, z_n^*)$, then we must have $\Sigma_i z_i \le \Sigma_i z_i^*$: an efficient level of production (improving upon the equilibrium) is not larger than the equilibrium level of production. In particular, suppose preferences are represented by quasi-linear utility functions (where the linearity can be with respect to the input or with respect to the output), so that the efficient level of production is (essentially) unique.[31] Then we conclude that, at the Nash equilibrium of the average-return mechanism, there is overproduction of the output unless the equilibrium is efficient. Exer-cise 6.20 explains the proof of these facts. Note that the same properties hold for the average-cost mechanism as well.

The voting mechanisms (Section 6.2) typically yield inefficient Nash equilibrium allocations, but for different reasons than the average-return (or average-cost) mechanism. In the former, inefficiency results

---

[31] It is determined by the maximization of the joint surplus. For instance, if utilities are linear in input, the efficient level of output maximizes $\Sigma_i u_i(y_i) - c(\Sigma_i y_i)$. It is unique if c is strictly convex and/or the utility functions u_i are strictly concave.

from differences in preferences (recall that the voting equilibrium is efficient if all preferences coincide), whereas in the latter, it results from variations in the marginal cost (recall that the average-return equilibrium is efficient if marginal cost is constant, i.e., if there are no production externalities). We turn to an example where both effects (varying marginal costs and different preferences) are present.

Example 6.7. A Fishing Model (Case [1979])

Each player sends a fishing fleet of size x_i. Total catch of fish, given a total fishing effort of size $x = \sum_i x_i$, is $f(x) = x(2b - x)$.[32] Here the production function decreases beyond $x = b$. In any Nash equilibrium of the average-return game, we must have $x \le b$. We assume linear utilities $u_i(y_i, x_i) = y_i - c_i x_i$, with c_i interpreted as the (constant) marginal cost of fishing.

Notice an alternative interpretation of the average-return game as a Cournot oligopoly with linear demand. Firm i supplies the quantity x_i (at marginal cost c_i) and the clearing price is $p(x) = 2b - x$; firm i's profit is then

$$\pi_i(x_1, \ldots, x_n) = x_i \cdot \left(2b - \sum_j x_j\right) - c_i x_i.$$

To fix ideas, we assume $n = 3$ and $b = 4$, $c_1 = c_2 = 2$, $c_3 = 6$. Total (efficient) surplus is

$$v(N) = \max_x \{x \cdot (8 - x) - 2x\} = 9 \quad \text{achieved for } x^* = 3.$$

In the average-return equilibrium, both agents 1 and 2 are active but agent 3 is not: $x_1 = x_2 > 0$, $x_3 = 0$. To compute x_1, we solve $\partial \pi_1 / \partial x_1 = 0$ (noticing that π_1 is a concave function of x_1) and find $x_1 = 2$. Then check that if $x_1 + x_2 = 4$, agent 3 does not want to enter (because $\pi_3(2, 2, x_3) = -2x_3 - x_3^2$). Thus the equilibrium entails 1 unit of overfishing. The corresponding surplus is $\pi_1 = \pi_2 = 4$, $\pi_3 = 0$.

Contrast this with the equilibrium of majority voting. The median player is agent 1 (or 2) and solves $\max_{x_1}\{x_1(8 - 3x_1) - 2x_1\} = 3$ achieved at $x_1 = 1$. The majority voting equilibrium has the correct fishing level $x = 3$, but inefficiently makes agent 3 active. In fact, the correspond-

---

[32] Case [1979] explains this formula by means of the Verhulst differential equation describing the evolution of the population z under a fishing effort x: $\dot{z} = z \cdot (2b - z) - x \cdot z$. Check that the steady state of this equation is $z(\infty) = 2b - x$, hence the function f.

ing net profit is 3 for agents 1 and 2 and −1 for agent 3; thus the average-return equilibrium is Pareto superior to the majority equilibrium.

The unanimity voting equilibrium is even worse: agent 3 imposes an inefficiently low level of fishing, namely,

$$\max_{x_3}\{x_3(8 - 3x_3) - 6x_3\} = \tfrac{1}{3} \quad \text{achieved at } x_3 = \tfrac{1}{3}.$$

The resulting net surplus is 3.66 (even lower than the net surplus of 5 at the majority equilibrium). Note that the unanimity equilibrium is not Pareto inferior to the average-return equilibrium because agent 3 prefers the former over the latter.

The profile $c_1 = c_2 = 2$, $c_3 = 6$ is one where preferences are quite different. Accordingly, unanimity voting brings less surplus than majority voting and less than average-return. By contrast, in the case $c_1 = c_2 = 2$, $c_3 = 3$, the reader will easily check that our three equilibria are as follows:

	x_1, x_2	x_3	π_1, π_2	π_3
average	1.75	0.75	3.1	0.6
majority	1	1	3	1
unanimity	0.83	0.83	2.9	2.1

We compare finally the average mechanism and the mechanism where agents contribute any amount of input they wish and divide the output in equal shares, irrespective of individual contributions. The latter mechanism, called the *equal-output-shares* mechanism, is formally equivalent to voluntary contribution to a public good (where x units of input bring $f(x)/n$ units of public good consumed by all agents). As already noted in Section 1.5, this mechanism is patently unfair, because rewards are not related to effort and therefore, whenever individual contributions differ, an agent with a largest contribution can complain in the name of the domination property (see Section 4.6).[33] Moreover, the equal-output-shares mechanism generally brings an equilibrium more inefficient than the average-return equilibrium. For instance, unlike average returns, the equal-output-shares mechanism is inefficient even when marginal costs are constant (the corollary to Lemma

---

[33] This unfairness is not a matter of concern in the case of a true public good because we have no choice but to let all agents consume the same amount of output.

6.4 holds for any cost function c).[34] Another major difference between the two mechanisms is the nature of the inefficiencies: equal-output-shares produces too little output (corollary to Lemma 6.4), whereas average-return produces too much or too little output depending on whether marginal costs increase or decrease (Lemma 6.5).

To illustrate further the advantages of the average-return mechanism, we look at one of our earlier numerical examples.

Example 6.8.   Example 5.4 Continued

The three agents have linear utilities

$$u_H = y_H - x_H; \qquad u_M = y_M - \tfrac{4}{3}x_M; \qquad u_L = y_L - 6x_L. \qquad (12)$$

The production function is $y = 6\sqrt{2x}$. In the average-return equilibrium, only agents H and M are active. The Nash equilibrium conditions for the profit functions

$$6\sqrt{2}\,\frac{x_i}{\sqrt{x_H + x_M}} - \mu_i \cdot x_i, \qquad i = H, M \quad (\text{and } \mu_H = 1, \ \mu_M = \tfrac{4}{3}),$$

yield the system of first-order conditions:

$$3\sqrt{2}\,\frac{x_i}{x} = 6\sqrt{2} - \mu_i \cdot \sqrt{x}, \qquad i = H, M.$$

The solution is $x_H = 21.25$, $x_M = 8.50$ (and output shares are $y_H = 33.1$, $y_M = 13.2$). The surplus distribution is

$$\pi_H = 11.80, \qquad \pi_M = 1.87, \qquad \pi_L = 0. \qquad (13)$$

Thus total equilibrium surplus is 13.2, or 74% of the efficient surplus 18 (efficient output level is $y^* = 36$, so that the average-return equilibrium involves nearly 30% of overproduction).

Consider now the equal-output-shares equilibrium: as computed in Example 6.6, only the High agent contributes voluntarily $x_H = 2$, and

---

[34] Note that if, moreover, preferences are identical or close to one another, the equal-output-shares equilibrium is also Pareto inferior to the (efficient) equilibrium of average return. However, if the equilibrium of equal output shares involves some "free-riding," it is entirely possible that a free rider prefers this mechanism over average return. See Example 6.8 as well as Exercise 6.14.

every agent enjoys 4 units of output. The resulting surplus distribution is

$$\pi_H = 2, \qquad \pi_M = 4, \qquad \pi_L = 4.[35]$$

Thus the average-return equilibrium brings significantly more surplus than the voluntary contribution one, but distributes it much more unequally.[36] Recall from Chapter 5 that in the common property regime, it is odd that a particular agent should not receive any surplus whatsoever; after all, the Low agent would derive 6 units of surplus by standing alone. A distinct advantage of the serial mechanism (discussed in Section 6.6) is to guarantee a positive share of the surplus to every participant who can benefit by standing alone; in this example, the Low agent receives 1 unit of surplus. See Example 6.13.

We give now an example where the average-return equilibrium yields a smaller total surplus than the equal-output-shares equilibrium. Overproduction in the former is more destructive than underproduction in the latter. This can happen even with just two agents and identical preferences. Consider the technology introduced in Example 6.7: $f(x) = x(20 - x)$, and the utilities

$$u_i(y_i, x_i) = y_i - \tfrac{1}{3}x_i^2 \qquad i = 1, 2. \tag{14}$$

The equal-output-shares mechanism has a unique equilibrium where both agents contribute $x_i = 3.75$ and receive a net benefit $u_i = 42.2$ (see Exercise 6.18 for a general argument about uniqueness when utilities are quasi-linear in the output). In the average-return mechanism, each agent faces the profit function

$$x_i(20 - x_1 - x_2) - \tfrac{1}{3} \cdot x_i^2,$$

and the first-order equilibrium conditions give

$$x_i = 5.45, \qquad u_i = 39.7$$

(note that efficiency here requires $x_i = 4.29$ and brings $u_i = 42.86$). Notice that if we change the utility functions to $u_i(x_i, y_i) = y_i - x_i^2$,

---

[35] These values differ from formula (9) because we measure surplus in output, whereas surplus is measured in input in Example 6.6.

[36] An even more striking example is the configuration of Example 5.3: same production function, $n = 2$, $u_H = y_H - x_H$, $u_L = y_L - 3x_L$. The average-return equilibrium is efficient but gives all 18 units of surplus to the High agent (the other agent staying out), whereas the equal-cost-shares equilibrium yields $\pi_H = 4.5$, $\pi_M = 9$.

the average-return equilibrium becomes Pareto superior to the equal-output-shares equilibrium.

We conclude this section by discussing the existence and uniqueness of the Nash equilibrium in the average-return mechanism. As in the previous sections, convexity of preferences and of the cost function play an essential role. Moreover, we will also use the property of binormality of preferences, a more demanding assumption than convexity.

We say that the (differentiable) utility function $u(y, x)$ represents a *binormal preference* is the slope dy/dx supporting the indifference contour of u at (y, x) is nondecreasing in x and nondecreasing in y.[37] Two important examples are utility functions quasi-linear in the input $(u_i(y_i, x_i) = v_i(y_i) - x_i)$ or in the output $(u_i(y_i, x_i) = y_i - w_i(x_i))$. Note that binormal preferences are in particular convex (when we move upward along an indifference contour, the slope dy/dx increases). The following result is stated for both the average-return and the average-cost mechanisms.

Theorem 6.2. *All statements apply indifferently to the average-return mechanism and to the average-cost mechanism.*

(*i*) *If preferences are convex and marginal cost is nondecreasing, the mechanism has at least one Nash equilibrium.*

(*ii*) *If preferences are binormal and marginal cost is increasing, the mechanism has a unique Nash equilibrium.*

(*iii*) *If preferences are nonconvex and/or marginal cost is decreasing, the mechanism may have no Nash equilibrium.*

The proof of statement (i) is a standard application of Nash's theorem (just as in Lemma 6.3). That of statement (ii) is due to Watts [1993]. Note that if marginal cost is nondecreasing, but not necessarily increasing, we still have a unique equilibrium utility vector. As for statement (iii), its proof is the subject of Exercise 6.19.

6.5. The Average-Cost Mechanism: A Lesser Tragedy

In this mechanism, every agent demands a quantity y_i of output, and must pay (per unit of demand) the average cost of total demand. If the

---

[37] A more general definition that does not use a differentiability assumption: Fix the relative price of input to output and assume our agent chooses his consumption (y_i, x_i) in the budget set $p_Y \cdot y_i - p_X \cdot x_i \leq r$; when his revenue r increases, he will neither demand less output y_i nor supply more input.

demand profile is $y_1, \ldots, y_n$, agent i's final utility is

$$u_i\left(y_i, \frac{y_i}{\Sigma_j \, y_j} \cdot c(\Sigma_j \, y_j)\right).$$

Examples include computer networks[38] or any kind of buyer's cooperative (where buyers pool their orders and are billed according to a (nonlinear) wholesale tariff).[39]

Obviously, the average-cost mechanism shares many properties of its "dual," the average-return mechanism. There are also some interesting differences that we discuss in some detail below. The parallel discussion of the two mechanisms is especially interesting when they can both be implemented: participants of a buyer's cooperative may place an order (and pay the unit cost determined by overall demand) or may send a check (and get a proportional share of whatever output can be purchased with the sum of all individual checks).

The first property common to both mechanisms is Theorem 6.2: the same convexity assumptions imply the existence of a Nash equilibrium and its uniqueness. In the same fashion, we have the exact analog of Lemma 6.5: if marginal costs increase, any Nash equilibrium of the average-cost mechanism entails overproduction (a small proportional reduction of y^* is Pareto improving); if marginal costs decrease, an equilibrium entails underproduction (a small proportional expansion of y^* is a Pareto improvement). The proof exactly parallels that of Lemma 6.5. See also Remark 6.7 and Exercise 6.20 for global statements about over- and underproduction.

There are some small families of economies where both mechanisms (average-return and average-cost) yield precisely the same equilibrium outcome. Three examples are the problems with constant marginal costs (Exercise: why?), the economy of Example 5.3,[40] and the economies where all agents have Leontief utility functions (see Exercise 6.8).

More interesting, and much more numerous, are the cases where both mechanisms have different equilibrium allocations. There, a somewhat unexpected finding is awaiting us; under increasing marginal cost, the equilibria of the average-cost mechanism result in less overproduction than those of the average-return mechanism. Consequently, the former equilibria tend to be less inefficient than the latter ones. Before

---

[38] Each demand represents a certain amount of computing services; the input is the waiting time necessary to fulfill these demands.

[39] See Roemer and Silvestre [1993].

[40] In the equilibrium of both mechanisms, the Low agent is inactive and the High agent stands alone and produces the efficient level of output. Thus the equilibrium is efficient.

stating this result formally, we illustrate the claim in (a slight generalization of) the common pasture model of Example 1.4, where the average-cost mechanism (unlike the average-return one) often avoids the tragedy of the commons altogether.

Example 6.8. A Technology with Two Marginal Costs

The example generalizes the commons model in Example 1.4: the marginal cost increases but takes only two values:

$$f(x) = \min\{x, \epsilon(x - 100) + 100\}$$

(marginal cost is 1 up to 100 units, and $1/\epsilon$ thereafter).

The positive number ϵ is smaller that $\frac{1}{2}$. When it becomes arbitrarily small, we approach the very technology of Example 1.4. All n agents, $n \geq 3$, have identical utilities:

$$u_i(y_i, x_i) = \tfrac{3}{2}y_i - x_i.$$

We compute first the equilibrium of the average-return game. It is unique by Theorem 6.2, hence symmetrical; each agent contributes the same amount x_i^* of input. Of course, x_i^* is at least $100/n$ (because the opportunity to produce with a unit marginal product is fully exploited by any agent standing alone). We check that it must be larger than $100/n$. Consider the utility of agent i:

$$\pi_i = \frac{3}{2}\frac{x_i}{x_N}f(x_N) - x_i, \quad \text{where } x_N = \sum_i x_i.$$

Assume $x_N = 100$ and compute the derivative of π_i when x_i increases:

$$\left(\frac{d\pi_i}{dx_i}\right)_+ = \frac{3}{2}(1 - \epsilon) \cdot \frac{x_{N\setminus i}}{100} - \left(1 - \frac{3}{2}\epsilon\right).$$

When all x_i are equal, we have

$$\frac{x_{N\setminus i}}{100} = \frac{n-1}{n} \Rightarrow \left(\frac{d\pi_i}{dx_i}\right)_+ > \frac{3}{2}\frac{n-1}{n} - 1 \geq 0,$$

so that every agent wishes to raise her input contribution. Therefore, the equilibrium inefficiently overproduces. Straightforward computations give

$$x_i^* = \frac{n-1}{n^2} \frac{3-3\epsilon}{2-3\epsilon} \, 100, \qquad u_i^* = \frac{1}{n^2} \frac{3-3\epsilon}{2} \, 100.$$

As in Example 1.4, total surplus collected at the average-return equilibrium is approximately $(3/n)$th of total surplus (with equality when $\varepsilon = 0$).

We turn to the average-cost mechanism, and show that when ε is small enough, its equilibrium is fully efficient. In other words, using this mechanism rather than average-return completely overcomes the tragedy of the commons! To see this, consider agent i's pay-off function in the average-cost mechanism:

$$\pi_i = \frac{3}{2} y_i - \frac{y_i}{y_N} \cdot c(y_N), \text{ where } c(y) = \max\left\{ y, \frac{1}{\varepsilon}(y - 100) + 100 \right\}.$$

Fix an outcome where $y_N = 100$ and compute

$$\left(\frac{d\pi_i}{dy_i} \right)_+ = \left(\frac{1}{\varepsilon} - 1 \right) \frac{y_{N \setminus i}}{100} - \left(\frac{1}{\varepsilon} - \frac{3}{2} \right).$$

Taking $y_i = 100/n$ for all i, we get

$$\left(\frac{d\pi_i}{dy_i} \right)_+ \leq 0 \Leftrightarrow \left(\frac{1}{\varepsilon} - 1 \right) \frac{n-1}{n} - \left(\frac{1}{\varepsilon} - \frac{3}{2} \right) \leq 0 \Leftrightarrow \varepsilon \leq \frac{2}{n+2}.$$

In other words, for $n = 3$, any value of ε below 0.4 guarantees an efficient equilibrium in the average-cost mechanism, whereas the equilibrium of average-return brings the fraction $(1 - \varepsilon)$ of total surplus. For $n = 9$, any value of ε not larger than 0.2 makes the equilibrium of average-cost efficient, whereas the equilibrium of average-return brings at most 38% of total surplus, and so on. Exercise 6.12 gives a couple of variants of this example.

Example 6.9 generalizes to all unanimous preference profiles.

Lemma 6.6

(*i*) *Suppose preferences are binormal (in particular, convex), and marginal costs are nondecreasing. Then in the average-return equilibrium, total production exceeds that of the average-cost equilibrium (the latter involves less overproduction than the former).*

(ii) *Therefore, the average-return equilibrium cannot be Pareto superior to the average-cost equilibrium.*

(iii) *In the case of unanimous preferences,the (symmetrical) average-cost equilibrium is Pareto superior to the (symmetrical) average-return equilibrium.*

The central idea of the proof is given in Exercise 6.15.[41]

Example 6.10.   Unanimous Linear Utilities, Quadratic Costs

The cost function is $c(y) = y^2/2$, and n agents have identical utilities $u_i(y_i, x_i) = y_i - \mu x_i$. As usual (see Examples 5.4 and 6.8), we measure surplus in the output good. In the average-return mechanism, the unique equilibrium is symmetrical. Assume every agent i contributes $x_i^* = \alpha$. Then agent 1 faces the profit function

$$\pi_1\left(x^* \,|^1 x_1\right) = \frac{\sqrt{2}\, x_1}{\sqrt{x_1 + (n-1)\alpha}} - \mu \cdot x_1$$

(where $(z \,|^i t_i)$ denotes the vector with ith coordinate t_i and jth coordinate z_j for all other j). This function reaches its maximum at $x_1 = \alpha$ if and only if

$$x_i^* = \alpha = \frac{2(n - \frac{1}{2})^2}{n^3 \mu^2}.$$

The corresponding equilibrium surplus is

$$\sum_i \pi_i(x^*) = \pi^r = \frac{n - \frac{1}{2}}{n^2} \cdot \frac{1}{\mu}. \tag{15}$$

In the average-cost mechanism, the unique equilibrium is symmetrical. If all agents $i = 2, \ldots, n$ demand $y_i^* = \beta$, agent 1 faces the profit function

$$\pi_1\left(y^* \,|^1 y_1\right) = y_1 - \frac{1}{2}\mu y_1 \cdot (y_1 + (n-1)\beta).$$

---

[41] See Moulin and Watts [1994] for a complete argument.

The equilibrium condition at $y_1 = \beta$ yields

$$y_i^* = \beta = \frac{2}{(n+1)\mu},$$

hence the total equilibrium surplus

$$\sum_i \pi_i(y^*) = \pi^c = \frac{2n}{(n+1)^2} \cdot \frac{1}{\mu}. \tag{16}$$

Comparing (15) and (16), we conclude that the ratio of total surplus at the average-return equilibrium and at the average-cost equilibrium decreases from 84% for $n = 2$, 74% for $n = 3$, to 57% for $n = 10$, and approaches 50% as n grows to infinity.

Example 6.11. Example 6.8 Continued

Costs are quadratic, namely, $c(y) = y^2/72$; utilities are still linear but not unanimous; see (12). In the average-cost equilibrium (as in the average-return one), only agents H and M are active. The profit function writes

$$y_i - \frac{\mu_i}{72} y_i \cdot (y_H + y_M), \qquad i = H, M,$$

hence the first-order conditions for Nash equilibrium:

$$1 = \frac{\mu_i}{72} (2y_i + y_j),$$

which gives $y_H = 30$, $y_M = 12$ (an overproduction of 17%) and the surplus shares

$$\pi_H = 12.5, \qquad \pi_M = 2.67, \qquad \pi_L = 0. \tag{17}$$

Both agents H and L are strictly better off than under the average-return equilibrium (and agent L is indifferent between the two mechanisms). Total equilibrium surplus rises to 84% of the efficient surplus (compare to 73% for the average-return equilibrium). Exercise 6.13 gives some general formulas for the equilibrium of our two average mechanisms when costs are quadratic and utilities are linear.

Our concluding example establishes the limits of Lemma 6.6: we cannot always expect that *all* active agents prefer the average-cost equilibrium over the average-return equilibrium (although statement (ii) of Lemma 6.6 tells us that at least one agent is not worse off under the former than under the latter). The example has two agents, both active at both equilibria, and disagreeing over their preferred equilibrium.

Example 6.12. A Technology with Two Marginal Costs

The production function is

$$f(x) = \min\{2x, x + 1\} \Leftrightarrow c(y) = \max\left\{\frac{y}{2}, y - 1\right\},$$

and the utilities of the two agents are linear:

$$u_1(y_1, x_1) = y_1 - (1.1)x_1, \qquad u_2 = y_2 - (1.6)x_2.$$

In the unique equilibrium of the average-cost equilibrium, the quantity of output demanded exceeds 2 (the critical level at which marginal costs jump up; compare with Example 6.9). It is easily computed (we omit the details):

$$y_1 = 1.727, \qquad x_1 = 0.922, \qquad \pi_1 = 0.713,$$

$$y_2 = 0.419, \qquad x_2 = 0.223, \qquad \pi_2 = 0.062.$$

Similar computations give the equilibrium of the average-return equilibrium:

$$y_1 = 2.08, \qquad x_1 = 1.224, \qquad \pi_1 = 0.734,$$

$$y_2 = 0.347, \qquad x_2 = 0.204, \qquad \pi_2 = 0.021.$$

Thus agent 1 prefers the average-return equilibrium to the average-cost one. Exercise 6.11 generalizes this example.

Remark 6.8

In the case of *decreasing marginal costs*, Lemma 6.6 cannot be adapted easily. Recall from Theorem 6.2 that the very existence of a Nash equilibrium is not guaranteed. Moreover, either or both of our mechanisms may have multiple equilibria, so that comparing the equilibrium surplus can be ambiguous.

Remark 6.9

An equilibrium concept with some similarities to the average-cost and average-return mechanisms on one hand, and the competitive equilibrium with equal incomes on the other hand, is the *average-cost competitive equilibrium* (ACCE) (Weitzman [1974]). Fix a cost function c, and a utility profile $u_1, \ldots, u_n$. For a given price p (1 unit of output costs p units of input), we denote agent i's demand by $\delta_i(p)$:

$$z_i = (y_i, x_i) \in \delta_i(p) \Leftrightarrow x_i = p \cdot y_i \quad \text{and}$$

$$\text{for all } z_i': \quad \{x_i' = p \cdot y_i'\} \Rightarrow \{u_i(z_i) \geq u_i(z_i')\}.$$

An average-cost competitive equilibrium consists of a price p and an allocation $(z_1, \ldots, z_n)$, such that

$$\sum_i x_i = c\left(\sum_i y_i\right) \quad \text{and} \quad z_i \in \delta_i(p), \quad \text{for all } i.$$

Unlike the CEEI (see Section 5.2), the ACCE concept makes no assumptions about the marginal cost at the equilibrium allocation; hence, the latter allocation is not Pareto optimal. Unlike the average-cost (or average-return) equilibrium, the ACCE does not result from a simple decentralized mechanism. Exercise 6.21 shows that, in general, the ACCE allocation is more inefficient than the average-cost (or even the average-return) equilibrium allocation.

6.6. SERIAL COST- (OR OUTPUT-) SHARING: IMPROVING UPON VOTING

Our last mechanisms to manage production externalities have the same simple message spaces as the average-cost (or average-return) mechanism. In the cost-sharing version of the serial mechanism, each agent demands (and obtains) a quantity y_i of output, the cost shares are computed according to a nonlinear pattern (that is to say, individual cost shares are not proportional to individual demands). In the output-sharing version of the mechanism, each agent contributes a quantity x_i of input, and output shares are computed according to a (the same) nonlinear pattern.

The easiest introduction to the serial pattern of cost- (or output-) sharing is as a refinement of unanimity voting. To fix ideas, suppose two agents share a technology with cost function c. In unanimity voting, agent i, $i = 1, 2$, selects an output level y_i expecting to pay $c(2y_i)/2$, and the smallest level, say y_1, is selected: every agent receives an identical allocation $(y_1, c(2y_1)/2)$. In the serial cost-sharing mechanism with the same messages (y_1, y_2), only agent 1 (with the lowest demand) gets the allocation $(y_1, c(2y_1)/2)$. Agents 2's higher demand is served, and agent 2 pays the balance $c(y_1 + y_2) - (c(2y_1)/2)$. In equilibrium, thus, agent 2 demands more than y_1 only if it is to his advantage to do so: both agents end up no worse than in the equilibrium of unanimity voting. However, as agent 2 has the option to demand any level of output higher than y_1 (provided he pays in full the incremental cost), he will typically be strictly better off than under unanimity voting.

Consider the numerical example 5.3. We have two agents (indexed H, L respectively for high and low demand of the output) with linear utilities and a quadratic cost function:

$$u_H(y_H, x_H) = y_H - x_H, \qquad u_L(y_L) = y_L - 3x_L, \qquad c(y) = \frac{y^2}{72}.$$

In unanimity voting, High demands $y_H = 18$ (that maximizes $y - c(2y)/2$), whereas Low demands $y_L^* = 6$. In the unanimity equilibrium, both agents consume $z_i = (6, 1)$ for the surplus shares $\pi_H = 5$, $\pi_L = 3$. Under serial cost-sharing, High can get any output level $y_H \geq 6$ provided she pays $c(y + 6) - 1$; therefore, she asks for $y_H^* = 30$ and ends up with the surplus $\pi_H = 13$ (Low's surplus is unchanged at $\pi_L = 3$). Thus total surplus is twice larger in the serial equilibrium than in unanimity voting. Figure 6.3 illustrates this example.

Contrast these equilibria with that of the average-cost mechanism (which in this case coincides with the average-return equilibrium): in the latter, Low must remain inactive (and gets zero surplus) because he must face the stiff average cost resulting from High's high demand. The average-cost equilibrium is $y_H = 36$, $y_L = 0$ with corresponding surplus $\pi_H = 18$, $\pi_L = 0$. The advantage of serial cost-sharing over average cost and average return is that an agent with a low demand is never "shut off" (as above, the Low agent): Low is protected from the high average cost imposed by the high-demand agent. See Lemma 6.7 for a general statement.

Note that both the serial and average-cost equilibrium outcomes produce the surplus-maximizing level of output, but only the average-cost equilibrium is truly efficient (given linear utilities, efficiency prevents Low from contributing any input). Thus total surplus is higher in the average-cost equilibrium than in the serial equilibrium. This configuration is not general, however; for instance, in Example 6.13, the inequality is reversed.

As an illustration of the serial formula (18), suppose 10 agents can demand 0, 1, or 2 units of output. If two agents demand $y_i = 0$ and

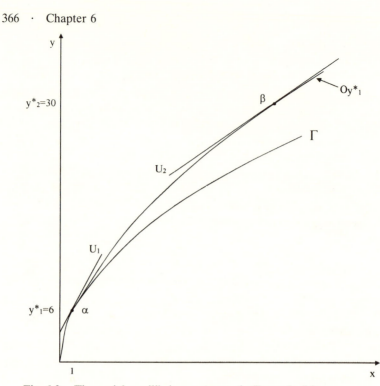

Fig. 6.3. The serial equilibrium outcome in Example 5.3.

eight agents demand $y_i = 1$, serial cost-sharing coincides with average cost-sharing: the eight active agents each pay $c(8)/8$ and two inactive agents pay nothing. Next, suppose two agents demand $y_i = 0$, five agents demand $y_j = 1$, and three agents demand $y_k = 2$. The serial formula splits total cost $c(11)$ as follows:

$$y_i = 0 \Rightarrow x_i = 0,$$

$$y_j = 1 \Rightarrow x_j = \frac{c(8)}{8},$$

$$y_k = 2 \Rightarrow x_k = \tfrac{1}{3}(c(11) - \tfrac{5}{8}c(8)).$$

Average cost-sharing would charge $c(11)/11$ to an agent demanding $y_j = 1$; serial cost-sharing, on the other hand, postulates that an agent demanding only one unit is not responsible for the second unit of demand by the last three agents.

Definition 6.3. The Serial Formula. Given a cost function and a demand profile $(y_1, \ldots, y_n)$, the *serial cost shares* are computed as follows. Arrange the individual demands in increasing order: to fix ideas, we assume $y_1 \le y_2 \le \cdots \le y_n$, and define

$$y^1 = ny_1, \qquad y^n = y_1 + \cdots + y_n$$

and

$$y^i = y_1 + \cdots + y_{i-1} + (n - i + 1)y_i \quad \text{for all } i,$$

$$x_1 = \frac{c(y^1)}{n}, x_2 = \frac{c(y^2)}{n - 1} - \frac{c(y^1)}{n(n - 1)}, \cdots$$

$$x_i = \frac{c(y^i)}{n - i + 1} - \sum_{k=1}^{i-1} \frac{c(y^k)}{(n - k + 1)(n - k)}, \qquad i = 1, \ldots, n.$$

(18)

In the general case, replace agent 1 by an agent with the lowest demand, agent 2 by an agent with the next to lowest demand, and so on.

Given a production function f and a profile of input contributions $(x_1, \ldots, x_n)$, the serial output shares are computed in exactly the same way (replacing y_i by x_i and c by f).

Serial cost-sharing is well defined whether the output is divisible or indivisible, but requires a divisible input (a symmetrical statement applies to serial output-sharing).

The serial formula is admittedly more complex than average cost (or average return). A simple mnemonics relies on two properties that together characterize the formula: (i) equal cost for equal demand (equal treatment of equals), and (ii) agent i's cost share does not depend upon demands higher than his own (provided they remain higher).

Fix a demand profile such that $y_1 \le y_2 \le \cdots \le y_n$. By property (ii), the cost share x_1 is unchanged if we lower y_i to y_1 for all $i \ge 2$. By (i), the cost shares at the profile $(y_1, \ldots, y_1)$ are all equal, hence $x_1 = c(ny_1)/n$. Next, we lower y_i to y_2 for all $i \ge 3$ (leaving y_1 unchanged) and invoke (ii) again at the profile $(y_1, y_2, \ldots, y_2)$: the cost shares x_1 and x_2 are unchanged, and property (i) says that agents $2, 3, \ldots, n$ all pay the same cost, hence,

$$c(y_1 + (n - 1)y_2) = x_1 + (n - 1)x_2,$$

which in turn yields (18) for $i = 2$, and so on.

In the remainder of this section, we describe the extraordinary strategic and normative properties of the two mechanisms, serial cost-sharing and serial output-sharing.[42] Because the formula (18) charges a different average cost to each participant, some coalition of agents may benefit by pooling, splitting, or redistributing individual demands. This poses a serious problem to the mechanism only if output and input are transferable and if actual consumptions of these goods cannot be monitored.[43] In some examples of cooperative production, this difficulty is enough to rule out the serial mechanism. When fishermen fish in a common lake, the context of free access (unregulated fishing) forces the average-return mechanism; using the serial formula to share the surplus requires monitoring of both the fishing effort and the catch. Similarly, the users of a copying machine can easily transfer output (pages copied) and input (money). On the other hand, in many examples of cooperative production, either the input and/or output is not transferable across agents, or individual consumptions of these goods can be monitored: think of the users of a computer network where output is the amount of computations demanded and input is the waiting time (neither of these two goods is easily transferable). Another example is a water system. Direct transfers of water across households, although conceivable, are easy to check; hence, a nonlinear pricing system if feasible. Similarly, the manager can often monitor the usage of the copying machine among her employees.

In the rest of this chapter, we shall rule out the possibility of transfers of either output or input among agents.

We describe the equilibrium of the serial mechanism in our familiar numerical example with linear utilities and quadratic costs.

Example 6.13. Examples 6.8 and 6.11 Continued

Call the three agents High, Medium, and Low; their linear utilities are given by (12). The cost function is $c(y) = y^2/72$. Consider first the unanimity voting equilibrium. The Low agent asks $y_L = 2$ (upon maximizing $y_L - 6c(3y_L)/3$), and other agents demand more. In the equilibrium of serial cost-sharing, Low gets the unanimity allocation $y_L = 2$,

---

[42] That is to say, each agent selects his demand of output (resp. his input contribution), and cost shares (resp. output shares) are computed by formula (18).

[43] Consider serial cost-sharing with three agents demanding $y_1 = 1$, $y_2 = 5$, $y_3 = 6$, and $c(y) = y^2$. Formula (18) assigns the cost shares $x_1 = 3$, $x_2 = 59$, $x_3 = 82$. Agents 1 and 2 now decide to demand $y_1' = y_2' = 3$ (agreeing that agent 1 will return 2 units to agent 2 afterwards), and pay $x_1' = x_2' = 27$ (so $x_3' = 90$), thus saving \$8. It is easy to show that the only cost-sharing formula where this kind of manipulation is never possible is average cost (see Moulin and Shenker [1992]).

$x_L = 0.17$. The remaining agents H, M now face a "residual" cost function

$$\bar{c}(y) = c(y + 2) - \tfrac{1}{6} \quad \text{for all } y \geq 2.$$

For this cost function, we compute the (two-person) unanimity voting equilibrium. The Medium agent maximizes

$$y_M - \frac{4}{3} \frac{\bar{c}(2y_M)}{2} = y_M - \frac{4}{3}\left(\frac{(y_M + 1)^2}{36} - \frac{1}{12} \right),$$

hence demands $y_M = 12.5$ (of course, High's unanimity demand is higher). Medium's serial equilibrium allocation is thus $y_M = 12.5$, $x_M = 4.98$. Finally, the High agent can demand any level of output y beyond y_M at a cost

$$c(y + y_M + y_L) - (4.98 + 0.17).$$

Therefore, High demands $y_H = 21.5$ and pays $x_H = 12.85$. Note that total production level is $y = 36$, precisely the efficient level, but the serial equilibrium inefficiently makes Medium and Low contribute some input. The surplus distribution is

$$\pi_H = 8.65, \qquad \pi_M = 5.86, \qquad \pi_L = 1, \tag{19}$$

for a total surplus 15.51, or 86% of maximal available surplus. Figure 6.4 illustrates this construction.

The serial equilibrium is Pareto superior to both the unanimity voting equilibrium (where all agents consume $y = 2$, $x = 0.17$) and to the majority voting equilibrium (where all agents consume $y = 4$, $x = 0.67$).[44] This is a general property explained in Lemma 6.7.

Now compare the serial equilibrium with the equilibria of the average-cost and average-return mechanisms (computed respectively in Examples 6.11 and 6.8). The corresponding surplus distributions ((17) and (13)) bring a smaller total surplus (although the difference is less than 2% of the efficient surplus in the case of average cost) with a more

[44] The corresponding surplus distributions are, respectively,

$$(\pi_H, \pi_M, \pi_L) = (1.83, 1.78, 1) \text{ and } (3.33, 3.11, 0).$$

Thus the serial equilibrium more than doubles total surplus. Note that in Examples 6.4 and 6.6, we arrived at different surplus figures because surplus was measured in the input (instead of output as we do now).

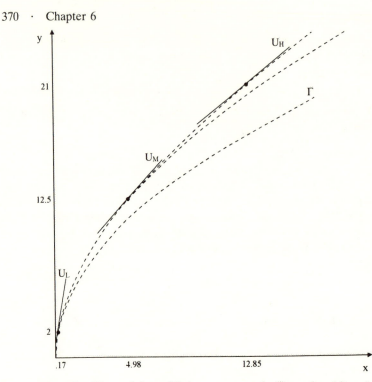

Fig. 6.4. The serial equilibrium outcome in Examples 6.8 and 6.11.

unequal distribution: Low gets nothing at all and High gets at least 80% of total surplus (she gets only 56% with the serial mechanism).

Theorem 6.3. (Moulin and Shenker [1992]). Assume that marginal cost increases and that preferences are convex and monotonic (alternatively, assume that marginal costs are nondecreasing and preferences are strictly convex). Then the serial cost-sharing mechanism (and the serial output-sharing mechanism) have a unique Nash equilibrium, the same for the cost-sharing and the output-sharing versions. This equilibrium is robust against coalition deviations: it is a strong equilibrium.[45] It also results from the successive elimination of strictly dominated strategies.

With convex preferences and nondecreasing marginal costs, the serial mechanism may have several equilibria, but one of them is Pareto superior.

---

[45] Namely, a Nash equilibrium robust against coalitional deviations as well. See, e.g., Moulin [1986].

Since both versions of the serial mechanism share the same equilibrium at all profiles of convex preferences, they are for all practical purposes the same mechanism. See Exercise 6.16 for a proof of this claim. Throughout the remainder of this section we always use the cost-sharing version of the mechanism in order to avoid confusion.

The intuition of Theorem 6.3 (of which the complete proof is omitted) is a generalization of the argument given above. Figure 6.3 illustrates the easy construction of the equilibrium in the case of two agents. Draw the unanimity cost function $c(2y)/2$. If the demands coincide, $y_1 = y_2 = y$, the individual cost shares are computed along the unanimity curve $\Gamma = \{y, (c(2y)/2\}$. Let y_1^*, y_2^* be the preferred demands of agents 1 and 2, respectively, when costs are computed on the unanimity curve. Assume $y_1^* \leq y_2^*$: we claim that y_1^* is the equilibrium demand of agent 1, and his equilibrium cost share is $c(2y_1^*)/2$.[46] Given the strategy y_1^*, agent 2's opportunity set is the curve $O_{y_1^*}$ on Figure 6.3, namely,

$$(y_2, x_2) \in O_{y_1^*} \quad \Leftrightarrow \quad \begin{aligned} & y_2 \leq y_1^* \quad \text{and} \quad x_2 = \frac{c(2y_2)}{2}, \\ & y_2 \geq y_1^* \quad \text{and} \quad x_2 = c(y_1^* + y_2) - \frac{c(2y_1^*)}{2}. \end{aligned}$$

Notice that $O_{y_1^*}$ coincides with Γ up to the point $\alpha = (y_1^*, c(2y_1^*)/2)$ and is above Γ afterwards. Moreover, $O_{y_1^*}$ and Γ have the same slope at α, namely, $c'(2y_1^*)$. Agent 2's preferred point on Γ is, by assumption, beyond α; by convexity of agent 2's preferences, this implies that agent 2's preferred point on $O_{y_1^*}$, call it β, is also beyond α. Denote by $\bar{y}_2$ the output level of β. As $\bar{y}_2 \geq y_1^*$, the opportunity set $O_{\bar{y}_2}$ coincides with Γ up to the point γ (see Figure 6.3); therefore, y_1^* is agent 1's best reply to agent 2's strategy $\bar{y}_2$. By construction of β, demand $\bar{y}_2$ is agent 2's best reply to y_1^*, and this completes the proof that $(y_1^*, \bar{y}_2)$ is a Nash equilibrium of the serial mechanism.

To see why it is the only Nash equilibrium is not difficult either. Note that agent i is guaranteed at least his unanimity utility by choosing strategy y_i^*; see Lemma 6.7. If (y_1, y_2) is a Nash equilibrium such that $y_2 < y_1$, this implies that $y_2 = y_2^*$; but the best reply of agent 1 to y_2^* is y_1^*! Similarly, if (y_1, y_2) is a Nash equilibrium such that $y_1 \leq y_2$, we must have $y_1 = y_1^*$, and we are back to the equilibrium just described.

---

[46] Recall that in unanimity voting, each agent would consume $(y_1^*, c(2y_1^*)/2)$ in equilibrium. In the equilibrium of the serial mechanism, agent 2 will do better than that.

The strategic properties of the serial mechanism are just as strong as those of the two voting mechanisms in Section 6.2. The equilibrium is unique and cannot be upset by any coalitional deviation. In particular, the equilibrium is "efficient" in the second-best sense: as long as they must allocate costs by means of the serial formula (18), the agents cannot cooperatively improve upon the outcome of the decentralized equilibrium. Of course, this outcome is typically inefficient within the set of allocations restricted only by the feasibility condition $\Sigma_i x_i = c(\Sigma_i y_i)$; see Example 6.13 and Lemma 6.8. The only difference between the serial and voting mechanisms is that the Nash equilibrium strategies in the serial mechanism are not dominant strategies. This difference is superficial: if is not difficult to alter the serial mechanism and turn it into a direct revelation mechanism (see Section 1.6), where truthful reporting of one's preferences is a dominant strategy for every participant at every profile of convex preferences.[47]

The construction of the equilibrium in Figure 6.3 (and in Example 6.13) explains why the serial equilibrium outcome is Pareto superior to the equilibrium outcome of any one of the voting mechanisms discussed in Section 6.2 (the unanimity and majority voting rules, as well as the k-rank mechanism of Remark 6.1): each agent can guarantee her unanimity utility by simply reporting her preferred demand under unanimity cost. On the other hand, in any voting rule, the allocation of any agent moves on the unanimity curve; hence, no agent can ever obtain more than her unanimity utility. Formally, we have the following.

Lemma 6.7. *Suppose marginal costs are nondecreasing. Then the cost shares given by* (18) *satisfy*

$$c(y_i) \le x_i \le \frac{c(ny_i)}{n}, \quad \text{for all } y_1, \ldots, y_n, \quad \text{all } i. \tag{20}$$

In every Nash equilibrium of the serial mechanism, every agent is guaranteed at least his unanimity utility and at most his stand alone utility:

$$\max_{y \ge 0} u_i(y, c(y)) \ge u_i(y_i, x_i) \ge \max_{y \ge 0} u_i\left(y, \frac{c(ny)}{n}\right). \tag{21}$$

Proof. Fix a demand profile $y_1 \le y_2 \le \cdots \le y_n$ and an agent i. Check first that the cost share x_i given by (18) does not decrease when

---

[47] The mechanism in question uses the reported preferences to compute the equilibrium of the serial mechanism as described in Figure 6.4. See Moulin and Shenker [1992] for details.

y_j increases, $j \le i$. Note that the term y_j appears in the quantities y^k for all k, $j \le k \le i$. When y_j increases, the numbers $c(y^k)$, $j + 1 \le k \le i - 1$, all increase but no more than $c(y^i)$ (because marginal costs are nondecreasing). Similarly, $c(y^j)$ increases at most $(n - j + 1)$ times faster than $c(y^i)$. Therefore, the equality

$$\frac{1}{n - i + 1} - \sum_{k=j+1}^{i-1} \frac{1}{(n - k + 1)(n - k)} - \frac{1}{n - j} = 0$$

implies the claim.

We noted earlier (see the discussion after Definition 6.3) that when we increase (or decrease) y_j, $j \ge i$, the cost share x_i does not move (as long as the move does not change the relative position of y_i and y_j). We conclude that x_i is nondecreasing in y_j for all j and all y_j. Now inequality (20) follows at once by comparing agent i's cost share at the initial profile and at the profiles $(0, \ldots, 0, y_i, 0, \ldots, 0)$ and $(y_i, \ldots, y_i, y_{i+1}, \ldots, y_n)$.

Inequality (21) follows at once from (20) and the Nash equilibrium property for agent i.   Q.E.D.

To put Lemma 6.7 in perspective, note that in the average-cost and in the average-return mechanisms, the low-demand (or low-contribution) agent never pays less (and usually pays more) than the unanimity cost.[48] Accordingly, the low-demand agent usually gets strictly less than his unanimity utility. See, for instance, Example 6.13.

Property (21) has the following important consequence. Whenever the vector of unanimity utilities is efficient, so is the serial equilibrium outcome. An example is when all individual preferences coincide (unanimous preferences): then the serial and voting equilibrium outcomes (for any reasonable voting rules) are equal and yield the highest feasible equal-utility vector. For instance, in Example 6.9, the serial equilibrium is fully efficient no matter what ϵ is.[49] Another circumstance where the serial mechanism yields a fully efficient equilibrium is when marginal costs are constant: then the serial equilibrium equals the average-cost

---

[48] Indeed, if y_i is the lowest demand, we have $n y_i \le y = \sum_j y_j$, implying $c(n y_i)/n y_i \le c(y)/y \Leftrightarrow c(n y_i)/n \le x_i$.

[49] For those values of ϵ such that the average-cost equilibrium is efficient (namely, $\epsilon \le 3/(n + 2)$), the serial mechanism has a much better strategic behavior: its equilibrium is unique and treats equals equally, whereas there is a continuum of equilibria to the average-cost mechanism, most of them treating equals unequally.

(or -return) equilibrium, with every agent achieving his or her stand alone utility.[50]

With nonunanimous preferences and nonconstant marginal costs, we expect that the serial equilibrium is not fully efficient. It entails over-production under increasing marginal cost and underproduction under decreasing marginal cost, just like the average-cost and average-return mechanisms.

Lemma 6.8

(*i*) *Suppose preferences are represented by linear utility functions and marginal costs are increasing. Then the serial equilibrium produces the efficient level of output, but allocates individual input contributions and output shares inefficiently.*

(*ii*) *Suppose preferences are convex and binormal (see Section 6.4), and assume nondecreasing marginal costs. Let* $(z_1^*, \ldots, z_n^*)$ *be the serial equilibrium outcome, and let* $(z_1, \ldots, z_n)$ *be a Pareto optimal outcome, also Pareto superior to the equilibrium outcome. Then we have* $\sum_{i=1}^n z_i \leq \sum_{i=1}^n z_i^*$.

(*iii*) *Assume nonincreasing marginal costs and otherwise the same premises as in statement (ii). Then we have* $\sum_{i=1}^n z_i^* \leq \sum_{i=1}^n z_i$.

Statement (i) is illustrated in Example 6.13 (see also Example 6.14). It follows directly from formula (18).[51]

Statements (ii) and (iii) (the proof of which is the subject of Exercise 6.20) have counterpart statements for the average-cost and -return mechanisms; see Remark 6.6. They are global statements about the type of inefficiency in the serial equilibrium; they should be contrasted with the local statements of Lemma 6.5. The latter describes a local Pareto improvement of the average-return or average-cost equilibrium, by means of a proportional reduction (or expansion) of the equilibrium strategies; this improvement is achieved within the constraints of the given mechanism. By contrast, in the serial mechanism, players cannot

---

[50] Check that for the cost function $c(y) = 3y$, the serial formula (18) gives $x_i = 3y_i$ for all i.

[51] Denote $u_i = y_i - a_i x_i$, with $a_1 \geq \cdots \geq a_n$. Then the serial equilibrium demands are ranked as $y_1 \leq \cdots \leq y_n$, and y_n maximizes

$$y_n - a_n \cdot \left(c\left(\sum_{i=1}^n y_i \right) + K \right),$$

where K does not depend on y_n. Therefore $1 = a_n \cdot c'(\sum_{i=1}^n y_i)$, which is precisely the first-order condition for maximizing joint surplus.

achieve a Pareto improvement without breaking away from the mechanism (indeed, the serial equilibrium is a strong equilibrium; see Theorem 6.3). This is illustrated by Example 6.13 and by the following example.

Example 6.14. Example 6.12 Continued

Compute unanimity demands. Agent i expects to pay $c(2y_i)/2$ for any demand y_i and maximizes his utility accordingly:

$$\mathrm{una}(u_1) = \max_{y_1 \geq 0} \left\{ y_1 - \frac{1.1}{2} \max(y_1, 2y_1 - 1) \right\}$$

$$= 0.45 \text{ achieved at } y_1 = 1,$$

$$\mathrm{una}(u_2) = \max_{y_2 \geq 0} \left\{ y_2 - \frac{1.6}{2} \max(y_1, 2y_1 - 1) \right\}$$

$$= 0.2 \text{ achieved at } y_2 = 1.$$

The serial equilibrium is therefore $y_1 = y_2 = 1$, $x_1 = x_2 = 0.5$ with corresponding surplus $\pi_1 = 0.45$, $\pi_2 = 0.2$. In this economy, efficiency commands producing also 2 units of output but letting agent 1 contribute the whole input ($x_1 = 1$). For instance, the ω-egalitarian-equivalent solution (Section 5.3) recommends a Pareto superior outcome, namely,

$$y_1 = 1.72, \quad y_2 = 0.28, \quad x_1 = 1, \quad x_2 = 0, \quad \pi_1 = 0.62, \quad \pi_2 = 0.28.$$

However, this requires giving some free output to agent 2, a configuration excluded by the serial formula (18).[52]

Comparing the overall surplus raised by the serial mechanism on one hand and the average-cost and surplus mechanisms on the other hand can go either way. In the case of a unanimity profile, the serial equilibrium is efficient and Pareto superior to the average-cost equilibrium. The same comparison (larger joint surplus at the serial equilibrium) holds in Example 6.13, as noted earlier. In the above example, by contrast, the joint surplus in the average-cost equilibrium and in the

---

[52] Notice that the ω-egalitarian-equivalent solution is not always Pareto superior to the serial equilibrium outcome. For instance, in Example 6.13, the ω-EE solution yields the surplus distribution $\pi_H = 8.6$, $\pi_M = 8.0$, $\pi_L = 1.4$ (computed at the end of Section 5.3), that only the Low and Medium agents prefer to (19).

average-return equilibrium (computed in Example 6.12) are, respectively, 0.77 and 0.75, both exceeding the joint surplus at the serial equilibrium (0.65) by at least 15%. See also the numerical Example 5.3 (discussed at the beginning of this section) as well as Example 6.7 (of which the serial equilibrium is discussed in Exercise 6.11).

The last normative property of the serial mechanism is not the least one: any equilibrium outcome of this mechanism passes the no envy test. Despite the fact that an agent demanding a higher quantity of output pays a greater "per-unit price" than an agent with a smaller demand, he would not want to swap his allocation with anyone else's: $u_i(y_i, x_i) \geq u_i(y_j, x_j)$ for all i, all j. Interestingly, the equilibrium outcomes of the average-cost and average-return mechanisms typically *do* generate envy (and yet everyone pays the same "unit price").

Consider the familiar example with quadratic costs and three agents with linear utilities analyzed successively in Examples 6.8, 6.11, and 6.13. Agent Medium envies High in both the average-return equilibrium [as $u_M(13.2, 8.5) = 1.87 < 4.77 = u_M(33.1, 21.25)$; see Example 6.8] and in the average-cost equilibrium [as $u_M(12, 7) = 2.67 < 6.67 = u_M(30, 17.5)$; see Example 6.11], but he is not envious at the serial equilibrium [as $u_M(12.5, 4.98) = 5.86 > 4.37 = u_M(21.5, 12.85)$; see Example 6.13], nor is any other agent envious in any of these three equilibria. The situation of Example 6.12 is similar, with agent 2 envying agent 1 in both the average-cost and average-return equilibrium. In fact, in *any* economy with linear utilities and increasing marginal costs, an active agent in the average-cost (or average-return) equilibrium is always envious of any agent demanding more than she does. (Exercise: prove this claim.)

By contrast, under nondecreasing marginal costs and convex preferences, the equilibrium of the serial mechanism always passes the no envy test. The construction of the serial equilibrium outcome in Figure 6.4 should make this property clear to the reader in the case of two agents. The proof for an arbitrary number of agents is omitted (see Moulin and Shenker [1992]).

To repeat the normative properties of the serial equilibrium outcome: it passes the unanimity, stand alone, and no envy tests (see Lemma 6.7); by contrast, the average-cost (or -return) equilibrium outcome passes only the stand alone test. Furthermore, the cost share of an arbitrary coalition S is never lower than its stand alone cost, a property also shared by the average-cost (or -return) equilibrium outcome. It can even be shown that when preferences are binormal, the serial (resp. average-cost, resp. average-return) equilibrium outcome is population-monotonic. (Of course, for an inefficient solution, population monotonicity does not imply the stand alone core property.)

Remark 6.10. The Case of Decreasing Marginal Costs

For brevity, we do not discuss in any detail the serial mechanism under *decreasing* marginal costs. Yet this mechanism has remarkable strategic and normative properties in that context, too. First of all, the existence of at least one Nash equilibrium is guaranteed under convex and binormal preferences (under the same assumptions, the average-cost and average-return mechanisms may have no equilibrium whatsoever). If the equilibrium may not be unique, at least there is a Nash equilibrium that Pareto dominates all other equilibria.

Moreover, every Nash equilibrium of the serial mechanism passes the no envy test, and stand alone test (a lower bound on individual utilities), and the unanimity test (*upper* bound on individual utilities). See Moulin [1993].

6.7. SERIAL COST-SHARING OF PARTIALLY EXCLUDABLE PUBLIC GOODS

A public good is *partially excludable* if different agents can enjoy it simultaneously at different levels. A patent excludes some agents from the benefit of the invention. When different radio stations are allowed to broadcast different fractions of the same program, we have partial exclusion. Other examples include cable TV and noncongested parks.

If agent i enjoys the excludable public good at the level y_i, the total cost to be covered is simply $c(\max_{i=1,\dots,n} y_i)$.

The serial mechanism is easily adapted to the present context. Say that agent i demands to consume the good at the level y_i, and assume (without loss of generality) $y_1 \le y_2 \le \cdots \le y_n$. Then the individual cost shares are computed as follows:

$$x_1 = \frac{c(y_1)}{n}, \qquad x_2 = \frac{c(y_2)}{n-1} - \frac{c(y_1)}{n(n-1)},$$

$$x_i = \frac{c(y_i)}{n-i+1} - \sum_{k=1}^{i-1} \frac{c(y_k)}{(n-k+1)(n-k)} \tag{22}$$

(note the formal analogy with formula (18)). There is also an output-sharing version of the mechanism, defined by precisely the same formula, where we switch the roles of c and f and those of y_i and x_i. Thus the agents select their input contributions and are allowed by the mechanism to consume only a fraction of the total public good produced. As in the precious section, this mechanism yields always the same equilibrium outcome as the cost-sharing one (see Exercise 6.16) and will be omitted from the discussion below.

Just as in the private good case, the above serial mechanism improves upon unanimity voting. Consider the case of a *binary* public good as in Example 6.1. Assume first the three agents have the following utilities for the public good:

$$b_1 = 2, \qquad b_2 = 6, \qquad b_3 = 7, \qquad c = 10.$$

The public good is not produced under unanimity voting but it will be under serial cost-sharing: agent 1 is excluded and agents 2 and 3 share equally its cost, thus raising 3 units of surplus. Note that majority voting efficiently raises 5 units of surplus in this example, but to do so it must force the participation of agent 1. The serial mechanism respects not only voluntary participation, but also the stand alone test. Consider the utility profile

$$b_1 = 2, \qquad b_2 = 4, \qquad b_3 = 12, \qquad c = 10.$$

This time, both agents 1 and 2 refuse to pay for the good in equilibrium (agent 2 would contribute at the cost $c/3$, but as agent 1 is out, he does not want to pay $c/2$). Under serial cost-sharing, they are excluded from the good, but agent 3 pays for his (own!) public good and enjoys 2 units of surplus. Here the serial equilibrium is Pareto superior to the majority equilibrium.[53]

In the probabilistic models of Example 6.2, we find that the expected surplus raised by the serial equilibrium given the distribution (1) is $(0.486)c$, or 180% of the expected surplus from unanimity voting (and 90% of the surplus from majority voting); we omit the details of this computation.

Figure 6.5 depicts the serial equilibrium outcome (as well as the unanimity and majority equilibrium outcomes) in a three-agent economy with a divisible public good and constant marginal cost. Here the serial outcome is Pareto superior to the unanimity outcome but not to the majority outcome.

We turn to our familiar example with three linear utility functions and quadratic costs.

---

[53] Clearly, in the binary example, the majority equilibrium cannot be Pareto superior to the serial equilibrium, although it may raise more surplus, as we just saw; this holds true whether the public good is divisible or not. (Exercise: why?)

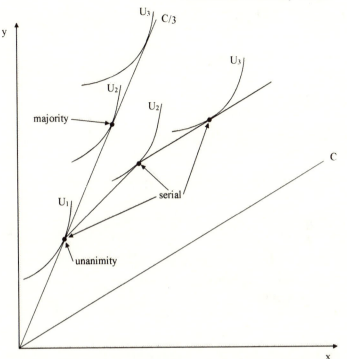

Fig. 6.5. The serial equilibrium outcome for a partially excludable public good.

Example 6.15. *Examples 5.8, 6.4, and 6.6 Continued*

The cost of producing the public good is $c(y) = y^2/8$, and the three utility functions are

$$u_1 = \tfrac{1}{6}y - x_1, \qquad u_2 = \tfrac{3}{4}y - x_2, \qquad u_3 = y - x_3.$$

Agent 1 has the lowest unanimity demand:

$$\max_{y \geq 0} \left(\frac{1}{6}y - \frac{1}{3} \cdot \frac{y^2}{8} \right) \Rightarrow y_1 = 2, \qquad x_1 = \frac{1}{6}.$$

Since agent 1 does not contribute anymore to the provision of $y \geq y_1$, agent 2's optimal contribution is computed as follows:

$$\max_{y \geq y_1} \left\{ \frac{3}{4}y - \frac{1}{2} \left(\frac{y^2}{8} - \frac{1}{6} \right) \right\} \Rightarrow y_2 = 6, \qquad x_2 = \frac{13}{6}.$$

Finally, agent 3 must pay the full incremental cost beyond 6 units of public good:

$$\max_{y \geq y_2} \left\{ y - \left(\frac{y^2}{8} - \frac{7}{3} \right) \right\} \Rightarrow y_3 = 6, \qquad x_3 = \frac{13}{6}.$$

The corresponding surplus distribution

$$\pi_1 = 0.17, \qquad \pi_2 = 2.33, \qquad \pi_3 = 3.83 \tag{23}$$

dominates (as always) the unanimity voting one ((6)) and is more equitable than the majority voting surplus distribution ((7)): total surplus is higher in the latter but agent 1's participation is not voluntary. The distribution (23) brings (slightly) more surplus than the voluntary contribution equilibrium ((19)) and seems fair too: agent 3 does get a larger share of surplus than agent 2.

Lemma 6.9. (Moulin [1994]). Suppose preferences are convex and marginal cost of the (excludable) public good is nondecreasing. Then the serial cost-sharing mechanism with partial exclusion has a unique strong equilibrium.[54] *The equilibrium outcome*

(i) never involves overproduction of the public good.

(ii) is Pareto superior (or Pareto indifferent) to the unanimity voting equilibrium (without exclusion).

(iii) passes the stand alone test (Section 5.6).

(iv) passes the unanimity test (see Lemma 5.7 in Section 5.8).

(v) passes the no envy test.

Once again, the serial equilibrium outcome passes the three equity test around which the whole normative discussion of first-best solutions (in Sections 5.6 to 5.8) is organized.

Note finally that the voluntary contribution mechanism may raise, in equilibrium, less surplus than serial cost-sharing (as in Example 6.14) or more surplus (as in the economy of Example 6.5; see Exercise 6.1). In fact, the serial equilibrium may be Pareto superior, but may not be Pareto inferior, to any voluntary contribution equilibrium. See Exercise 6.6.

---

[54] See note 45.

APPENDIX TO CHAPTER 6

A6.1. Strategy-proof Voting in the Single-Peaked Context (Moulin [1980], Sprumont [1991])

The set A of possible outcomes is *either* $[0, 1]$ *or* a finite subset of $[0, 1]$ containing 0 and 1. The set of single-peaked preferences is denoted SP(A) (Definition 6.1). Consider first a one-person voting rule. To each u in SP(A), it associates an outcome $x = f(u)$ in A. The strategy-proofness property reads

$$u(f(u)) \geq u(f(u')) \quad \text{for all } u, u' \in SP(A).$$

Denote by A_f the range of f. Strategy-proofness amounts to

$$\text{for all } u, \quad \text{all } a \in A_f: \quad u(f(u)) \geq u(a).$$

From now on, we shall restrict ourselves to voting rules of which the range is an interval of A. Given a single-peaked preference u with peak a, and an interval $A_f = [\alpha, \beta]$, the best outcome in A_f for u is simply the projection of a on $[\alpha, \beta]$, namely,

$$\text{proj}(a; [\alpha, \beta]) = \min\{\beta, \max(a, \alpha)\}.$$

Summarizing, a one-person strategy-proof voting rule on SP(A), of which the range is an interval, elicits only the peak of one's preference and enforces the projection of this peak on the said interval. Note that if A_f is not an interval, knowing the peak of u is not enough to determine its maximum on A_f.

We turn to two-person voting rules. Let f be strategy-proof and its range be an interval. Fix a preference u_2 in SP(A): the one-person voting rule $u_1 \to f(u_1, u_2)$ is clearly strategy-proof. It can be proven that its range must be an interval $[\alpha(u_2), \beta(u_2)]$ (see Sprumont [1991]). From the above discussion, f takes the form

$$f(u_1, u_2) = \min\{\beta(u_2), \max(a(u_1), \alpha(u_2))\}, \qquad (24)$$

where $a(u_1)$ is the peak of u_1 and where $\alpha(u_2) \leq \beta(u_2)$. A similar argument exchanging the roles of u_1 and u_2 yields

$$f(u_1, u_2) = \min\{\beta'(u_1), \max(a(u_2), \alpha'(u_1))\}, \qquad (25)$$

with $\alpha'(u_1) \le \beta'(u_1)$ for all u_1. Equating the two dual forms of f for the cases $a(u_2) = 1$ and $a(u_2) = 0$, respectively, gives

$$a(u_2) = 1 \Rightarrow \beta'(u_1) = \min\{\beta(1), \max(a(u_1), \alpha(1))\},$$
$$a(u_2) = 0 \Rightarrow \alpha'(u_1) = \min\{\beta(0), \max(a(u_1), \alpha(0))\},$$
(26)

with the abuse of notation identifying the preference with peak at 1 (resp. 0) and the outcome 1 (resp. 0).

As $\alpha' \le \beta'$, we get $\alpha(0) \le \alpha(1)$ and $\beta(0) \le \beta(1)$. Combining (25) and (26) and using the distributivity of max with respect to min, we get for all u_1, u_2:

$$f(u_1, u_2) = \min\{\beta(1), \max(a(u_1), \alpha(1)),$$

$$\max(a(u_2), \beta(0)), \max(a(u_1), a(u_2), \alpha(0))\}. \quad (27)$$

The above formula where the four parameters like $\alpha(0)$ are chosen arbitrarily in A, except for $\alpha(0) \le \beta(0)$, $\alpha(1) \le \beta(1)$, is the fully general form of strategy-proof two-person voting rules of SP(A) such that their range is an interval.

To interpret this result, we add the mild assumption of nonimposition. Call a voting rule "nonimposed" if its range equals A (for instance, any voting rule respecting the unanimous preferences of the agents—choosing a if a is the common peak of all agents—is nonimposed). In formula (27), nonimposition means $\beta(1) = 1$ and $\alpha(0) = 0$, so the rule takes a simpler form

$$f(u_1, u_2) = \min$$

$$\{\max(a(u_1), \alpha(1)), \max(a(u_2), \beta(0)), \max(a(u_1), a(u_2))\}. \quad (28)$$

This rule respects the unanimous preference of the agents (if $a(u_1) = a(u_2) = a$, then $f(u_1, u_2) = a$). On the other hand, agent 1 can guarantee an outcome not higher than $\alpha(1)$ by reporting $\alpha(1)$ or less, and not smaller than $\beta(0)$ by reporting $\beta(0)$ or more. If we further require that the voting rule treats both agents equally (anonymity), then $\alpha(1) = \beta(0) = \alpha$ and the above rule boils down to

$$f(u_1, u_2) = \text{median}\{a(u_1), a(u_2), \alpha\}. \quad (29)$$

Note that for $\alpha = 0$, the rule is $\min\{a(u_1), a(u_2)\}$ or simply unanimity voting (where the rule is biased in favor of lower levels of public good), and for $\alpha = 1$, we get the rule $\max\{a(u_1), a(u_2)\}$.

Exercise 6.22 analyzes the family of strategy-proof voting rules on SP(A) for an arbitrary number of agents n (generalizing the formulas (27), (28), and (29)).

A6.2. The Gibbard–Satterthwaite Theorem

For a finite set of outcomes A, denote by $L(A)$ the set of linear orderings of A, (i.e., preference orderings without indifferences). Given the set N of agents, a voting rule associates to each preference profile $(u_1, \ldots, u_n)$ an outcome $f(u_1, \ldots, u_n) = a$. The voting rule is strategy-proof if, for all profile $u = (u_1, \ldots, u_n)$ in $L(A)^N$, all agent i, and all preference u'_i in $L(A)$, we have

$$u_i(f(u)) \geq u_i\left(f\left(u \mid^i u'_i\right)\right).$$

*Theorem. (Gibbard [1973], Satterthwaite [1975]). Suppose the voting rule f contains at least three outcomes in its range. Then f is strategy-proof (over the domain $L(A)^N$) if and only if f is **dictatorial**. We say that f is dictatorial if there exists an agent i such that the voting rule selects agent i's top outcome in A at all profiles.*

An easy proof in the case of two agents is in Barbera and Peleg [1980], who also present a topological version of this result. For an exposition of the Gibbard–Satterthwaite theorem emphasizing the connections with Arrow's theorem, see Moulin [1988].

A6.3. Strategy-Proof Voting and Condorcet Winners: The Case of Multiple Public Goods

1) SEPARABLE PREFERENCES

The main restrictive assumption in the single-peaked context is the fact that outcomes are distributed along a single dimension. Examples include the choice of the drinking age or of the level of public spending on a specific public good. Yet most real-life voting situations deal with several public goods at a time (as when congressmen vote over the budget, or when citizens elect representatives, who will represent them on more than one issue).

Suppose that the outcome y to be decided upon is a vector where each component y_k is a real number (think of K different public goods, and assume that the total cost of y is divided equally among participants). Say that agents i's preferences over the range of y are *separable* if his preferences between two levels y_k and y'_k of the kth coordinate do not depend upon the level of production of other goods.[55] Then, if all individual preferences are separable and single-peaked (in each coordinate), the coordinatewise majority voting rule (enforcing the median peaks in each coordinate) is strategy-proof, and so is the (coordinatewise) k-rank voting, or indeed any mixed mechanism such as majority voting in some coordinates and unanimity voting in some others. It is even possible to define a larger preference domain including some nonseparable preferences where coordinatewise strategy-proofness is logically equivalent to overall strategy-proofness. See Exercise 6.22 reporting on the work of Barbera, Gül, and Stacchetti [1993].

Yet in the case of separable and single-peaked preferences, the core stability property of majority voting (or, indeed, of any strategy-proof voting rule) is lost. In other words, the coordinatewise Condorcet winner is not an overall Condorcet winner; hence, direct majority voting is strategy-proof against individual, but not coalitional, deviations. In our terminology, majority voting (or any combination of coordinatewise strategy-proof mechanisms) works in the decentralized mode but not in the direct agreement mode. This important limitation is illustrated by an example with two public goods and separable preferences.

The two goods are produced by a (separable) linear technology $c(y_1, y_2) = y_1 + y_2$. Let a be a strictly concave utility function such that $a(0) = 0$, $a(6) = 3$, $a'(6) = \frac{1}{3}$, (e.g., $a(\lambda) = (3\lambda^2/4)^{1/3}$), and consider a society of three agents with the following separable, concave, and quasi-linear utilities for the public goods:

$$u_1(y_1, y_2) = a(y_1),$$

$$u_2(y_1, y_2) = a(y_2),$$

$$u_3(y_1, y_2) = a(y_1) + a(y_2).$$

Consider majority voting over the first public good only (assume a fixed level y_2 of the second public good; by separability of costs and of utilities, the particular value of y_2 is irrelevant). Agents 1 and 3 vote for the level $y_1 = 6$ (by construction of a, the function $a(y_1) - y_1/3$ peaks at $y_1 = 6$) and agent 2 votes for $y_1 = 0$ (he does not care for public good 1): the majority winner is $y_1 = 6$. Similarly, voting for the second public good only (at any fixed level y_1) yields a majority winner $y_2 = 6$

---

[55] This means: for all outcomes z and z', agent i prefers $(z \, |^k y_k)$ to $(z \, |^k y'_k)$ if and only if he prefers $(z' \, |^k y_k)$ to $(z' \, |^k y'_k)$.

(supported by agents 2 and 3). Yet $y = (6, 6)$ is not a Condorcet winner, as agents 1 and 2 both prefer no production at all:

$$u_i(6, 6) - \frac{c(6, 6)}{3} = a(6) - 4 = -1 < u_i(0, 0) \quad \text{for } i = 1, 2.$$

In fact, the core of majority voting is empty at this profile. (Exercise: why?)

2) SPATIAL VOTING

We maintain the assumption that an outcome is a vector $y = (y_1, \ldots, y_K)$. The coordinate y_k is interpreted as the level of provision of a public good or, more generally, an issue in the political debate (see Riker [1982]). Individual preferences take a very specific form: each voter has a certain ideal point (a certain outcome y^i) and uses the distance between the actual outcome and his ideal point to measure the disutility of the outcome in question. A key simplifying assumption is that each voter uses the same distance. Note that if this distance is the euclidean metric, individual preferences are also separable.

For instance, take $K = 2$, so an outcome is a point in the plane. Suppose that four voters use the euclidean distance, and denote agent i's ideal point as a_i. Figure 6.6 is an example with four voters where the Condorcet winner x is simply the intersection of $[a_1, a_3]$ and $[a_2, a_4]$. To

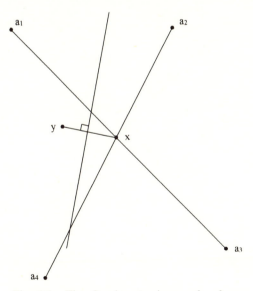

Fig. 6.6. The Condorcet winner of a four-person spatial voting game.

see this, observe that every line through x defines two half-spaces (excluding the line itself) that never contain more than half of the ideal points. If x were dominated by another outcome y, the median between x and y (the line perpendicular to $[x, y]$ through its midpoint) would leave three or more ideal points in the half-space containing y, hence the line parallel to the median through x would leave three or more ideal points in one of its half-spaces.

By contrast, Figure 6.7 is a typical five-voter situation with *no* Condorcet winner. To see this, consider a point x outside the line a_1a_2 and draw the parallel to this line through x: surely one of the half-spaces that it defines contains three or more ideal points. Then pick a point y on the perpendicular through x in the direction of the "densely popu-lated" half-space: if y is close enough to x, it will surely majority dominate x. Thus a Condorcet winner must be on the a_1a_2 line, and by a similar argument, it must be on the a_3a_4 line as well; but the intersection z of these two lines is dominated by a_5.

The situation of Figures 6.6 and 6.7 generalizes. In a two-dimensional outcome space, a Condorcet winner may or may not exist if we have an *even* number of voters, but such a core outcome will typically not exist with an *odd* number of voters. For instance, in Figure 6.7, we would need a_5 to coincide with z, but such a configuration is exceptional (it is destroyed by an arbitrary small perturbation of any one ideal point;

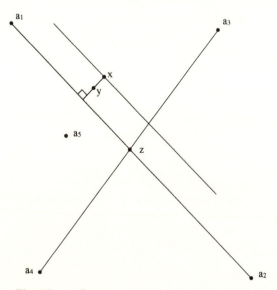

Fig. 6.7. A five-person spatial voting game with no Condorcet winner.

contrast with Figure 6.6, where the existence of a Condorcet winner is robust to small perturbations of the ideal points). General conditions relating the number of voters and the dimension of the outcome space to the existence of a Condorcet winner and its "exceptional" character (called structural instability) are discussed in McKelvey and Schofield [1987], Schofield [1985], and Banks [1994].

In the above model, the choice of the euclidean metric is fairly arbitrary. By far that most restrictive assumption is that all voters share the same distance to compare the candidate's platform to their ideal point. Most of the negative results are preserved when this assumption is relaxed.

EXERCISES ON CHAPTER 6

Exercise 6.1.   Numerical Examples of the (Divisible) Public Good Provision Problem

(a) In Example 6.5, compute the serial equilibrium outcome (assuming the good is partially excludable) and compare its surplus distribution to that of the voluntary contribution mechanism, and to that of the ratio equilibrium, of the PGE solution (Definition 5.5), and of the ω-EE solution (Lemma 5.7).

(b) Answer the same questions for Example 6.13.

(c) Answer the same questions for the following numerical specifications of Example 5.12. Five agents and the a_i-profile is successively:

$$(2, 1.5, 0.8, 0.4, 0.1),$$

$$(1.2, 0.8, 0.4, 0.3, 0.3),$$

$$(0.9, 0.8, 0.8, 0.3, 0.2).$$

Include the equilibria of unanimity voting and of majority voting in the overall comparison of surplus distributions.

Exercise 6.2.   A Public Good Problem with Explicit Formulas

The cost of the public good is quadratic: $c(y) = y^2/2$. Utilities are quasi-linear in the input:

$$u_i(y, x_i) = 2\sqrt{\lambda_i y} - x_i,$$

and we assume $0 < \lambda_1 \leq \lambda_2 \leq \cdots \leq \lambda_n$. Compute the equilibrium outcome and corresponding surplus distribution for the three mechanisms: unanimity voting, majority voting, and voluntary contribution.

Exercise 6.3.   Example 6.3 Continued (public good)

(a) In the increasing marginal cost version of Example 6.3 (cost function given by (4), utilities by (3)), compute the equilibria of the voluntary contribution mechanism. Show in particular that there is a unique efficient equilibrium outcome, as well as many inefficient ones. Check that if we reduce $u_2(2)$ by an arbitrarily small amount, the efficient equilibrium outcome disappears.

(b) Compare these equilibria with the equilibrium outcomes in majority voting, in unanimity voting, and in the serial mechanism (the latter assuming that the public good is excludable).

(c) Answer the same questions as in (a) and (b) for the decreasing marginal cost version of Example 6.3 (cost function given by (5)).

Exercise 6.4.   Binary Input Contributions to a Public Good

Each agent contributes zero or one unit of input. Agent i, $i = 1, \ldots, n$, has a disutility $v_i = i$ for contributing one unit (and zero for contributing nothing). If t agents contribute, the level $y = f(t)$ of public good results, and agent i's final utility is $f(t) - i$ if he contributed and $f(t)$ otherwise.

We assume throughout the exercise that the production function f increases, that its marginal product $\partial f(t) = f(t) - f(t - 1)$ is non-increasing, and that $\partial f(1) > 1$, $\partial f(n) < n$.

Show that in the unique equilibrium of the voluntary contribution mechanism, the first t^* agents contribute, where t^* is the largest integer such that $t \le \partial f(t)$ (for simplicity, assume throughout the exercise that an agent who is indifferent between contributing or not always contributes; thus agents are altruistic once their own utility is maximized).

(b) Show that if t^* is such that $f(t^*) < f(n) - n$, then the unanimity (as well as the majority) voting outcome is Pareto superior to the voluntary contribution outcome. Give a numerical example. Conversely, the voluntary contribution outcome dominates both voting outcomes when these voting equilibrium outcomes involves zero production. Give a numerical example.

(c) Assume now that the public good is excludable and consider the serial mechanism (Section 6.7): agent i's final utility is $f(t) - i$ if she contributes and zero otherwise, Show that this mechanism has a Nash equilibrium of which the outcome is Pareto superior to every other Nash equilibrium outcome. In the Pareto superior equilibrium, the first $\bar{t}$ agents contribute, where $\bar{t}$ is the largest integer such that $t \le f(t)$. Show that if $f(n) \ge n$, the (Pareto superior) serial equilibrium, the unanimity, and the majority equilibrium all coincide.

(d) Assume $f(n) < n$ (so that the unanimity voting yields zero production). Assume also $f(n) > 1 + (n/2)$ (so that majority voting yields the production of $f(n)$ units of public good). Denote by $\sigma(\theta)$ the total surplus generated by the mechanism θ. Show by numerical examples that the inequality $\sigma(\theta) > \sigma(\theta')$ is possible, when θ, θ' are any two mechanisms among serial, voluntary contribution, and majority.

Exercise 6.5. Example 6.6 Continued (public good)

The public good is produced by a quadratic cost function and utilities are linear. We use the notations of Example 6.6, and set $a = 1$ without loss of generality.

(a) Show that $\sigma^{\text{una}} \leq \sigma^{\text{vc}}$ holds true always.

(b) Show that $\sigma^{\text{vc}} - \sigma^{\text{una}}$ has the same sign as $\lambda_n - n\lambda_1$ (in particular they are both zero or both nonzero).

(c) Denote the median by $\lambda_{n'}$ (where $n = 2n' - 1$). Show the following properties:

$$\sigma^{\text{vc}} \leq \sigma^{\text{maj}} \Leftrightarrow \frac{\lambda_n}{n} \leq \lambda_{n'},$$

Hint: the following inequality holds true

$$\frac{1}{2}(\lambda_n + n\lambda_{n'}) \leq \lambda_N$$

(d) Show that if $(\lambda_n/n) < (\lambda_1/2)$, the unanimity equilibrium is Pareto superior to the voluntary contribution equilibrium.

(e) Show that if $2\lambda_1 < (\lambda_n/n)$, the voluntary contribution equilibrium is Pareto superior to the unanimity equilibrium. Show that, in fact, the slightly weaker constraint

$$\left(n + \sqrt{n^2 - n}\right) \cdot \lambda_1 < \lambda_n$$

implies the announced Pareto comparison.

(f) Show that if $2\lambda_{n'} < (\lambda_n/n)$, the voluntary contribution equilibrium is Pareto superior to the majority equilibrium and to the unanimity equilibrium as well.

(g) Show that if $(\lambda_n/n) < (\lambda_{n'}/2)$, the majority equilibrium is Pareto superior to the voluntary contribution equilibrium. In fact, this property holds if we have

$$\lambda_n < \tfrac{1}{2}\left(n + \sqrt{n^2 - 2n} \right) \cdot \lambda_{n'}.$$

Exercise 6.6. Pareto Ranking of the Voluntary Contribution and Serial Mechanisms (public good)

Assume the public good is divisible and produced with nondecreasing marginal costs. Assume individual preferences are strictly convex.[56]

(a) Consider a unanimity profile where the (common) utility is $u_i(y, x_i) = y - v(x_i)$ and $c(y) = y$. Show that the serial equilibrium outcome (Section 6.7) is Pareto superior to the unique voluntary contribution equilibrium outcome (uniqueness of this equilibrium is the subject of question (a) in Exercise 6.18). Give an example of a unanimity profile with $u_i(y, x_i) = v(y) - x_i$ where the serial equilibrium outcome is Pareto superior to all voluntary contribution equilibrium outcomes (described in question (b) of Exercise 6.18).

(b) With a nonanonymous preference profile, show that a voluntary contribution equilibrium outcome may not be Pareto inferior to the serial equilibrium outcomes. However, a voluntary contribution outcome cannot be Pareto superior to the serial equilibrium outcome.

Exercise 6.7. Public Good Production with Leontief Utilities

In this exercise and the following, a Leontief utility is characterized by a continuous decreasing function h with domain $[0, \bar{x}]$ and range $[0, h(0)]$ (where both $\bar{x}$ and h are positive). To such a function h we associate the Leontief utility function

$$u(y, x) = \min\{y, h(x)\}, \quad \text{for all } x \in [0, \bar{x}], \quad \text{all } y \geq 0.$$

We denote by k the inverse function of h (with domain $[0, h(0)]$ and range $[0, \bar{x}]$). Fix the cost function $c(y) = y$ and a profile of n Leontief utilities. Throughout the exercise, it will be useful to draw figures.

(a) Show that the unanimity demand of agent i is the (unique) solution of $y_i = h_i(y_i/n)$.

(b) Consider the unique solution y^* of the equation $\sum_{i=1}^{n} k_i(y) = y$. Show that $x_i = k_i(y^*)$ (for all i) is the unique equilibrium of the voluntary contribution mechanism and results in the production of y^* units of public good. Show that this equilibrium is fully efficient. Show

---

[56] Alternatively, we may assume convex preferences and make an equilibrium selection assumption (e.g., all agents choose the highest level of public good when indifferent).

that it is Pareto superior to the unanimity equilibrium. Give a three-agent example where the voluntary contribution equilibrium is Pareto superior to the majority equilibrium. (*Hint:* Make sure more public good is produced in the former than in the latter.) Give an example where none of these equilibria is Pareto superior to the others and yet all three equilibrium outcomes are different. Show that the majority equilibrium cannot be Pareto superior to the voluntary contribution equilibrium.

(c) Assume the public good is partially excludable. Show that the serial equilibrium can be neither Pareto superior nor Pareto inferior to the voluntary contribution equilibrium.

Exercise 6.8. Private Good Production with Leontief Utilities

Fix a continuous cost function c with $c(0) = 0$ and $c(\infty) = \infty$. Note that we place no restriction on the marginal cost. Fix a profile of n Leontief utilities (as defined in the previous exercise). As in Exercise 6.7, most arguments can be made with the help of figures.

(a) Show that the equilibrium outcomes of the average-return and average-cost mechanisms coincide and are fully efficient. (*Hint:* Denote as $x_i = \theta_i(\lambda)$ the unique solution of $\lambda x = h_i(x)$ for all $\lambda \geq 0$.) Show that $\theta_i(\lambda)$ is continuous and decreasing in λ and that $\lambda \cdot \theta_i(\lambda)$ is increasing in λ. Then solve

$$c\left(\sum_{i=1}^{n} \lambda \cdot \theta_i(\lambda) \right) = \sum_{i=1}^{n} \theta_i(\lambda).$$

(b) Compute the average-return (-cost) equilibrium as well as the serial equilibrium in the six numerical specifications of questions (a) and (b) in Exercise 5.9. Compare these equilibria with the first-best solutions computed in Exercise 5.9.

(c) Show that the serial equilibrium outcome is always fully efficient.

Exercise 6.8. Free Access to a Public Bad (Remark 6.5)

Given a private good output Y and a public bad input X, we denote by g the technology to produce the public bad: if total output consumed is y, each agent must consume $x = g(y)$ units of public bad. In the free-access mechanism, each agent selects the amount of output y_i she wishes to consume and the resulting public bad $g = (\sum_i y_i)$ is inflicted upon all agents.

(a) Suppose g is quadratic, $g(y) = y^2/2$, and utilities are linear, $u_i(y_i, x) = y_i - \mu_i x$. Compute the equilibrium of the free-access mechanism: show in particular that (if the parameters μ_i are all different) only one agent consumes a positive quantity of output, and all other agents suffer a net loss. Characterize the profiles of preferences such that the total equilibrium surplus is negative.

(b) Suppose g has increasing marginal cost and that all agents have Leontief preferences (see Exercise 6.7). Show that the equilibrium of the free-access mechanism is efficient and describe it.

Exercise 6.10. Two Examples with Indivisible Private Output

(a) Consider Example 5.1 where four agents share an increasing marginal cost technology for producing an indivisible private output:

$$\partial c(y) = 3 \quad 1 \le y \le 4,$$
$$\partial c(y) = 7 \quad 5 \le y \le 8,$$
$$\partial c(y) = 11 \quad 9 \le y \le 12.$$

The utilities are given in Example 5.1. Compute the outcome and surplus distribution in the following mechanisms:
  (i) unanimity voting,
  (ii) majority voting where the largest Condorcet winner prevails,
  (iii) average cost-sharing,
  (iv) serial cost-sharing.
Check the following Pareto rankings: serial outcome dominates unanimity, majority, and average-cost outcomes; serial outcome is dominated by the CEEI and equilibrium surplus is (approximately) 80% of total efficient surplus under unanimity or majority, 93% under average-cost, and 97% under serial. What about the equilibrium of the equal-cost-shares mechanism (where each agent chooses a demand and total cost is shared equally) discussed in Remark 6.5 and Exercise 6.9?

(b) Our next example has three agents and the following cost function:

$$\partial c(y) = 1 \quad \text{for } y = 1, 2, 3,$$
$$\partial c(y) = 2 \quad \text{for } y = 4, 5, 6,$$
$$\partial c(y) = 2.5 \quad \text{for } y = 7, 8,$$
$$\partial c(y) = 3 \quad \text{for } y = 9, 10,$$
$$\partial c(y) = \frac{y}{2} - \frac{3}{2} \quad \text{for } y \ge 11.$$

The (quasi-linear) utilities are

y	1	2	3	4	5	6	7	8	9
∂u_1	6	5	4	3	2.5	2.5	2	1	0
∂u_2	5	4	3	2.5	2.5	2	1	0	0
∂u_3	4	3	2.5	2.5	2	1	0	0	0

Answer the same questions as in (a). Moreover, compute the CEEI solution and list all Pareto comparisons between the various outcomes.

Exercise 6.11. Average-Cost (-Return) Equilibrium in Two Increasing Marginal Cost Examples (private goods)

(a) Consider the fishing model of Example 6.7, with the numerical specification $n = 3$, $b = 4$, $c_1 = c_2 = 2$, $c_3 = 6$. Compute the equilibrium outcome of the average-return mechanism. Use $c(y) = 4 - \sqrt{16 - y}$ for $y \le 16$, and $+\infty$ if $y > 16$. Compute that of the serial mechanism. Compare the surplus distributions in all three mechanisms (serial, average-cost, and average-return). Check that total surplus is highest for average-cost (98.5% of efficient surplus) and lowest for serial. (*Hint:* To compute the average-cost equilibrium, write the first-order equilibrium conditions. Find a linear combination of these equations giving a simple equation in total output y.)

Compute the CEEI solution, the ω-EE, and the CRE solutions, and compare them to the three equilibrium outcomes.

(b) We generalize Example 6.12, keeping the same technology and two agents with arbitrary linear utilities $u_i = y_i - a_i \cdot x_i$. We assume $1 < a_i < 2$, $i = 1, 2$. Show that the average-cost mechanism has a unique equilibrium such that $y_1 + y_2 > 2$ if and only if $(1/a_1) + (1/a_2) > 3/2$. Similarly, the average-return mechanism has a unique equilibrium such that $x_1 + x_2 > 1$ if and only if $a_1 + a_2 < 3$. Assuming that both of these inequalities hold, compute total equilibrium surplus for both mechanisms (as a function of $s = a_1 + a_2$ and $p = a_1 + a_2$) and check that it is always higher for the average-cost mechanism, and yet the average-cost equilibrium is not always Pareto superior to the average-return equilibrium.

Exercise 6.12. A Variant of Example 6.9 (private good)

The production function is now

$$f(x) = \min\left\{\frac{x}{\alpha}, \epsilon(x - \alpha) + 1\right\}.$$

All n agents, $n \geq 3$, have the same utility $u_i = \lambda \cdot y_i - x_i$, where $\alpha < \lambda < 1/\epsilon$. Show that the average-return equilibrium involves overproduction ($x_N > \alpha$) if and only if

$$\lambda > \frac{n\alpha}{n - 1 + \alpha\epsilon},$$

and yields the total surplus $\pi_N = (1 - \alpha\epsilon) \cdot \lambda/n$, to be compared with the efficient surplus $v(N) = \lambda - \alpha$.

Show that the average-cost equilibrium(s) involve the efficient level of production ($x_N = \alpha$) if and only if

$$\lambda \leq \frac{1}{n\epsilon} + \frac{n - 1}{n} \alpha.$$

When this inequality is violated, compute the total surplus of the (unique) average-cost equilibrium and compare it with that of average-return.

Exercise 6.13. Generalization of Examples 6.10 and 6.11 (private goods)

The cost function is quadratic, $c(y) = y^2/2$, and utilities are linear:

$$u_i = y_i - \mu_i \cdot x_i, \qquad i = 1, \ldots, n.$$

(a) Show that all agents are active ($x_i > 0$) in the average-return equilibrium if and only if

$$(n - \tfrac{1}{2}) \cdot \mu_i < \mu_N \quad \text{for all } i = 1, \ldots, n.$$

In this case, show that the equilibrium outcome is given by the formulas

$$x_N = 2 \frac{(n - \tfrac{1}{2})^2}{\mu_N^2}, \qquad \frac{x_i}{x_N} = 2\left(1 - (n - \tfrac{1}{2})\frac{\mu_i}{\mu_N}\right), \qquad \frac{y_i}{x_i} = \frac{\mu_N}{(n - \tfrac{1}{2})},$$

and total surplus is given by

$$\pi_N = 4 \frac{(n - \tfrac{1}{2})}{\mu_N} \cdot \left(\left(n - \frac{1}{2}\right)^2 \cdot \frac{\sigma}{(\mu_N)^2} - (n - 1)\right), \quad \text{where } \sigma = \sum_i \mu_i^2.$$

(b) Show that all agents are active $(y_i > 0)$ in the average-cost equilibrium if and only if

$$\frac{\lambda_N}{n+1} < \lambda_i \quad \text{for all } i, \quad \text{where } \lambda_i = \frac{1}{\mu_i}.$$

In this case, show that the equilibrium outcome is given by the formulas

$$y_N = \frac{2}{n+1}\lambda_N, \quad y_i = 2\left(\lambda_i - \frac{\lambda_N}{n+1}\right), \quad \frac{x_i}{y_i} = \frac{\lambda_N}{n+1}.$$

Compute the corresponding total surplus and compare it to that of the average-return equilibrium.

(c) In the case $n = 2$, show that total surplus under average-cost is never smaller that total surplus under average-return (prove this for *all* pairs of nonnegative numbers). Show that this property does not hold anymore for $n \geq 3$.

Exercise 6.14. Comparing Average-Return and Equal-Output-Shares (private good)

The private good is produced by the linear technology $c(y) = y$ and individual utilities are quasi-linear in output:

$$u_i(y_i, x_i) = y_i - \frac{1}{2\lambda_i}x_i^2.$$

We assume $0 < \lambda_1 \leq \lambda_2 \leq \cdots \leq \lambda_n$.

Compute the unique equilibrium outcome of the average-return mechanism, and that of the equal-output-shares mechanism (Section 6.4). Show that the former always brings more joint surplus than the latter. Show that the former is Pareto superior to the latter if we have

$$\frac{2n}{n^2+1} \cdot \lambda_e \leq \lambda_1, \quad \text{where } \lambda_e = \frac{\lambda_N}{n}.$$

Thus, with $n = 10$, we need that any coefficient λ_i be no smaller than one-fifth of average, with $n = 100$ no smaller than one-twentieth of average, and so on.

Exercise 6.15. Sketch of the Proof of Lemma 6.6 (private good)

The production function f and its inverse, the cost function c, are differentiable, their derivative is strictly positive, and marginal cost is increasing. We fix the number of agents. For a given allocation $z = (y, x)$, we denote by Γ_z the opportunity set through z in the average-return mechanism. Denote by δ the unique solution of

$$y = \frac{x}{x + \delta} \cdot f(x + \delta).$$

The Γ_z is defined by

$$z' = (y', x') \in \Gamma_z \Leftrightarrow y' = \frac{x'}{x' + \delta} \cdot f(x' + \delta).$$

Define similarly by Δ_z the opportunity set through z in the average-cost mechanism.

(a) Show that the slope (dy/dx) of Γ_z at z is larger than the slope of Δ_z at z.

(b) Show statement (iii) in Lemma 6.6: if preferences are (binormal and) unanimous, the average-cost equilibrium is Pareto superior to the average-return equilibrium.

Exercise 6.16. The Two Serial Mechanisms Have the Same Equilibrium Outcome (private good)

Consider the serial output-sharing mechanism among two agents and denote by Γ_z the corresponding opportunity set through z. For a given allocation $z = (y, z)$, show that there is at most one number δ such that, if the two agents contribute x and δ respectively, agent 1 receives y units of output. Then Γ_z is defined as the set of allocation feasible for agent 1 as long as agent 2's strategy δ is fixed. A similar definition yields the opportunity set Δ_z through z in the serial cost-sharing mechanism.

Fix an increasing, continuous, and invertible cost function c and an allocation z. Show that only two cases are possible:

(i) None of the opportunity sets Γ_z, Δ_z can be defined.

(ii) Both can be defined and they coincide.

Deduce that both mechanisms have precisely the same equilibrium outcomes. Note that this property holds for arbitrary preferences and arbitrary marginal cost (increasing, decreasing, or whatever).

Exercise 6.17. *Voluntary Contribution Games with No Nash Equilibrium (public goods)*

(a) CASE OF INDIVISIBLE PUBLIC GOODS

The public good can be produced at the level 0, 1, or 2 at a cost of 0, 1, or 2, respectively. Two agents have quasi-linear utilities for the good:

Public Good Level	u_1	u_2
0	0.0	0
1	0.3	2
2	1.6	2.5

Show that the voluntary contribution game has no Nash equilibrium. (*Hint:* Without loss of generality, we can restrict individual strategies to $x_i = 0$, 1, or 2.) Check that the same property happens if agent 1's preferences satisfy

$$u_1(2,1) > u_1(1,0) > u_1(0,0) > u_1(1,1), u_1(2,2).$$

Thus the lack of equilibrium could occur even when both agents have convex preferences.

(b) CASE OF DIVISIBLE GOODS

Fix the technology $f(x) = x$ and consider the voluntary contribution game among two agents. Construct agent 1's preferences in such a way that his best-reply strategy to the contribution x_2 by agent 2 is $br_1(x_2) = x_2$. This can be done with convex preferences for agent 1. Construct agent 2's preferences (necessarily nonconvex) in such a way that his best reply "jumps down":

$$br_2(x_1) = 10 \text{ if } x_1 < 8, \quad br_2(8) = \{10, 0\}, \quad br_2(x_1) = 0 \text{ if } x_1 > 8.$$

Exercise 6.18. *Unique Equilibrium in the Voluntary Contribution Mechanism (public good)*

The production function f is concave and increasing (with $f(0) = 0$). We discuss several assumptions on individual utilities.

(a) Suppose utilities are quasi-linear in the public good: $u_i = y - v_i(x_i)$ for all i, where v_i is strictly convex and increasing. Assume v_i and f are differentiable and define the function h_i as follows:

$$x_i = h_i(t) \Leftrightarrow v_i'(x_i) = f'(t).$$

For a level of contribution x_{-i} by agents other than i, denote by $\mathrm{br}_i(x_{-i})$ the best-reply contribution for agent i (namely, his best-reply strategy in the voluntary contribution game). Show that

$$x_i = \mathrm{br}_i(x_{-i}) \Leftrightarrow x_i = h_i(x_{-i} + x_i).$$

Show that the equation $\sum_i h_i(t) = t$ has a unique solution and deduce that the voluntary contribution game has a unique Nash equilibrium.

(b) Suppose utilities are quasi-linear in the private good: $u_i = v_i(y) - x_i$ for all i, where v_i is strictly concave. Check that the stand alone contribution x_i^* of agent i (his contribution if no one else contributes) is unique. Show that the voluntary contribution game has a unique equilibrium if the largest stand alone contribution corresponds to a single agent (e.g., $x_1^* > \max_{i \geq 2} x_i^*$): in this equilibrium, only one agent (namely, agent 1) is active. Show that if the largest stand alone contribution is positive and is achieved by two or more agents, the voluntary contribution game has a continuum of equilibria. In any equilibrium, only the agents with the largest stand alone contributions are active, and they share the cost $\max x_i^*$ in arbitrary nonnegative shares.

(c) Suppose individual preferences are binormal (see Theorem 6.2) and strictly convex: the slope dy/dx supporting the indifference contour at (y, x) is increasing in x and increasing in y. Define the best-reply function br_i as in question (a) and show (i) that br_i is decreasing, (ii) that $\mathrm{br}_i(t) + t$ is increasing. Then define a function h_i by the property

$$x_i = h_i(t) \Leftrightarrow x_i = \mathrm{br}_i(t - x_i),$$

and check that h_i is decreasing. Conclude as in question (a) that the equilibrium is unique.

Exercise 6.19.   Average-Cost (-Return) Game with No Equilibrium Under Decreasing Marginal Cost (private good)

The output comes in indivisible units and the two agents have quasi-linear utilities:

y	1	2	3	4	5	6	7	8	9	10
$c(y)$	140	168	189	203	210	217	224	231	238	245
u_1	175	217	245	263	279	279	...	...	...	...
u_2	175	202	226.5	248.5	248.5	...	...	...	...	...

Compute the best-reply correspondences in the average-cost game:

y_i	0	1	2	3	4	5
br_1	5	5	{5,3}	3	3	3
br_2	4	1	1	1	{1,4}	4

and conclude.

Construct a similar example for the average-return game with indivisible units of input.

Exercise 6.20. Overproduction Under Increasing Marginal Cost (private good)

The cost function c is strictly convex and differentiable. All agents have convex and binormal preferences.

(a) Denote by $(y_1^*, \ldots, y_n^*)$ the equilibrium (unique by Theorem 6.2) of the average-cost mechanism. We show by contradiction that the equilibrium involves overproduction. Suppose (y_i, x_i), $i = 1, \ldots, n$, is a Pareto optimal allocation Pareto superior to the equilibrium allocation and such that $\sum_i y_i = y_N > y_N^* = \sum_i y_i^*$. Denote by σ_i^* the slope (dx/dy) of the indifference contour of agent i through (y_i^*, x_i^*), and by σ that of his indifference contour through (y_i, x_i). Show $\sigma_i < c'(y_N^*)$ and $\sigma_i < \sigma$ for all i. Use binormality of preferences and the assumption $u_i(y_i, x_i) \geq u_i(y_i^*, x_i^*)$ to deduce

$$\frac{y_i}{x_i} \geq \frac{y_i^*}{x_i^*},$$

and derive a contradiction.

(b) Show the same property holds for the average-return mechanism as well as the serial mechanism.

Exercise 6.21. The Average-Cost Competitive Equilibrium (Remark 6.9) (private good)

The cost function is convex and differentiable. Assume first that all agents have quasi-linear utility $u_i = v_i(y_i) - x_i$ with v_i strictly concave and differentiable. Denote by θ_i the inverse of agents i's marginal utility for the private good:

$$y_i = \theta_i(\lambda) \Leftrightarrow \lambda = u_i'(y_i).$$

(a) Check that the efficient level of production y^* is the unique solution of

$$\sum_i \theta_i(c'(y)) = y.$$

Show that the ACCE allocation is determined by solving

$$\sum_i \theta_i\left(\frac{c(y)}{y}\right) = y.$$

If $\bar{y}$ is the (unique) solution of the above equation, the price is $p = c(\bar{y})/\bar{y}$ and agent i's demand is $y_i = \theta_i(p)$.

Show that the ACCE allocation overproduces more than the average-return (and the average-cost) equilibrium. (*Hint:* At the average-return equilibrium allocation (y_i, x_i), we have $u_i'(y_i) \geq c(y)/y$.)

(b) Show that the conclusion of question (a) (the ACCE overproduces more than the average-return or the average-cost equilibrium) holds true for arbitrary convex preferences.

Exercise 6.22. Single-Peaked Preferences on the Line and on a Lattice (Moulin [1980], Barbera, Gül, and Stachetti [1993])

(a) Suppose A is a (finite or infinite) subset of the real line and suppose the n agents are endowed with preferences $R_1, \ldots, R_n$ in SP(A). The peak of such a preference R_i is denoted by a_i. For every coalition S, $S \subseteq \{1, \ldots, n\}$, we fix an element α_S in A. We define a voting rule f as follows:

$$f(R_1, \ldots, R_n) = \min_S \left\{ \max\left(\alpha_S, \max_{i \in S} a_i \right) \right\},$$

$$\text{for all } R_1, \ldots, R_n \in \text{SP}(A). \quad (29)$$

Show that f is strategy-proof (even in the coalitional sense). Conversely, Moulin [1980] shows that every strategy-proof voting rule on SP(A) takes the form (29).

(b) Same assumptions as in question (a). Show that a strategy-proof voting rule f as described in (a) is *anonymous* (is unaffected when we permute the preferences R_i among agents) if and only if it takes the form

$$f(R_1, \ldots, R_n) = \text{median}\{a_1, \ldots, a_n; \beta_0, \beta_1, \ldots, \beta_n\}, \quad (30)$$

where β_j, $j = 0, \ldots, n$, are $(n + 1)$ fixed elements (not necessarily distinct) of A. Without loss of generality, we assume $\beta_0 \leq \beta_1 \leq \cdots \leq \beta_n$. Assume f given by (30) is *nonimposed*: f is *onto* A [i.e., $f(\mathrm{SP}(A)^n) = A$]. Then show that f writes

$$f(R_1, \ldots, R_n) = \mathrm{median}\{a_1, \ldots, a_n; \beta_1, \ldots, \beta_{n-1}\}.$$

(c) In this question, the set A is a sublattice of $\mathbb{R}^K$, namely, a closed subset stable by the coordinatewise supremum operation and by the coordinatewise infimum operation, denoted $x \vee y$ and $x \wedge y$, respectively. Denote by R a preference on A with peak a. Denote by $[x, y]$ the interval $x \wedge y \leq z \leq x \vee y$, where $\leq$ is the partial ordering of $\mathbb{R}^K$, and consider the following property of R:

$$\text{for all } x, y \in A: \qquad x \in [a, y] \Rightarrow xRy. \tag{31}$$

Finally, we say that R is *single-peaked on A* if it has a unique peak element a in A and satisfies (31). Denote by $\mathrm{SP}(A)$ the set of such preferences. Suppose f is a voting rule on $\mathrm{SP}(A)$, i.e., a mapping from $\mathrm{SP}(A)^n$ into A, denoted

$$f(R_1, \ldots, R_n) = (f_k(R_1, \ldots, R_n))_{k=1, \ldots, K}.$$

Say that f is decomposable if each mapping f_k depends only upon the kth coordinates a_{ki} of the peak of preference R_i, $i = 1, \ldots, n$. Suppose that f is a decomposable voting rule such that each voting rule f_k is strategy-proof on the kth coordinate; show that f is strategy-proof on $\mathrm{SP}(A)^n$. (*Hint:* Use the representation (29) to show that each mapping f_k satisfies $f_k(R_i) \in [a_{ki}, f_k(R_i')]$ for all R_i, R_i' in $\mathrm{SP}(A)$ (where a_{ki} is the kth coordinate of the peak of R_i).) Conversely, show that if f is strategy-proof on $\mathrm{SP}(A)^n$, and if f depends only upon the peaks a_i, $i = 1, \ldots, n$, then f must be decomposable and each partial rule f_k must be strategy-proof.

Cooperative Games

7.1. Games in Characteristic Function Form

This chapter is a brief overview of the two main concepts of the theory of cooperative games, namely, the core and the Shapley value. The technical level of the discussion is higher than in the previous chapters: our attention is focused on the main mathematical results about core existence and on the axiomatic characterization of the Shapley value. Many important (yet, in my view, secondary) concepts are not discussed at all. The most notable omissions are the second-order stability concepts (such as the Von Neumann–Morgenstern stable sets or the bargaining set) and the alternative concepts of value (such as the nucleolus or the τ-value). We refer the interested reader to Owen [1982] or Shubik [1984] for a more comprehensive survey of cooperative game theory.

The model of games in characteristic function form, originally proposed by Von Neumann and Morgenstern [1947], is the central analytical tool of cooperation by direct agreements. This should be already clear from the discussion of Chapters 2, 3 and 5: the surplus available to a coalition of agents from trading its own resources is the key to the core analysis.

Formally, a game in characteristic function form is a pair (N, v), where N is the set of agents (or players), and v is the characteristic function associating to each coalition S in N (including the "grand coalition" N itself) a set $v(S)$ of utility vectors for the members of S. In the *transferable utility* version of the model, $v(S)$ is a real number and represents the set of utility vectors $(u_i)_{i \in S}$ such that $\sum_{i \in S} u_i \le v(S)$; the interpretation is that agents have quasi-linear utilities and $v(S)$ represents a quantity of the numeraire good (e.g., money) with respect to which utilities are linear. In the general (nontransferable utility) context, $v(S)$ represents an arbitrary subset of R^S.

The concept of games in characteristic function form, commonly called cooperative games (on this terminology, see our comments below) makes two essential contributions to the cooperative methodology outlined in Chapter 1. First, it gives a mathematical tool to study the existence (and other properties) of the core, namely, the set of stable direct agreements (Section 1.4). Second, it provides a reduced model for the discussion of fair division of a surplus (or a cost), applicable to all specific resource allocation problems discussed in previous chapters: the Shapley value formula (or any other "value" concept) is a very general method for equitable division. In the former case, the set $v(S)$ must be interpreted as the set of utility vectors feasible for coalition S when its members rely on their own resources (exercise their own property rights). In the latter, the set $v(S)$ may or may not represent utilities that coalition S can secure on its own; it may simply describe a reference set of utilities that plays a focal role in the normative discussion (as when the stand alone test is used in a normative context; see Section 5.3). See the discussion after Definition 7.3 in Section 7.2.

We comply with the (universal) usage by calling the model of games in characteristic function form a "cooperative game," but our interpretation of the model is qualified in two important ways. First, our cooperative games do not model a situation involving binding agreements between the players (as opposed to noncooperative games where direct communication between the players, if any, must be nonbinding).[1] Indeed, the very essence of core stability (as of any other notion of strategic stability) is that agreements are self-enforcing (I am free to trade with you or with another trader: if our deal corresponds to a core outcome, I will stick to it of my own will) but not binding.

Second, the model of cooperative games is a mathematical reduction of a full-fledged allocation problem. It is a useful framework to discuss general properties of the core and to derive general formulas like the Shapley value. However, this is only one of many models fit to discuss the justice mode of cooperation (Section 1.5). For instance, a cooperative production problem (Chapter 5) can be reduced to a cooperative game if we choose to summarize all information about utilities and marginal cost into the list of stand alone surplus levels, but this erases a considerable amount of information about the allocation process: e.g., we cannot talk anymore about no envy and egalitarian-equivalence.

Section 7.2 defines the core of cooperative games. Three classes of games where the core is nonempty are the subject of the next three sections: games where only some coalitions are effective (Section 7.3), supermodular games (also called convex games; Section 7.4), and balanced games (Section 7.5). The Shapley value of a transferable utility game is defined in Section 7.6. It is compared to the core in Section 7.7.

7.2. The Core: Definition

Definition 7.1. Given a finite set N of agents, a *cooperative game* (or game in characteristic function form) is a pair (N, v) where the function v associates to every nonempty coalition S, $S \subseteq N$, a comprehensive

---

[1] See Myerson [1991] or Moulin [1986].

subset $v(S)$ of $\mathbb{R}^S$. A cooperative game with transferable utility (in short, a TU cooperative game) is a pair (N, v) where for all nonempty S, $v(S)$ is a real number.

In the *positive* interpretation of the cooperative game model, $v(S)$ represents the stand alone utility set of coalition S, namely, the set of utility vectors that S can achieve by combining the resources of its members. Similarly, in the transferable utility case, $v(S)$ represents the joint surplus (or joint utility) that coalition S secures by utilizing optimally its resources. In this case, the cooperative game has the following property.

Definition 7.2. The TU cooperative game (N, v) *is superadditive* if we have

$$\text{for all } S, T \subseteq N: \qquad S \cap T = \varnothing \Rightarrow v(S) + v(T) \leq v(S \cup T). \quad (1)$$

The cooperative game (N, v) is superadditive if we have

for all $S, T \subseteq N$:

$$\{S \cap T = \varnothing : u_S \in v(S), u_T \in v(T)\} \Rightarrow \{(u_S, u_T) \in v(S \cup T)\}.$$

All exchange and production economies in Chapters 2 and 3 yield superadditive games: if a certain trade within S (resp. T) brings the utility vector u_S (resp. u_T), then the coalition $S \cup T$ can achieve (u_S, u_T) by performing these two trades separately. In the production economies of Chapter 5, the free access of each coalition to the technology yields a superadditive game under decreasing marginal costs (or in the provision of a public good) but not under increasing marginal costs.

In a superadditive cooperative game, the cooperative opportunities of the grand coalition N exceed (or at least, are not smaller than) those of any partition of N into subcoalitions. Accordingly, all efficient (Pareto optimal) utility vectors can be achieved by the grand coalition N. The next definition describes the utility profiles in $v(N)$ corresponding to stable direct agreements.

In line with the notations of earlier chapters, we speak of agent i's surplus share π_i in the case of TU cooperative games, and of his utility u_i in the general case where utility is not transferable.

Definition 7.3. Given a cooperative game (N, v), a utility vector $u = (u_i)_{i \in N}$ is in the *core* if we have

$u \in v(N)$ and for all $S \subseteq N$:

$$\{u'_S \in v(S), u_i \leq u'_i \text{ for all } i \in S\} \Rightarrow \{u_i = u'_i \text{ for all } i \in S\}.$$

For a TU cooperative game, the core stability property of the surplus vector $(\pi_i)_{i \in N}$ takes one of two equivalent formulations:

$$\sum_{i \in N} \pi_i = v(N), \quad \text{and for all } S \subsetneq N: \qquad v(S) \leq \sum_{i \in S} \pi_i, \qquad (2)$$

$$\sum_{i \in N} \pi_i = v(N), \quad \text{and for all } S \subsetneq N: \qquad \sum_{i \in S} \pi_i \leq v(N) - v(N \backslash S).$$

Clearly, if a cooperative game has a nonempty core, it must satisfy a consequence of the superadditivity property, namely,

$$\text{for all partition } S_1, \ldots, S_K \text{ of } N: \qquad \sum_{k=1}^{K} v(S_k) \leq v(N) \qquad (4)$$

(and a similar property in the general—nontransferable utility—context). The converse is not true: a superadditive cooperative game may have an empty core. This fundamental observation has been amply illustrated in Chapters 2, 3 and 5 by many exchange or production economies with an empty core; see, in particular, Section 2.8 and many exercises in Chapters 2 and 3. It motivates the analysis of the next three sections.

In some cases, it is convenient to think of the number $v(S)$ as a loss that must be distributed among the participants; the most familiar example is cost-sharing (see Example 7.4 in Section 7.4). The individual share π_i of the total loss $v(N)$ is then a *disutility* that agent i tries to minimize, and the relevant (positive) stability concept is the core$_-$. The core$_-$ of the TU game (N, v) is the set of utility vectors $(u_i)_{i \in N}$ such that

$$\sum_{i \in N} \pi_i = v(N) \quad \text{and for all } S \subset N: \qquad \sum_{i \in S} \pi_i \leq v(S).$$

Similarly, the core$_-$ of a (non-TU) cooperative game (N, v) is defined by the two properties:

(i) $u \in v(N)$ and for all $u' \in v(N)$: $\{u_i \leq u'_i$ for all $i\} \Rightarrow \{u_i = u'_i$ for all $i\}$,
(ii) for all $S \subsetneq N$, there exist $u'_S \in v(S)$ such that $u_i \leq u'_i$ for all $i \in S$.

Recall that in the *positive* interpretation, $v(S)$ is feasible for coalition S merely by exercising its property rights: in this case, the game (N, v) will be superadditive (see the discussion before Definition 7.2; alternatively, the cost-sharing game will be subadditive) and the core describes stable direct agreements. In the *normative* interpretation, the number $v(S)^2$ represents a reference level of surplus that is not actually feasible

---

[2] In the TU case; or the set $v(S)$ in the general case.

for coalition S but is nevertheless useful in the normative discussion. In some interesting cases, the level $v(S)$ is used as an upper bound on the joint utility of coalition S, and the relevant concept is that of core_. See the cooperative production games of Chapter 5 when marginal costs increase (there the stand alone test is an upper bound on utilities) and the output-sharing games when the production function has decreasing marginal product (Example 7.4). See also fair division with money (Example 7.8).

7.3. Universally Stable Families of Coalitions

In this section, we look for restrictions in the pattern of coalition formation that make it easy to decide if the core of a game is empty or not. Specifically, we rule out certain coalitions as "ineffective." In some cases, this implies that the core is nonempty if and only if the value function v is superadditive.

Given the set $N = \{1, 2, \ldots, n\}$ of agents, we suppose that only a subset $\mathscr{F}$ of (nonempty) coalitions ($\mathscr{F}$ is a subset of $2^N \setminus \varnothing$) can effectively cooperate. We assume that $\mathscr{F}$ contains all one-agent coalitions (also called the singleton coalitions) and some (but not all) "real" coalitions as well. We fix arbitrarily the surplus $v(S)$ achievable by a coalition S in $\mathscr{F}$ (to fix ideas, we work in the TU context). To express the fact that coalitions outside $\mathscr{F}$ are not effective, we extend the definition of v to all coalitions as follows:

$$\text{for all } S \subseteq N: \qquad v^*(S) = \max \sum_{k=1}^{K} v(S_k), \qquad (5)$$

where the maximum bears over all partitions of S such that $S_k \in \mathscr{F}$ for all $k = 1, \ldots, K$. This maximum is well defined because $\mathscr{F}$ contains at least the singletons.[3]

Note that v^* coincides with v on $\mathscr{F}$ if (and only if) the function v is superadditive on $\mathscr{F}$. Moreover, an allocation (π_i) is in the core of (N, v^*) if and only if (a) $\sum_N \pi_i = v^*(N)$ and (b) $\sum_S \pi_i \geq v(S)$ for all S in $\mathscr{F}$.

Definition 7.4. Given are N and a subset of coalitions $\mathscr{F}$ containing the singleton coalitions. We say that $\mathscr{F}$ is *universally stable* if for all

---

[3] Similarly, in the nontransferable utility case, the extension v^* of v is defined as

$$v^*(S) = \bigcup \operatorname*{X}_{k=1}^{K} v(S_k).$$

(real-valued) superadditive function v defined on $\mathcal{F}$, the cooperative game (N, v^*) defined by (5) has a nonempty core.

The above definition is given for TU games; its extension to general cooperative games is straightforward (using note 3).

Example 7.1. Games with a Veto Player

Let $\mathcal{F}$ contain all singleton coalitions and all coalitions containing player 1. Up to renormalization, we choose a value function v such that $v(i) = 0$ for each singleton coalition $\{i\}$. Then v is superadditive on $\mathcal{F}$ if and only if it is monotonic (namely, $1 \in S \subseteq T \Rightarrow v(S) \leq v(T)$) Clearly, the allocation giving all the surplus to player 1 is in the core.[4] There are other core allocations as well: e.g., for $n = 3$, check that $\pi_1 = v(12)$, $\pi_2 = 0$, $\pi_3 = v(123) - v(12)$ is a core allocation.

Example 7.2. Connected Coalitions on a Line

Fix an ordering of the agents, say $N = \{1, 2, \ldots, n\}$, and consider the interval coalitions

$$S \in \mathcal{F} \Leftrightarrow \{S = \{i, i+1, \ldots, i'\} \quad \text{for some } i, i', i \leq i'\}.$$

Interpretation: the agents are located along a road and cooperation within S requires setting a cable connecting all members of S. Agreement of agent i is necessary to go through location i.

We fix a (TU) superadditive function v with domain $\mathcal{F}$. The greedy algorithm (introduced in Section 5.5) yields a core allocation:

$$\pi_1 = v(1), \qquad \pi_2 = v(1, 2) - v(1), \ldots,$$

$$\pi_i = v(1, 2, \ldots, i) - v(1, \ldots, i-1), \ldots,$$

$$\pi_n = v(N) - v(N \setminus n).$$

To check the core property, fix an interval $S = \{i, \ldots, i'\}$ and compute (with the help of superadditivity):

$$\sum_{j=i}^{i'} \pi_j = v(1, 2, \ldots, i') - v(1, 2, \ldots, (i-1))$$

$$\geq v(i, i+1, \ldots, i').$$

Of course, the greedy algorithm is easily adapted in the nontransferable utility case: u_1 is the largest utility in $v(1)$; u_2 is the largest feasible

---

[4] If v is not normalized, check that $\pi_1 = v(N) - \sum_{i \geq 2} v(i)$ and $\pi_i = v(i)$ for all $i \geq 2$ yields a core allocation.

utility in $v(12)$ given that agent 1 gets u_2; and so on. This algorithm picks a core utility vector if v is superadditive. (Exercise: why?)

Exercise 7.16 generalizes Example 7.2 to the case of connected coalitions along a tree.

Example 7.3. Bilateral Coalitions

Suppose N is partitioned as $N = M \cup W$ and define the set $\mathscr{F}$ as follows:

$$S \in \mathscr{F} \Leftrightarrow \{S \text{ is a singleton}\} \text{ or } \{S = \{i, j\} \text{ for some } i \in M \text{ and } j \in W\}.$$

The bilateral assignment market (Section 3.4) and the marriage market (Section 3.3) are two important examples, one in the transferable utility context, the other in the nontransferable utility case. Thus the nonemptiness of the core in these models is not caused by the particular assumptions on preferences but by the fact that only the pairs containing one man and one woman cooperate effectively.

A general result (stated in Section 7.5 as a corollary to Theorem 7.3) characterizes all the universally stable families of coalitions and proves the claim of Example 7.3.

7.4. CONVEX (SUPERMODULAR) GAMES

We only discuss the case of TU cooperative games, although the notion of convex games can be extended to the nontransferable utility case and some of the results below can be preserved (see Vilkov [1977], Greenberg [1985], and Milgrom and Shannon [1994]).

Given a set $N = \{1, 2, \ldots, n\}$ of agents, an ordering of N is described by a one-to-one mapping σ from N into itself: the player $\sigma(1)$ comes first, followed by player $\sigma(2)$, and so on. Given a function v, we denote by π^σ the marginal contribution vector corresponding to the ordering σ:

$$\pi^\sigma_{\sigma(1)} = v(\sigma(1)), \qquad \pi^\sigma_{\sigma(2)} = v(\sigma(1), \sigma(2)) - v(\sigma(1)), \ldots,$$

$$\pi^\sigma_{\sigma(i)} = v(\sigma(1), \ldots, \sigma(i)) - v(\sigma(1), \ldots, \sigma(i-1)), \quad \text{for all } i.$$

The marginal contribution vector is the allocation of total surplus $v(N)$ in the greedy algorithm where player $\sigma(1)$ is served first, followed by player $\sigma(2)$, and so on.

Theorem 7.1. (Shapley [1971], Ichiishi [1981]). *For any TU coopera-tive game* (N, v), *the four following statements are equivalent:*

(a) *For all orderings* σ *of* N, *the marginal contribution vector* π^{σ} *is in the core.*

(b) *The core equals the convex hull of the marginal contribution vectors.*

(c) *For all* i *and all coalitions* S, T:

$$\{S \subseteq T \text{ and } i \notin T\} \Rightarrow \{v(S \cup \{i\}) - v(S) \le v(T \cup \{i\}) - v(T)\}. \quad (6)$$

The marginal contribution of i *to coalition* S *increases with* S.

(d) *The function* v *is supermodular:*

$$v(S) + v(T) \le v(S \cup T) + v(S \cap T) \quad \text{for all } S, T \subseteq N \quad (7)$$

(with the convention $v(\varnothing) = 0$).

In a convex game, the greedy algorithm selects a core allocation for every conceivable ordering of the players. Moreover, every core alloca-tion obtains as a convex combination of the marginal contribution vectors.[5] In particular, fix a coalition S and consider an ordering of N where the players in S come first: in the corresponding marginal contribution vector, coalition S receives exactly $v(S)$. Symmetrically, if the players in S come last, coalition S receives exactly $v(N) - v(N \setminus S)$. Therefore, in the core of a convex game, both inequalities (2) and (3) can be equalities for any coalition S. We give a few examples of convex games.

Example 7.4. *Output-Sharing and Cost-Sharing Games*

In an output-sharing game, a production function f is given and each one of the n agents has an inelastic supply of input x_i. The stand alone game (N, v) is defined as follows:

$$v(S) = f\left(\sum_{i \in S} x_i\right) \quad \text{for all } S \subseteq N. \quad (8)$$

If f is superadditive $(f(x) + f(x') \le f(x + x'))$, efficiency commands using the inputs in a single machine, and the stand alone core repre-sents the distribution of output not threatened by the secession of any coalition of agents (in the positive interpretation where each one owns a

---

[5] Note that for *any* TU cooperative game, the (possibly empty) core is contained in the convex hull of the marginal contribution vectors; see Exercise 7.8.

copy of f), or simply the distributions where no coalition S subsidizes the complement coalition (in the normative interpretation where a single machine is the common property of all agents; see Section 5.4): $\Sigma_S y_i \geq v(S)$ for all S. If f is convex (decreasing marginal costs), this TU game is convex; indeed, property (6) reads

$$f(x_S + x_i) - f(x_S) \leq f(x_T + x_i) - f(x_T),$$

and follows from $x_S \leq x_T$. Thus Theorem 7.1 applies.

In a cost-sharing game, the cost function c is given and each agent i has an inelastic demand of output y_i. The game $(N, \tilde{c})$ is defined by

$$\tilde{c}(S) = c\left(\sum_{i \in S} y_i\right) \quad \text{for all } S \subseteq N. \tag{9}$$

If c is subadditive, the stand alone core is defined as the core_. If c is concave (decreasing marginal costs), the TU game $(N, \tilde{c})$ is *concave*: it satisfies the opposite inequalities of (6) and (7). Theorem 7.1 reads in this case: the core_ of a concave TU game is the convex hull of its marginal contribution vectors. Exercise 7.4 generalizes the above discussion to the case of production (or cost) functions with multiple inputs or outputs.

Notice an interesting variant of the cost-sharing game where the cost function takes the form $c(y_1, \ldots, y_n) = c_0(\max_i y_i)$. The corresponding TU game is concave for any nondecreasing function c_0. (Exercise: why?) We discuss this game as Example 7.10 (Section 7.6).

Example 7.5. Cooperative Production Games

If we introduce variable demands of output (and variable input contributions) in the cost-sharing game, we find ourselves in the context of Chapter 5. Assume that utilities over input and output are quasi-linear in the input: $u_i(y_i, x_i) = w_i(y_i) - x_i$. Then, for a given cost function c, the following TU cooperative game obtains:

$$v(S) = \max_{y_i} \left\{ \sum_{i \in S} u_i(y_i) - c\left(\sum_{i \in S} y_i\right) \right\} \quad \text{for all } S. \tag{10}$$

It was shown in Section 5.4 that this game is convex if marginal costs are nonincreasing and the functions u_i, $i = 1, \ldots, n$, are nondecreasing (no assumption on marginal utilities is needed). Similarly, consider the provision of a public good problem. There we get the following TU game:

$$v(S) = \max_{y \geq 0} \left\{ \sum_{i \in S} u_i(y) - c(y) \right\}.$$

In Section 5.6, we showed that this game is convex as soon as the functions u_i are nondecreasing.

The next result is a very useful property of convex games. It allows us to select a core allocation respecting an exogenous lower bound and/or an upper bound on individual utilities. By way of motivating this property, we look at a three-person game, $N = \{1, 2, 3\}$. Suppose the core of the game is nonempty. By Theorem 7.3, this is equivalent to

$$v(i) + v(jk) \leq v(123), \qquad v(12) + v(23) + v(13) \leq 2v(123).$$

Consider a surplus vector $(\tilde{\pi}_1, \tilde{\pi}_2, \tilde{\pi}_3)$ and look for an allocation (π_1, π_2, π_3) in the core and bounded above by $\tilde{\pi}$ ($\pi_i \leq \tilde{\pi}_i$ for all i). Necessary conditions for the existence of π are

$$\tilde{\pi}_i \geq v(i) \quad \text{for all } i, \qquad \tilde{\pi}_i + \tilde{\pi}_j \geq v(ij) \quad \text{for all } j. \tag{11}$$

However, the above system does not guarantee the existence of π. Suppose $\tilde{\pi}_3 = v(3)$ and $\tilde{\pi}_1, \tilde{\pi}_2$ are both very large (so as to meet inequalities (11)). Then the desired allocation π must be as follows:

$$\pi_3 = v(3), \quad \pi_1 \geq v(13) - v(3),$$

$$\pi_2 \geq v(23) - v(3), \quad \pi_1 + \pi_2 = v(123) - v(3).$$

Thus it is not possible to find π unless the inequality $v(13) + v(23) \leq v(123) + v(3)$ is satisfied. This is precisely the supermodularity inequality (7) for three-person games.

Theorem 7.2. (Sharkey [1982a], Ichiishi [1990]). Let (N, v) be a convex TU cooperative game and consider a surplus distribution $\tilde{\pi} = (\tilde{\pi}_1, \ldots, \tilde{\pi}_n)$ such that

$$\text{for all } S \subseteq N: \quad v(S) \leq \sum_S \tilde{\pi}_i. \tag{12}$$

Then there exists a core allocation π such that $\pi_i \leq \tilde{\pi}_i$ for all i. Consider next a surplus distribution $\tilde{\tilde{\pi}} = (\tilde{\tilde{\pi}}_1, \ldots, \tilde{\tilde{\pi}}_n)$ such that

$$\text{for all } S \subseteq N: \quad \sum_S \tilde{\tilde{\pi}}_i \leq v(N) - v(N \setminus S) \quad (\text{with } v(\emptyset) = 0). \tag{13}$$

Then there exists a core allocation π such that $\overset{=}{\pi}_i \leq \pi_i$. Finally, if $\tilde{\pi}$ and $\overset{=}{\pi}$ satisfy (12) and (13) respectively, and if $\overset{=}{\pi}_i \leq \tilde{\pi}_i$ for all i, there exists a core allocation π such that $\overset{=}{\pi}_i \leq \pi_i \leq \tilde{\pi}_i$.

The proof of Theorem 7.2 is constructive and uses an interesting descending algorithm; see Appendix 7.1.[6] A typical application of this result is to the provision of a public good. Consider the unanimity utility of a given player i:

$$\mathrm{una}(u_i) = \max_{y \geq 0} \left\{ u_i(y) - \frac{c(y)}{n} \right\}.$$

The vector $(\mathrm{una}(u_1), \ldots, \mathrm{una}(u_n))$ satisfies inequality (12). Indeed, for any public good level y and any coalition S, we have

$$\sum_S u_i(y) - c(y) \leq \sum_S \left(u_i(y) - \frac{c(y)}{n} \right) \leq \sum_S \mathrm{una}(u_i).$$

Therefore, Theorem 7.2 establishes the existence of an allocation in the stand alone core meeting the unanimity upper bound as well.

Next, consider the surplus distribution associated with a Nash equilibrium of the voluntary contribution mechanism (Section 6.3). It is shown in Exercise 7.9 that this vector satisfies the system (13); thus there exists a stand alone core allocation Pareto superior to the equilibrium allocation in question. See also in Exercise 7.9 an application of Theorem 7.2 to the cooperative production of private goods.

7.5. BALANCED GAMES

The Bondareva–Scarf theorem (Theorem 7.3) gives a necessary and sufficient condition for the nonemptiness of the core in a TU cooperative game.

The superadditivity inequality (4) is a necessary (but not sufficient) condition for a nonempty core. More necessary conditions follow from applying the upper bound (3). Fix a coalition S and observe that if π is

---

[6] Note that if a TU cooperative game (N, v) is not convex, then at least one of the reduced games (S, v) must violate the above property, so that Theorem 7.2 can be viewed as yet another characteristic property of convex games. See Moulin [1990] for details.

a core allocation, we have

$$\pi_i \leq v(N) - v(N \setminus i) \quad \text{for all } i \in S \quad \text{and} \quad v(S) \leq \sum_S \pi_i.$$

Combining these inequalities gives another necessary condition for nonemptiness of the core:

$$v(S) + \sum_{i \in S} v(N \setminus i) \leq |S| \cdot v(N). \tag{14}$$

Consider a three-person game. Inequality (14) applied to any two-person coalition S (or to $S = \{123\}$) yields

$$v(12) + v(23) + v(13) \leq 2v(123), \tag{15}$$

In turn, (15) and the four superadditivity inequalities like $v(12) + v(3) \leq v(123)$ and $v(1) + v(2) + v(3) \leq v(123)$ characterize the nonemptiness of the core.

Consider a four-person game. Applying (14) to any three-person coalition S (or to $S = \{1234\}$) yields

$$v(123) + v(124) + v(134) + v(234) \leq 3v(1234). \tag{16}$$

Next, (14) applied to the two-person coalition $S = \{12\}$ yields

$$v(12) + v(134) + v(234) \leq 2v(1234). \tag{17}$$

There are six distinct inequalities like (17) corresponding to each two-person coalition. Theorem 7.3 implies that a superadditive four-person game (property (1)) has a nonempty core if and only if it satisfies inequality (16) and six additional inequalities like (17) (obtained by permuting the agents).

Definition 7.5. Given the set N of agents, a balanced family of coalitions is a subset $\mathcal{B}$ of $2^N \setminus \{N\}$ such that there exists, for each S in $\mathcal{B}$, a "weight" δ_S, $0 \leq \delta_S \leq 1$, satisfying

$$\text{for all } i \in N: \quad \sum_{S \in \mathcal{B}_i} \delta_S = 1, \quad \text{where } \mathcal{B}_i = \{S \in \mathcal{B} \mid i \in S\}. \tag{18}$$

A mapping δ satisfying the above equations is called a vector of balanced weights.

Theorem 7.3. (Bondareva [1963], Scarf [1967])

(a) *Let* (N, v) *be a TU cooperative game. Its core is nonempty if and only if, for every vector of balanced weights* δ, *we have*

$$\sum_{S \subsetneq N} \delta_S \cdot v(S) \leq v(N). \tag{19}$$

(b) *Let* (N, v) *be a (nontransferable utility) game such that, for all* S, *the set* $v(S)$ *is closed in* $\mathbb{R}^S$. *If the game* (N, v) *satisfies the following property:*

for all balanced family of coalition $\mathfrak{B}$ *and for all utility vector u in* $\mathbb{R}^N$:

$$\{u_S \in v(S) \text{ for all } S \in \mathfrak{B}\} \Rightarrow \{u \in v(N)\}, \tag{20}$$

then the core of the game is nonempty.

We say that a TU cooperative game (resp. a general cooperative game) is *balanced* if it satisfies property (19) (resp. property (20)). Thus balancedness of the game is a necessary and sufficient condition for the existence of the core in the TU case; it is only a sufficient condition in the general case.

Any partition of N in nonempty coalitions (as well as any union of partitions) is a balanced family of coalitions (with associated weights $\delta_S = 1$ for all S). Therefore, the balancedness inequality (19) implies in particular property (4).

A general vector of balanced weights represents a pattern of coalition formation more involved than (but similar to) a partition of the set N. Agent i joins coalition S for a fraction δ_S of his time; all members of S devote the same fraction of their time to this particular coalition; coalition S produces $\delta_S \cdot v(S)$ units of surplus. Inequality (19) says that in any such coalitional pattern, the overall surplus does not exceed $v(N)$. Inequalities (15) (for three players) and (17) (for four players) are two examples. (Exercise: for what vector of balanced weights?)

In principle, it is enough to check inequality (19) for a finite number of vectors δ; indeed, the set of vectors δ satisfying (18) (as well as $0 \leq \delta_S \leq 1$) is a convex polyhedron in a euclidean space and the system of inequalities (19) is linear in δ. Therefore, we only need to check (19) at every extreme point of this polyhedron. However, the number of extreme points grows rapidly with n.[7]

---

[7] They are 5 for $n = 3$, and 23 for $n = 4$; no systematic count of these points is known.

Corollary to Theorem 7.3. (Kaneko and Wooders [1982]). The set $\mathcal{F}$ of coalitions in N is universally stable for all TU cooperative games if and only if it is universally stable for all (nontransferable utility) cooperative games. This occurs if and only if every balanced family of coalitions in $\mathcal{F}$ contains a partition.

See Exercise 7.16 and Example 7.3 for applications of this result. Now to a few important applications of the Bondareva–Scarf theorem, starting with (TU) games.

Example 7.6. Example 7.4 Continued

Consider output-sharing first. Given the production function f, the game (8) is balanced for all n and all input profiles $(x_1, \ldots, x_n)$ if and only if the average return $f(x)/x$ is nondecreasing.

To prove the claim, we suppose first that f is given such that the game (8) is balanced for all input profiles. Fix $x \geq 0$ and two integers p,q such that $p < q$. Denote $x' = px/q$ and $x^* = x/q$. Consider the output-sharing game with q players each supplying x^*. In this game, the set of all coalitions of size p is balanced for the weights

$$\delta_S = 1 \bigg/ \binom{q-1}{p-1}.$$

Applying (19) gives

$$\sum_{S: |S|=p} \delta_S \cdot f(px^*) \leq f(qx^*) \Leftrightarrow \frac{q}{p} \cdot f(x') \leq f(x),$$

therefore the inequality $f(x')/x' \leq f(x)/x$ holds for any x' smaller than x and such that x'/x is a rational number. Extending this inequality to the case where x'/x is irrational is easy because f is nondecreasing.

Conversely, assume that f is given with nondecreasing average returns. One can check directly that all resulting output-sharing games (8) are balanced. Alternatively, we note that the proportional output allocation $y_i = (x_i/x_N) \cdot f(x_N)$ is in the stand alone core.

The same result applies to the cost-sharing games (9). Their core_ is nonempty for all demand profiles if and only if the average cost $c(y)/y$ is nonincreasing. Notice that the core_ of a TU game is nonempty if and only if, for every vector of balanced weights δ, the opposite inequality of (19) holds.[8]

Example 7.7. Exchange Economies

In the Arrow–Debreu economies of Section 3.6, the nonemptiness of the core under convex continuous preferences is a consequence of Scarf's theorem. Let N be fixed, ω_i be agent i's initial endowment, and u_i be a concave utility function representing his preferences.[9] Pick a balanced family of coalitions $\mathcal{B}$ with weights δ, and a vector π in $\mathbb{R}^N$. The trading opportunities within coalition S are captured by the utility set $v(S)$:

$$\pi \in v(S) \Leftrightarrow \text{for all } i \in S: \quad \text{there exists } z_i \in \mathbb{R}_+^K:$$

$$\sum_S z_i = \sum_S \omega_i \quad \text{and } \pi_i \le u_i(z_i), \quad \text{all } i \in S.$$

Let the vector π be in $v(S)$ for all S in $\mathcal{B}$. Denote by z^S the corresponding trade within coalition S. Define an allocation $(y_1, \ldots, y_n)$ as follows:

$$y_i = \sum_{\mathcal{B}_i} \delta_S \cdot z_i^S, \quad \text{where } \mathcal{B}_i = \{S \in \mathcal{B} \mid i \in S\}.$$

By property (18), we have

$$\sum_N y_i = \sum_{\mathcal{B}} \delta_S \left(\sum_S z_i^S \right) = \sum_{\mathcal{B}} \delta_S \left(\sum_S \omega_i \right) = \sum_N \omega_i,$$

therefore y is a feasible allocation (for the grand coalition N). Concavity of u_i, property (18), and our assumption on π imply

$$u_i(y_i) \ge \sum_{\mathcal{B}_i} \delta_S u_i(z_i^S) \ge \sum_{\mathcal{B}_i} \delta_S \cdot \pi_i = \pi_i,$$

so we conclude that π belongs to $v(N)$, as was to be proved.

Note that the existence of a core allocation can also be deduced from that of a competitive equilibrium allocation (by Lemma 3.6), or even of

---

[8] The result of Example 7.6 has been extended to the cases of production functions with multiple inputs and of cost functions with multiple outputs by Scarf [1986] and others. See Moulin [1988], Section 4.3 for a survey of these results.

[9] The assumption that preferences can be represented by a concave utility function is only slightly stronger than convexity of their upper contour sets.

a constrained competitive equilibrium (see Remark 3.4). As shown in Chapters 2 and 3, the lack of convexity of preferences can easily result in an empty core.

In the context of cooperative production games (Example 7.5), empty cores are also very easy to achieve; see the discussion of Section 5.4. Exercises 7.5 and 7.6 give some more examples in the context of cost-sharing games (Example 7.4). Our last example belongs to the normative interpretation of the core; it is a rich family of (subadditive) TU cooperative games with a nonempty core.

Example 7.8. Fair Division with Money

The n agents divide a bundle of (divisible) goods $\omega \in \mathbb{R}_+^K$ and money is the $(K + 1)$th good. Utilities are quasi-linear in money: if agent i receives a share z_i of the "manna" ω ($z_i \in \mathbb{R}_+^K$) and a net transfer t_i of money, his final utility is $u_i(z_i) + t_i$. The stand alone surplus of coalition S obtains when S consumes all the resources ω:

$$v(S) = \max\left\{\sum_S u_i(z_i) \,\middle|\, \sum_S z_i = \omega, \quad z_i \geq 0 \quad \text{all } i \in S\right\}.$$

Clearly, v is subadditive ($v(S \cup T) \leq v(S) + v(T)$), so that the relevant normative concept is the core _. Actually, if the utilities u_i are monotonic (and continuous), the TU game (N, v) is balanced_ (i.e., the opposite inequality of (19) holds for any vector of balanced weights). Therefore, by Theorem 7.3, the stand alone core_ is never empty.

To check the balancedness, call $(z_1, \ldots, z_n)$ an allocation of ω such that $\sum_N u_i(z_i) = v(N)$, and pick a vector of balanced weights δ. Note that, for all S, we have $\sum_S u_i(z_i) \leq v(S)$ by the monotonicity of utilities. Hence

$$\sum_N u_i(z_i) = \sum_N \left(\sum_{\ni i} \delta_S\right) u_i(z_i) = \sum_{\ni} \delta_S \left(\sum_S u_i(z_i)\right) \leq \sum_{\ni} \delta_S v(S),$$

as was to be proved.

7.6. THE SHAPLEY VALUE: DEFINITION

For the rest of the chapter, we restrict attention to TU cooperative games.[10]

---

[10] Several extensions of the TU Shapley value to nontransferable utility games have been proposed; their discussion is beyond the scope of this volume. See Harsanyi [1963], Shapley [1969], Kalai and Samet [1985], and Hart [1985].

The Shapley value is a normative solution for any TU cooperative games. Given any game (N, v), the Shapley value proposes a distribution $(\pi_1, \ldots, \pi_n)$ of the surplus $v(N)$. Because so many problems of fair allocation can be reduced to such a cooperative game (e.g., all fair-division models of Chapter 4 and all cooperative production models of Chapter 5, provided utilities are quasi-linear; see also the power index of voting committees: Example 7.9 below), the formula has a wide gamut of applications. Its limit is the reductionist nature of the cooperative game model, extracting only scant information about coalitional surplus.

A common feature of the core and of the Shapley value is the dummy axiom. In the game (N, v), we call player i a *dummy* if the incremental value of adding i to any coalition is zero:

$$v(S \cup i) = v(S) \quad \text{for all } S.$$

Any core allocation gives a zero share of surplus to a dummy player (in view of inequality $v(i) \leq x_i \leq v(N) - v(N \setminus i)$). This is the dummy axiom: a dummy player should get zero.

Let us look for solutions that satisfy the dummy axiom and compute agent i's share π_i by way of a formula *additive* in the numbers $v(S)$, $S \subseteq N$. Consider a fixed ordering σ of the players: the marginal contribution vector π^σ is such a solution. Indeed, the share π_i^σ takes the form $\pi_i^\sigma = v(S \cup i) - v(S)$ for a coalition S independent of v. Clearly, any convex combination of marginal contribution vectors (with fixed coefficients independent of the particular choice of v) is also such a solution (the dummy axiom and the additivity property are both preserved by the convex combination). It turns out that the *only* solutions satisfying the dummy and additivity properties are the convex combinations of marginal contribution vectors.[11] The most interesting among these solutions is the convex combination giving equal weight $1/n!$ to every ordering σ, because this solution treats equals equally.

Definition 7.6. Given a game (N, v), the Shapley value is the arithmetic average of the marginal contribution vectors:

$$\pi_i = \frac{1}{n!} \sum_\sigma \pi_i^\sigma = \sum_{s=0}^{n-1} \frac{s!(n-s-1)!}{n!} \sum_{\mathfrak{Z}_{-i}(s)} \{v(S \cup i) - v(S)\},$$

$$\text{where } \mathfrak{Z}_{-i}(s) = \{S \mid i \notin S \text{ and } |S| = s\}. \tag{21}$$

---

[11] This result is due to Weber [1988]. See Exercise 7.8.

Theorem 7.4. (Shapley [1953]). *The Shapley value is the only solution satisfying the three axioms*:

(*i*) **Dummy:** $\pi_i = 0$ *if i is a dummy player.*

(*ii*) **Additivity:** *for all* v_1, v_2: $\pi(v_1 + v_2) = \pi(v_1) + \pi(v_2)$.

(*iii*) **Equal treatment of equals:** *for all game* (N, v) *and any two players* i, j,

$$\{v(S \cup i) = v(S \cup j) \quad \text{for all } S \subseteq N \setminus \{ij\}\} \Rightarrow \{\pi_i(v) = \pi_j(v)\}.$$

The fourth axiom of efficiency ($\sum_i \pi_i = v(N)$) is often listed explicitly in the statement of Shapley's theorem. We only discuss efficient solutions; therefore, we do not state the property explicitly.

The proof of Theorem 7.4 is explained after Lemma 7.1.

Formula (21) gives the probablistic interpretation of the Shapley value. Agent i receives his expected marginal contribution $v(S \cup i) - v(S)$ when an ordering is drawn at random (with uniform probability on all orderings) and S is the (random) coalition of these agents preceding i. The last part of formula (21) develops the coefficients of the probabilistic computation. The other major interpretation of the Shapley value, the incremental interpretation, is discussed after Example 7.10.

An important application of the Shapley value is the power index of a committee (Shapley and Shubik [1954]). Consider a simple game among the set N of agents: every coalition S is either winning (in which case $v(S) = 1$) or losing (in which case $v(S) = 0$). The former occurs if the members of S can control the outcome of the voting process (by coordinating their votes), the latter if they cannot. We also assume that the complement of a losing coalition is winning and vice versa ($v(S) = 1 \Leftrightarrow v(N \setminus S) = 0$). Moreover, v is monotonic, namely, if $S \subset T$ and S is winning, then T is winning as well. The Shapley value offers a measure of the relative share of decision power held by each committee member, based on the notion of pivotal player. Suppose a certain binary choice (yea or nay to a particular motion) is to be voted upon, and order the voters by the intensity of their support of the motion, say as $1, 2, \ldots, n$. This means that agent 1 is the fiercest supporter of the motion, and agent n its fiercest opponent. The *pivotal* player is the smallest i such that the coalition $\{1, \ldots, i\}$ is winning. Therefore, both coalitions $\{1, \ldots, i\}$ and $\{i, i + 1, \ldots, n\}$ are winning: if voter i is convinced to support the motion, all voters with a lower index will be convinced as well and the motion passes; if voter i opposes the motion, so do all voters with a higher index and the motion fails. The pivotal role of player i is reflected in the fact that he receives the full surplus ($\pi_i^\sigma = 1$) in the marginal contribution vector corresponding to this particular ordering. Therefore, the Shapley value gives to each agent her probability of being pivotal, when all orderings by decreasing support of the motion are equally probable.

Example 7.9. Quota Games

Given are N, the set of players, and a set of convex weights $p_1, \ldots, p_n$. We assume

$$p_i \geq 0 \quad \text{for all } i; \qquad \sum_N p_i = 1 \quad \text{and for no coalition } S: \qquad \sum_S p_i = \tfrac{1}{2}.$$

A coalition S is winning if its total weight exceeds one-half and losing otherwise. Hence the simple cooperative game:

$$v(S) = 1 \quad \text{if } \sum_S p_i \geq \tfrac{1}{2}$$

$$v(S) = 0 \quad \text{otherwise}.$$

The core of a quota game is empty unless there is a *veto player*, namely, a player without whom no coalition can win.[12]

Consider the 11-player quota game: $p_1 = 1/3$ and $p_i = 1/15$ for $i = 2, \ldots, 11$. The big player 1 has one-third of the voting weight, but the Shapley value gives her an even bigger share of power, reflecting the fact that she is pivotal more than one-third of the time. Specifically, player 1 is pivotal if at least three and at most seven other players precede her. As all ranks are equiprobable, this gives $\pi_1 = 5/11$, therefore $\pi_i = 6/110$ for $i = 2, \ldots, 11$. When we increase the number of small players while keeping one big player with $p_1 = 1/3$ (so that $p_i = 2/3(n-1)$ for $i = 2, \ldots, n$), the Shapley value gives nearly one-half to the big player (because player 1 is pivotal if at least $(n-1)/4$ and at most $3(n-1)/4$ small players precede him).

On the other hand, consider the 12-player quota games $p_1 = p_2 = 1/3$, $p_i = 1/30$, $i = 3, \ldots, 12$. Here player 1 is pivotal if he is preceded by *either* at least five small players and not player 2 *or* player 2 and at most four small players. Hence $\pi_1 = \pi_2 = 0.21$, or only one-fifth of the

---

[12] Say agent i is a veto player if $p_i > \tfrac{1}{2}$. Exercise: prove the claim.

surplus, because the competition between the two big players reduces their voting power.

Our next example is a popular application of the Shapley value, and illustrates an important decomposition technique.

Example 7.10. Sharing the Cost of a Capacity (Littlechild and Owen [1973])

In this cost-sharing game (Example 7.4), the cost of serving the demand profile $(y_1, \ldots, y_n)$ is $c_0(\max_i y_i)$. Examples include the "airport game" where airline i needs a runway of size y_i (determined by the size of its airplanes), the cost-sharing of an elevator in an apartment building (where agent i lives on the y_i-th floor), and the excludable public goods of Section 6.7. To compute the Shapley value of the stand alone cost game, we order the fixed demand profile as $y_1 \leq y_2 \leq \cdots \leq y_n$, and denote $c_i = c_0(y_i)$. Then, for any coalition T, we define an auxiliary (simple) game γ_T:

$$\gamma_T(S) = 1 \quad \text{if } S \cap T \neq \varnothing$$

$$\gamma_T(S) = 0 \quad \text{otherwise.}$$

(22)

Observe that the stand alone cost-sharing game (N, v) can be decomposed as follows:

$$v = c_1 \cdot \gamma_N + (c_2 - c_1) \cdot \gamma_{N \setminus 1} + \cdots + (c_i - c_{i-1}) \cdot \gamma_{\{i, \ldots, n\}}$$

$$+ \cdots + (c_n - c_{n-1}) \cdot \gamma_{\{n\}}. \quad (23)$$

Indeed, for any coalition S where the highest player is player i, we have $\gamma_{\{j, \ldots, n\}}(S) = 1$ if and only if $j \leq i$; hence, the right-hand game yields $c_1 + (c_2 - c_1) + \cdots + (c_i - c_{i-1}) = c_i$ at coalition S.

Now the Shapley value of the game γ_S is easy to compute, since all players of $N \setminus S$ are dummies and all players of S are equal. We must give $1/|S|$ units of surplus to each agent in S and zero to each agent in $N \setminus S$. In view of the additivity property, we derive the Shapley value π of game v:

$$\pi_i = \frac{c_1}{n} + \frac{c_2 - c_1}{n - 1} + \cdots + \frac{c_i - c_{i-1}}{n - i + 1}, \quad \text{for all } i. \quad (24)$$

The incremental cost of extending the capacity from c_{i-1} to c_i is shared equally among all agents using a capacity greater than c_{i-1}.[13] Exercise 7.6 (question (c)) generalizes this formula to a large class of cost-sharing games.

The additive decomposition (23) generalizes to any game: the set of games γ_T, where T can be any coalition of N, is a linear basis of the set of TU cooperative games v (where v is viewed as a vector in a euclidean space of dimension $2^n - 1$). A slightly more convenient linear basis is the set of so-called *unanimity games* δ_T, namely,[14]

$$\delta_T(S) = 1 \quad \text{if } T \subseteq S$$

$$\delta_T(S) = 0 \quad \text{otherwise.} \tag{25}$$

We can compute explicitly the coordinates of any game (N, v) in the basis of unanimity games.

Lemma 7.1. Fix a TU game (N, v) and define α_T for all coalition T:

$$\alpha_T = \sum_{S \subseteq T} (-1)^{t-s} \cdot v(S), \quad \text{where } t = |T|, \ s = |S|. \tag{26}$$

Then we have

$$v = \sum_{T \subseteq N} \alpha_T \cdot \delta_T. \tag{27}$$

---

[13] Interestingly, formula (24) appears in the talmudic literature (Rabbi Ibn Ezra, 1140 AD) in the dual context of an output-sharing problem. An estate of 120 units must be divided among four brothers: Reuben claims 120, Simon claims 60, Levi claims 40, and Judah claims 30. Interpret agent i's claim as an input x_i producing c_i units of surplus (where c_i is the face value of the claim), and the stand alone output of a coalition S as $\max_S c_i$. Here is Ibn Ezra's solution: "In accordance with the view of the Jewish Sages, the three older brothers say to Judah, "Your claim is only on 30, but all of us have an equal claim on them. Therefore, take $7\frac{1}{2}$, which is one-quarter and depart." Each one of the brothers takes a similar amount. Then Reuben says to Levi, "Your claim is only on 40. You have already received your share of the 30 which all four of us claimed; therefore take $\frac{1}{3}$ of the remaining 10 and go." Thus Levi's share is $10\frac{5}{6}$. Reuben also says to Simeon, "Your claim is for only half of the estate which is 60, while the remaining half is mine. Now you have already received your share of the 40, so that the amount at issue between us is 20—take half of that and depart." Thus Simeon's share is $20\frac{5}{6}$ and Reuben's share is $80\frac{5}{6}$. (See O'Neill [1982] and references therein.)

[14] In view of the identity $\delta_T(S) + \gamma_T(N \setminus S) = 1$, it should be clear that both linear basis $\{\delta_T\}$ and $\{\gamma_T\}$ yield dual decomposition formulas. See Exercise 7.13 for details.

The easy proof is omitted. The proof of Theorem 7.4 follows at once. If a solution satisfies dummy and equal treatment of equals, it solves the game $\alpha_T \cdot \delta_T$ by giving $\alpha_T/|T|$ to every agent in T and zero to every agent outside T. Therefore, by additivity, is solves the game v as follows:

$$\pi_i = \sum_{\mathfrak{I}_i} \frac{\alpha_T}{|T|}, \quad \text{where } \mathfrak{I}_i = \{T \subseteq N \mid i \in T\}. \tag{28}$$

Thus there is a unique solution satisfying the three required axioms, namely, the Shapley value.

In view of (27), we can interpret the coefficient α_T as the *dividend* of coalition T is game v: this is the fraction of surplus generated exclusively by the cooperation of all members of T. The Shapley value (28) divides equally these dividends among all coalition members. Another useful consequence of (28) follows.

Lemma 7.2. (Myerson [1977]). Fix a game (N, v) and for every coalition S, $S \subseteq N$, denote by π^S the Shapley value of the reduced game (S, v) (where agents in $N \setminus S$ are eliminated, along with their cooperative opportunities). Then we have, for all distinct i, j in N,

$$\pi_i^N - \pi_i^{N \setminus j} = \pi_j^N - \pi_j^{N \setminus i}. \tag{29}$$

In other words, the net effect on i of adding j to the set of players equals the net effect on j of adding i. To prove this, use (28) to compute

$$\pi_i^N - \pi_i^{N \setminus j} = \sum_{\mathfrak{I}_i(N)} \frac{\alpha_T}{|T|} - \sum_{\mathfrak{I}_i(N \setminus j)} \frac{\alpha_T}{|T|} = \sum_{\mathfrak{I}_{ij}(N)} \frac{\alpha_T}{|T|},$$

$$\text{where } \mathfrak{I}_{ij} = \{T \mid i, j \in T\}.$$

The huge literature on the Shapley value offers several alternative axiomatic characterizations. One possibility due to Myerson [1977] is to combine property (29) with the requirement that the solution coincides with the standard "equal split of nonseparable surplus" formula for two-person games (namely, $\pi_i = v(i) + (v(12) - v(1) - v(2))/2$).[15] A related characterization (due to Hart and Mas-Colell [1992]) relies on

---

[15] Clearly, formula (29) together with efficiency determines π^N uniquely if $\pi^{N \setminus k}$ is known for all k. Therefore, the characterization follows an obvious induction argument.

the observation that formula (28) can be written as

$$\pi_i^N = \phi(N) - \phi(N \setminus i), \quad \text{where } \phi(T) = \sum_{S \subseteq T} \frac{\alpha_S}{|S|}.$$

Thus the number π_i is interpreted as the derivative with respect to i of the "potential" function ϕ. The above formula easily implies (29); hence, the Shapley value is characterized by two properties: (i) it is the derivative of some potential function, and (ii) it is "standard" for two-person games.

Other characterizations of the value are related to Shapley's original result (Theorem 7.4). The most interesting axiom is the marginalism property, requiring that agent i's surplus share π_i be only a function of the vector of his marginal contributions $(v(S \cup i) - v(S))$ when S runs over all coalitions not containing i. Together with equal treatment of equals, the marginalism axiom characterizes the Shapley value (Young [1985b]). On this result and a related characterization by Chun [1989], see Exercise 7.14.

Remark 7.1

Several noncooperative bargaining games to implement the Shapley value have been proposed. The simplest one relies on a sequence of ultimatum games and a random device. In the first round, a player i is drawn at random (with uniform probability distribution); he proposes a division of $v(N)$ among all players that the other players must take or leave. If a single player refuses the offer, the proposal is cancelled and a random device determines with a small probability δ whether player i (the author of the unsuccessful offer) should be eliminated or not (the latter occurring with probability $(1 - \delta)$). Then we go to round 2, where we play the same game again (starting with the random selection of a proposer) among the surviving agents: a player is selected to propose a division of $v(S)$, where S is the set of surviving agents (thus $S = N$ or $N \setminus i$ depending on the fate of the first-round proposer), and so on. Hart and Mas-Colell [1992] show that, for all δ, this game has a unique subgame perfect equilibrium in stationary strategies, and the equilibrium allocation converges to the Shapley value as δ goes to zero.

7.7. THE SHAPLEY VALUE AND THE CORE

Lemma 7.3. Suppose the game (N, v) is superadditive (resp. subadditive). Then the Shapley value π satisfies $v(i) \leq \pi_i$ (resp. $\pi_i \leq v(i)$).

In the marginal contribution vector π^σ, player i receives $v(S \cup i) - v(S)$ (where S is the coalition of players preceding her in σ), hence at least $v(i)$ (if v is superadditive). As this holds for all vectors π^σ, it holds for any convex combination of these, hence the announced inequality.

Despite this encouraging result, in general the Shapley value is not a core allocation (even when the core of the game is nonempty). Consider the case of the three-player game, $N = \{1, 2, 3\}$, where we assume for simplicity $v(i) = 0$, $i = 1, 2, 3$, and $v(N) = 1$. The core of this game is nonempty if we have (see (15)):

$$v(ij) \leq 1 \quad \text{all } i, j \quad \text{and} \quad v(12) + v(13) + v(23) \leq 2. \quad (30)$$

The Shapley value is

$$\pi_3 = \frac{1}{3}\left(1 + \frac{v(13) + v(23)}{2} - v(12)\right).$$

Therefore, the core constraint $\pi_1 + \pi_2 \geq v(12) \Leftrightarrow \pi_3 \leq 1 - v(12)$ is met if and only if

$$v(12) + \frac{v(13) + v(23)}{4} \leq 1. \quad (31)$$

Clearly, inequality (31) can be violated, whereas (30) holds; e.g., $v(12) = 0.8$, $v(13) = v(23) = 0.5$.

In the important class of convex (resp. concave) games, the Shapley value is a core (resp. core_) allocation. This follows at once from Theorem 7.1 (stating that each marginal contribution vector π^σ is a core allocation) and the definition of the Shapley value as the arithmetic average of all vectors π^σ. In fact, the vectors π^σ are the extreme points of the core, and the Shapley value occupies a central position in the core. Thus it is a natural normative selection from the core.

In Example 7.10, for instance, the Shapley allocation (24) satisfies the core inequalities $\sum_S \pi_i \leq \max_S c_i$. Note that the equally simple allocation of the cost c_n in proportion to c_i (π_i/c_i is independent of i) does not satisfy these inequalities. (Exercise: why?)

Example 7.11.   Fair Assignment with Money

The n agents share a finite set of indivisible objects and can transfer money. Each agent wants to consume at most one object and utilities are quasi-linear. Using the notations of Section 4.3, a fair assignment problem is described by N, a set of houses H, and a matrix of utilities

$(u_i(h))$, $i \in N$, $h \in H$. As in the more general model of fair division with money (Example 7.8), the stand alone surplus $v(S)$ of a coalition S is the maximal surplus of the coalition helping itself in the set H. It sets an upper bound on the joint surplus of S.

$$v(S) = \max_S \sum u_i(h_i),$$

where the maximum bears on all one-to-one mappings $i \to h_i$ from S into H (we assume $|H| \geq |N|$). It turns out that this cooperative game is always concave. A proof of this fact is the subject of Exercise 7.11. Thus the Shapley value is a natural core selection. In the numerical example (5) in Chapter 4, compute the stand alone game:

$$v(1) = 12, \quad v(2) = 9, \quad v(3) = 9, \quad v(N) = 27,$$

$$v(12) = 21, \quad v(13) = 21, \quad v(23) = 15,$$

of which the Shapley value is $\pi_1 = 12$, $\pi_2 = \pi_3 = 7.5$. This corresponds to the optimal assignment (1 gets h_2, 2 gets h_3, 3 gets h_1) and to the transfers $t_1 = 0$, $t_2 = 1.5$, $t_3 = -1.5$. Interestingly, this allocation is envy-free. In fact, it is precisely the allocation recommended by the Aragones algorithm (see Appendix 4.1) and minimizing the largest out-of-pocket payment within the envy-free set.

In general, the Shapley value of the fair-assignment problem does not recommend an envy-free allocation. Consider, for instance, estate division (Example 1.1), the particular case of fair assignment with a single house to allocate. The Shapley value is then given by a formula similar to (24).[16] We showed in Section 4.2 that the stand alone test by itself is not always compatible with the no envy test.

In every convex (resp. concave) game, the Shapley value is *population-monotonic*: agents i's surplus share in the game (N, v) is nondecreasing (resp. nonincreasing) with respect to the set N of agents. To check this, it is enough to show that a marginal contribution solution π^σ (with a fixed ordering of σ of the agents) is population-monotonic. Consider two coalitions T_1, T_2 such that $T_1 \subseteq T_2$. Then we have, for any given player i,

$$\pi_i^{\sigma, T_\epsilon} = v(S_\epsilon \cup i) - v(S_\epsilon),$$

where S_ϵ is the set of players in T_ϵ preceding i for σ.

---

[16] Namely, if the valuations are $u_1 \geq u_2 \geq \cdots \geq u_n$, then $\pi_n = u_n/n$, $\pi_{n-1} = u_n/n + (u_{n-1} - u_n)/(n-1)$, and so on. Exercise: why?

As $S_1 \subseteq S_2$, property (c) in Theorem 7.1 implies $\pi_i^{\sigma, T_1} \le \pi_i^{\sigma, T_2}$, as desired.

Population monotonicity is a normative strengthening of the core property (already discussed in Section 5.5) that reinforces the appeal of the Shapley value in convex (concave) games. However, in these games, the competition for the most appealing core selection is severe; e.g., if there are a natural upper and lower bound on individual utilities, we may use the descending algorithm (see Appendix 7.1) to select a core allocation respecting these bounds.[17]

We conclude this section by mentioning the numerous (nonconvex) games where the Shapley value is typically outside the core. Most exchange economies (with transferable utility) are in this case. We start with an excessively simple example of the Böhm–Bawerk horse market (Section 2.3).

Example 7.12. *Horse Market with Identical Buyers and Identical Sellers (Shapley and Shubik [1969a]).*

We have m identical buyers with $u_i = 1$, all i, and n identical sellers with $v_j = 0$, all j. The core contains a single (competitive) allocation if $m \ne n$: if $m > n$, (the price is 1 and) each seller gains \$1, each buyer breaks even; if $n > m$, (the price is 0 and) all surplus goes to the buyers.

By contrast, the Shapley value distributes a positive surplus share to every player. Suppose, for instance, that we have two buyers and three sellers. We compute the joint surplus share $\beta_1 + \beta_2$ of the buyers at the various marginal contribution vectors:

Ordering	b	b	b	b	s	s	s	s	s	s
	b	s	s	s	b	b	b	s	s	s
	s	b	s	s	b	s	s	b	b	s
	s	s	b	s	s	b	s	b	s	b
	s	s	s	b	s	s	b	s	b	b
$\beta_1 + \beta_2$	0	0	1	1	1	2	2	2	2	2

The short side of the market (the buyers) receives 65% of total surplus in the Shapley value (each buyer gets 0.65, each seller gets 0.117), so the sellers are far from totally exploited. If we have one buyer and five

---

[17] Dutta and Ray [1989] argued for a selection of the core equalizing the surplus shares π_i as much as possible. They show that the descending algorithm with $\bar{\bar{\pi}} = 0$ and $\bar{\pi} = v(N) \cdot (1, \ldots, 1)$ picks exactly such a selection. See also the discussion of the public good provision problem after Theorem 7.2.

sellers, the Shapley value gives 0.833 to the buyer and 0.033 to each seller: the total share of the sellers is still 20% of the buyer's surplus.[18]

The example generalizes. In a typical Böhm–Bawerk market, some agents (buyers or sellers) who are inactive in the efficient allocation (and therefore get no surplus in any core allocation) do receive a positive share of surplus from the Shapley value. Take a utility profile

$$u_1 \geq u_2 \geq \cdots \geq u_m \quad \text{for buyers,}$$

$$v_1 \leq v_2 \leq \cdots \leq v_n \quad \text{for sellers.}$$

Then any seller s_j such that $v_j < u_1$ gets at least $(u_1 - v_j)/(m + n)^2$, because he gets $(u_1 - v_j)$ in the ordering where b_1 is first and s_j is second. Similarly, a buyer b_i such that $u_i > v_1$ gets at least $(u_i - v_1)/(m + n)^2$.

The Shapley value tends to spread out the surplus among the agents much more than the core does. For instance, in the segregated markets with decreasing marginal costs and identical firms of Lemma 2.5, each firm would get a positive share of surplus. Moreover, the surplus distribution is very sensitive to the number of potential firms. Therefore, its application to these economies is of limited interest.

APPENDIX TO CHAPTER 7: PROOF OF THEOREM 7.2

We are given a convex game (N, v) and two utility vectors $\tilde{\pi}$ and $\tilde{\tilde{\pi}}$ satisfying (12) and (13), respectively, and $\tilde{\tilde{\pi}} \leq \tilde{\pi}$. The descending algorithm follows the line from $\tilde{\pi}$ to $\tilde{\tilde{\pi}}$ and stops whenever one of the core constraints becomes tight. If $\tilde{\tilde{\pi}}$ and/or $\tilde{\pi}$ is a feasible allocation (e.g., $\sum_N \tilde{\pi}_i = v(N)$), Theorem 7.2 is trivial and the algorithm chooses this allocation. From now on, we assume

$$\sum_N \tilde{\tilde{\pi}}_i < v(N) < \sum_N \tilde{\pi}_i. \tag{32}$$

---

[18] The general formula is as follows, assuming $m \leq n$:

$$\beta_i = \frac{1}{2} + \frac{n - m}{2m} \cdot \sum_{k=0}^{n} \frac{m! \, n!}{(m + k)!(n - k)!} \quad \text{for a buyer,}$$

$$\sigma_j = \frac{1}{2} - \frac{n - m}{2n} \cdot \sum_{k=0}^{m} \frac{m! \, n!}{(m - k)!(n + k)!} \quad \text{for a seller.}$$

See Shapley and Shubik [1969a].

For all λ, $0 \leq \lambda \leq 1$, define $x(\lambda) = \lambda\overset{\leftrightarrow}{\pi} + (1 - \lambda)\bar{\pi}$. Let λ_1 be the smallest number such that one of the inequalities $\Sigma_S x_i(\lambda) \geq v(S)$ is an equality. In view of (32), this number is well defined and $0 \leq \lambda_1 < 1$. We claim that the set of coalitions S such that $\Sigma_S x_i(\lambda_1) = v(S)$ has a largest element, denoted S_1. More precisely, we show that if S_1, S_2 are two such coalitions, so is $S_1 \cup S_2$ (and incidentally, $S_1 \cap S_2$). To see this, note that by definition of λ_1, we have

$$v(S_i) = \lambda_1\overset{\leftrightarrow}{\pi}_{S_i} + (1 - \lambda_1)\bar{\pi}_{S_i}, \qquad i = 1, 2$$

$$\text{and} \qquad v(S) \leq \lambda_1\overset{\leftrightarrow}{\pi}_S + (1 - \lambda_1)\bar{\pi}_S, \quad \text{all } S \subseteq N. \quad (33)$$

Therefore, if we write $\delta_i = \bar{\pi}_i - \overset{\leftrightarrow}{\pi}_i$ and $w(S) = v(S) - \overset{\leftrightarrow}{\pi}_S$, the ratio $w(S)/\delta_S$ is maximal at S_i, $i = 1, 2$. (Note that if $\delta_i = 0$, then $\lambda_1 = 0$ and the computations below can be adapted.) Combining this with the convexity of v (hence of w), we compute

$$w(S_1 \cup S_2) + \delta_{S_1 \cap S_2} \cdot \frac{w(S_1)}{\delta_{S_1}} \geq w(S_1 \cup S_2) + w(S_1 \cap S_2)$$

$$\geq w(S_1) + w(S_2)$$

$$= w(S_1) + \frac{\delta_{S_2}}{\delta_{S_1}} \cdot w(S_1) \Rightarrow \frac{w(S_1 \cup S_2)}{\delta_{S_1 \cup S_2}} \geq \frac{w(S_1)}{\delta_{S_1}},$$

proving the claim. We denote by T_1 the largest coalition such that $x_S(\lambda_1) = v(S)$. The T_1-coordinates of our desired core allocation π will be $x_i(\lambda_1)$.

The algorithm continues by considering the following reduced game $(N \setminus T_1, v_1)$:

$$v_1(S) = v(S \cup T_1) - v(T_1) \quad \text{for all } S \subseteq N \setminus T_1.$$

Note that for all $S \subseteq N \setminus T_1$, we have, by (33),

$$v(S \cup T_1) \leq x_{S \cup T_1}(\lambda_1) = v(T_1) + x_S(\lambda_1) \Rightarrow v_1(S) \leq x_S(\lambda_1) \leq \bar{\pi}_S,$$

$$v_1(N \setminus T_1) - v_1(N \setminus (T_1 \cup S)) = v(N) - v(N \setminus S) \geq \overset{\leftrightarrow}{\pi}_S.$$

Moreover, the game $(N \setminus T_1, v_1)$ is convex as well. Therefore, we can apply the same construction to the reduced game as we did to the original game.

Define λ_2 as the smallest number such that one of the inequalities $x_S(\lambda) \geq v_1(S)$, $S \subseteq N \setminus T_1$, is an equality, and T_2 as the largest coalition in $N \setminus T_1$ where equality holds. By construction, we have $\lambda_1 < \lambda_2 < 1$, and we set the T_2-coordinates of our desired core allocation as $x_i(\lambda_2)$, and so on: define the game $(N \setminus (T_1 \cup T_2), v_2)$ by $v_2(S) = v_1(S \cup T_2) - v_1(T_2)$, or $v_2(S) = v(S \cup T_1 \cup T_2) - v(T_1 \cup T_2)$, and proceed to define λ_3. The algorithm stops whenever a coalition T_k equals $N \setminus (T_1 \cup \cdots \cup T_{k-1})$. At that point, the allocation π is entirely defined and $\tilde{\tilde{\pi}} \leq \pi \leq \tilde{\pi}$ holds because, for all i, $\pi_i = x_i(\lambda_t)$ for some λ_t, $0 \leq \lambda_t \leq 1$ (of course, if i belongs to T_α, the definition of $x_i(\lambda)$ is adapted to the game v_α).

It remains to check that π is a core allocation. Clearly, its restriction to T_k is in the core of (T_k, v_{k-1}) by $v_{k-1}(T_k) = x_{T_k}(\lambda_k) = \pi_{T_k}$ and property (33). Check that its restriction to $T_{k-1} \cup T_k$ is in the core of $(T_{k-1} \cup T_k, v_{k-2})$. Fix a coalition S in $T_{k-1} \cup T_k$ and denote $S_k = S \cap T_k$, $S_{k-1} = S \cap T_{k-1}$. Using (33) and the convexity of v_{k-2}, we have

$$\pi_S = x_{S_{k-1}}(\lambda_{k-1}) + x_{S_k}(\lambda_k) \geq v_{k-2}(S_{k-1}) + v_{k-1}(S_k)$$

$$= v_{k-2}(S_{k-1}) + v_{k-2}(S_k \cup T_{k-1}) - v_{k-2}(T_{k-1})$$

$$\geq v_{k-2}(S_k \cup S_{k-1}) = v_{k-2}(S),$$

as desired. An induction argument shows that the restriction of π to $T_t \cup \cdots \cup T_k$ is in the core of v_{t-1}, and the proof is complete.

EXERCISES ON CHAPTER 7

Exercise 7.1. The Garbage Game (Shapley and Shubik [1969b])

Player i has a bag of trash of size $b_i > 0$. Each player can dump his bag in the backyard of the player of her choice. The disutility to agent i of a pile of trash of size b in his backyard is b. She gets no disutility from a pile of trash in someone else's backyard.

(a) Show that the cooperative opportunities are described by the following cooperative game:

$$v(N) = -\sum_N b_i, \qquad v(S) = -\sum_{N \setminus S} b_i, \quad \text{for all } S \subsetneq N, \quad S \neq \varnothing.$$

Check that this game is superadditive and that its core is empty for $n \geq 3$.

(b) Compute the Shapley value of this game.

Exercise 7.2. A Game of Information Sharing (Muto [1986])

The set N of agents contains a subset I of "informed" agents. An informed agent is able to secure a certain profit r_i, $r_i \geq 0$, independently of what other agents do. In any coalition, if one agent is informed, he can pass the information at no cost. Hence the following characteristic function game:

$$v(S) = \sum_{i \in S} r_i \quad \text{if } S \cap I \neq \varnothing, \qquad v(S) = 0 \quad \text{if } S \cap I = \varnothing.$$

An alternative interpretation is that an agent in I holds a patent and can share it with any other member of a coalition of which he is a member.

(a) Show that the above game is convex if I contains a single agent. Show that the game is not convex if I contains two or more agents and at least one r_i, $i \in N \setminus I$, is positive.

(b) Show that the core contains a single allocation if $|I| \geq 2$ but is "big" if $|I| = 1$.

(c) Show that the Shapley value is

$$\pi_i = \frac{p}{p+1} \cdot r_i \qquad \text{if } i \notin I, \quad \text{where } p = |I|,$$

$$\pi_i = r_i + \frac{1}{p+1} \sum_{N \setminus I} r_j \quad \text{if } i \in I.$$

Exercise 7.3. Pairwise Cooperation Games

For every pair of players i, j, the number v_{ij} measures the (positive or negative) surplus generated by the interaction of i and j. We consider the following game (N, v):

$$v(S) = \sum_{\{i, j\} \subseteq S} v_{ij}, \qquad \text{where the sum bears over the } s(s-1)/2 \text{ pairs of } S.$$

(a) Show that this game is convex if and only if v_{ij} is nonnegative for all i, j.

(b) Show that the game is balanced only if there is no *negative cut* of N. A negative cut is a partition of N as $N_1 \cup N_2$, such that

$$\sum_{i \in N_1, \, j \in N_2} v_{ij} < 0.$$

(*Hint:* If there is a negative cut, show that the game is not super-additive.)

Note that the converse property (the game is balanced if there is no negative cut) is also true (Kalai and Zamel [1982]).

Exercise 7.4. Output-Sharing and Cooperative Production Games

(a) We generalize Example 7.4 to the case of multi-input production functions. Agent i supplies (inelastically) a vector x_i of inputs ($x_i \in \mathbb{R}_+^K$) and the single output production function f has domain $\mathbb{R}_+^K$ and range $\mathbb{R}$. Assume that f is differentiable. The stand alone game is defined by (8). Show that this game is convex (resp. concave) if and only if we have

$$\frac{\partial^2 f}{\partial x^k \, \partial x^{k'}}(x) \geq 0 \quad \text{for all } k, k' = 1, \dots, K \quad \text{and} \quad \text{all } x \in \mathbb{R}_+^K$$

(resp. the opposite inequality).

(b) Consider a cooperative production game (Example 7.5) with a single input, a single output, and a convex cost function c. Show that the TU game (10) is concave (the function v is submodular).

Exercise 7.5. The Traveling Lecturer Game (Tamir [1989])

A lecturer located at 0 is invited to lecture in n universities located at $\{1, 2, \dots, n\}$, respectively. The cost of traveling between any two locations is c_{ij} (where $i, j = 0, 1, \dots, n$ and $c_{ij} = c_{ji}$). The cost $c(S)$ of visiting a subset S of $N = \{1, \dots, n\}$ is defined by the least expensive tour starting at 0 and visiting all universities in S.

(a) Show that c is subadditive.

(b) Consider the following example:

$c_{ij} = 1$ for the following pairs:

$$(0, 1) \ (0, 2) \ (0, 3) \ (1, 4) \ (2, 5) \ (3, 6) \ (4, 5) \ (4, 6) \ (5, 6),$$

$c_{ij} = 2$ for every other pair.

Draw a figure of the transportation network. Show that the core_ of the corresponding cost-sharing game is empty. (*Hint:* Use the symmetries of the game: if the core_ is nonempty, it must contain an allocation where 1, 2, and 3 get the same share and so do 4, 5, and 6.)

(c) Remove player 6 in the above game. Show that the reduced five-player game has a nonempty core_.

(d) Remove player 1: is the core_ of the reduced game empty or not? (Note that the lecturer may still travel through the city formally occupied by player 1.)

Exercise 7.6. Minimal Cost-Spanning Tree Games (Granot and Huberman [1981], Sharkey [1993])

Let N be a set of players and 0 be a source. Given is a symmetrical matrix (c_{ij}), $i, j = 0, 1, \ldots, n$, where c_{ij} represents the cost of connecting i and j. We interpret 0 as the source of a (telephone) network and N as the users os the network. The cost $c(S)$ of serving the consumers of S is defined by the least expensive graph with nodes $S \cup \{0\}$ connecting the source and all consumers in S (note that coalition S is not allowed to use the nodes occupied by players outside S).

(a) Show that the least expensive graph serving S is always a tree (namely, a connected graph without cycles) with nodes $S \cup \{0\}$ and that the cost function c is subadditive. Check that the cost-sharing game is concave in the following example with three players:

1	2	3	
10	16	17	0
	9	11	1
		18	2

and compute its core_.

(b) Consider the following example with $N = \{1, 2, 3, 4\}$:

$$c_{20} = c_{30} = c_{24} = 2, \qquad c_{ij} = 1 \quad \text{for all other pair } ij.$$

Show that this game is not concave; show it is balanced_.

(c) Given are a fixed tree T^* with a set of nodes $N \cup \{0\}$ (each player and the source occupy exactly one node), and a cost c_{ij}^* for each edge ij of the tree. To (T^*, c^*) we associate the minimal cost-spanning tree game where c_{ij} is the total cost of the path connecting i to j on T^*. Show that this game is concave. Note that the game of Example 7.10 is a particular case where the graph T^* is a line, with the source at one end.

Taking 0 as the root of the tree, define $p(i)$, the predecessor of agent $i \in N$ on T^*, as the node adjacent to i on the path connecting 0 to i on T^* (between any two nodes of a tree, there is a unique path along the edges of the tree). Note that $p(i) \in N \cup \{0\}$. Denote by $S(i)$ the set of successors of i, namely, those $j \in N$ such that the path from 0 to j

visits i. Note that $i \in S(i)$. Show that the Shapley value x is as follows:

$$x_i = \sum_j \frac{c_{jp(j)}}{|S(j)|}, \quad \text{where the sum bears on all nodes}$$

visited in the path from 0 to i.

This formula generalizes (24).

(d)* We are back in the general context of questions (a) and (b). Show that every minimal cost-spanning tree game is balanced_ . Fix N and the matrix (c_{ij}). Pick a tree T with cost $c(N)$ (i.e., a minimal cost-spanning tree for the grand coalition). Taking 0 as the root of the tree, show that the allocation $x_i = c_{ip(i)}$ is in the core_ (where the predecessor $p(i)$ is defined in the previous question).

(e) We now generalize the minimal cost-spanning tree game model by allowing coalitions of customers to add more nodes to their connecting tree if this allows them to reduce total cost. We think of a minimal cost-spanning tree game in which certain nodes (other than the source) are vacant. For example, consider the initial minimal cost-spanning tree game with six players:

$$c_{0i} = 1, \quad \text{if } i = 4,5,6, \qquad c_{0j} = 2, \qquad \text{if } j = 1,2,3,$$

$$c_{ij} = 1, \quad \text{if } i = 4,5,6, \qquad j = 1,2,3 \quad \text{and } i + j \neq 7,$$

$$c_{34} = c_{25} = c_{16} = 3.$$

(*Hint:* This is conveniently depicted by a triangular configuration with 0 in the center of the triangle, 1, 2, and 3 as vertices, and 4, 5, and 6 as the middle points of the edges opposed to 3, 2, and 1.) Describe the symmetrical allocations in the core_. Next, remove players 4, 5, and 6 and consider the cost-sharing game among 1, 2, and 3, where a coalition can build a network visiting the nodes vacated by the players 4, 5, and 6 (at the same cost as before). Show that the core_ is empty.

Exercise 7.7. A Property of Convex Games

Say that the TU game (N, v) is *inessential* if we have $v(S) = \sum_S v(i)$ for all S. Clearly, the core of an inessential game contains the single allocation $\pi_i = v(i)$. Show that if (N, v) is a convex game, its core contains a single allocation (if and) only if the game is inessential. Give an example of a (three-player) noninessential game with a single element in its core.

Exercise 7.8. *A General Property of Marginal Contribution Vectors (Weber [1988])*

For *any* TU cooperative game (N, v), show that the core is contained in the convex hull of the marginal contribution vectors. *Hint:* Suppose not. Use a separation argument to show the existence of a vector p in $\mathbb{R}^N$ and a core allocation π^* such that

$$\text{for all ordering } \sigma: \qquad p \cdot \pi^\sigma > p \cdot \pi^*.$$

Choose an ordering σ such that

$$p_{\sigma(1)} \geq p_{\sigma(2)} \geq \cdots \geq p_{\sigma(n)},$$

and compute

$$p \cdot \pi^\sigma = p_{\sigma(n)} \cdot v(N) + \sum_{1}^{n-1} (p_{\sigma(i)} - p_{\sigma(i+1)}) \cdot v(S_i),$$

where $S_i = \{\sigma(1), \ldots, \sigma(i)\}$. Deduce a contradiction.

Exercise 7.9. *Applying Theorem 7.2*

(a) Consider the public good provision problem with utilities quasi-linear in the private good, $u_i = v_i(y) - x_i$, and strictly concave v_i. Assume also nondecreasing marginal cost.

We pick a Nash equilibrium $(x_1, \ldots, x_n)$ of the voluntary contribution mechanism as described in Exercise 6.18 (question (b)). Without loss, assume that $x_i > 0$ for all $i = 1, \ldots, m$ and $x_j = 0$ for $j = m + 1, \ldots, n$. Denote by $\bar{y}$ the equilibrium level of public good. We wish to show the existence of an allocation in the stand alone core and Pareto superior to the above equilibrium allocation. By Theorem 7.2, this amounts to showing that for all coalition S, we have

$$\sum_{S} (u_i(\bar{y}) - x_i) \leq v(N) - v(N \setminus S). \tag{34}$$

Fix S and denote by y^* the stand alone public good level of $N \setminus S$. Distinguish two cases. If $N \setminus S$ contains at least one agent i active in equilibrium $(i \leq m)$, show that $\bar{y} \leq y^*$ and deduce (34). If all active agents are in S, show that $\sum_S x_i = c(\bar{y})$ and deduce (34) again.

(b) Consider a concave multi-output production function c such that $c(0) = 0$ (in particular, c is subadditive). Each one of the n agents is endowed with a quasi-linear utility $u_i(y_i) - x_i$ (here, $y_i \in \mathbb{R}_+^K$), where u_i

is continuous and monotonic but otherwise arbitrary. Assume that the surplus is maximized at the consumption profile $(y_1^*, \ldots, y_n^*)$:

$$\sum_N u_i(y_i^*) - c(y^*) = v(N), \quad \text{where } y^* = \sum_N y_i^*.$$

Show the existence of a cost-sharing $(x_1, \ldots, x_n)$ of $c(y^*)$ satisfying all three following properties:

(i) Stand alone core on costs:

$$\sum_S x_i \leq c\left(\sum_S y_i^*\right), \quad \text{for all } S.$$

(ii) Stand alone test on utilities:

$$\max_y \{u_i(y) - c(y)\} \leq u_i(y_i^*) - x_i.$$

(iii) Unanimity test on utilities:

$$u_i(y_i^*) - x_i \leq \max_y \left\{ u_i(y) - \frac{c(ny)}{y} \right\}.$$

Exercise 7.10.   Shapley Value in the Model of Exercise 4.5

In the fair-assignment problem with multiplicative utilities of Exercise 4.5, compute the Shapley value. Recall that agent i's utility for house h is $\alpha_i \cdot \beta_h$, where we assume the same number of agents and of houses, and $\alpha_1 \geq \alpha_2 \geq \cdots \geq \alpha_n \geq 0$, $\beta_1 \geq \beta_2 \geq \cdots \geq \beta_n \geq 0$. Denote $\gamma_i = (\sum_1^i \beta_k)/i$. Show that the Shapley value is given by the formula

$$\pi_i = \sum_{j=i}^n (\alpha_j - \alpha_{j+1}) \cdot \gamma_j \quad \text{for all } i, \text{ with the convention } \alpha_{n+1} = 0.$$

Exercise 7.11.   Shapley Value in the Fair-Assignment Model (Moulin [1992])

Given a fair-assignment problem (N, H, u) (Section 4.3), we denote by $v(S, K)$ the stand alone surplus of coalition S, $S \subseteq N$, when it can allocate the houses in K, $K \subseteq H$, among its members.

(a) Show that for all $K \subseteq H$, the game $(N, v(\cdot, K))$ is concave. Hence the Shapley value is population-monotonic: $\pi_i(S, K)$ is nonincreasing in S (see Section 7.7).

(b) Show that for all $S \subseteq N$, all $K \subseteq H$, we have the inequality:

$$v(S \cup i, K) + v(S, K \cup h) \leq v(S \cup i, K \cup h) + v(S, K)$$

$$\text{for all } i \in N, \quad \text{all } h \in H.$$

Deduce that the Shapley value is resource-monotonic: $\pi_i(S, K)$ is nondecreasing in K $(\pi_i(S, K) \leq \pi_i(S, K \cup h)$ for all h, all $i)$.

(c) Show that the Shapley value satisfies the fair share guaranteed lower bound (4) in Chapter 4.

Exercise 7.12.   A Generalization of Example 7.10

(a) A facility is jointly used by n users. The facility can provide a (finite) set A of services (each service is an indivisible public good; all goods are different). Player i uses a certain subset A_i of services $(A_i \subseteq A)$. The cost of service is c_a, $c_a \geq 0$, and that of a subset B of A is $\tilde{c}(B) = \Sigma_B \, c_a$. The cost-sharing game (N, c) is defined by

$$c(S) = \tilde{c}\left(\bigcup_{i \in S} A_i \right) \quad \text{for all } S \subseteq N. \tag{35}$$

Show that this game is concave. Show that c can be written as

$$c = \sum_{T \subseteq N} \beta_T \cdot \gamma_T, \quad \text{where } \gamma_T \text{ is defined by (22),}$$

and

$$\beta_T = \tilde{c}\left(\bigcap_N B_T(i) \right), \quad \text{where } B_T(i) = A_i \quad \text{if } i \in T,$$

$$= A \setminus A_i \quad \text{if } i \notin T.$$

Conversely, every cost function that is a positive linear combination of the games γ_T can be written as a game (35).

(b) Consider now the game

$$v(S) = \tilde{c}\left(\bigcap_{i \in N \setminus S} A_i \right) \quad \text{for all } S \subseteq N. \tag{36}$$

Show that this game is convex and can be written as

$$v = \sum_{T \subseteq N} \alpha_T \delta_T, \quad \text{where } \delta_T \text{ is defined by (25),}$$

for some nonnegative coefficients α_T. Conversely, every positive linear combination of the games δ_T can be written as a game (36).

Exercise 7.13. A Dual Form of Lemma 7.1

Given a TU game (N, v), we denote by $(N, \tilde{v})$ the dual game: $\tilde{v}(S) = v(N) - v(N \setminus S)$.
   (a) Check that $\tilde{\gamma}_T = \delta_T$, where γ_T, δ_T are defined by (22) and (25).
   (b) Show that formula (27) implies

$$v = \sum_{T \subseteq N} \alpha_T(\tilde{v}) \cdot \gamma_T,$$

where $\alpha_T(v)$ is given by (26).
   (c) Compute:

$$\alpha_T(\tilde{v}) = (-1)^{t+1} v(N) + \sum_{N \setminus T \subseteq S'} (-1)^{s'+1-(n-t)} v(S'),$$

$$\text{where } t = |T|, \quad s' = |S'|.$$

Exercise 7.14. Two Characterizations of the Shapley Value (Young [1985b], Chun [1989])

Given N, we denote by ϕ a solution for TU cooperative games on N, namely, a mapping associating to every game (N, v) an allocation $\phi(v)$ of $v(N)$: $\sum_N \phi_i(v) = v(N)$. We define two axioms:
Marginalism (Young).   For any games v_1, v_2 and any agent i:

$$\{v_1(S \cup i) - v_1(S) = v_2(S \cup i) - v_2(S)$$

$$\text{for all } S \subseteq N \setminus i\} \Rightarrow \phi_i(v_1) = \phi_i(v_2).$$

   The Chun Axiom.   For any game v, any coalition T, and any number λ:

$$\phi_i(v + \lambda \delta_T) = \phi_i(v) \quad \text{for all } i \in N \setminus T.$$

(a) Show that for any solution ϕ, marginalism implies the Chun axiom. Show that the combination additivity plus dummy implies the Chun axiom.

(b) Show that the Shapley value is the *only* solution satisfying the Chun axiom and equal treatment of equals. (*Hint:* Take ϕ satisfying these two axioms and show that ϕ coincides with the Shapley value for all games $\lambda \cdot \delta_T$, next for all games $\lambda \delta_T + \mu \delta_T$, and so on.)

Exercise 7.15. A Property of the Shapley Value in the Public Good Provision Problem

The cost of the public good is increasing and continuous. Individual utilities are quasi-linear in money and increasing in the public good: $u_i = u_i(y) - x_i$. We assume that the efficient level of public good y^* is positive.

Show that in the Shapley value allocation of the stand alone surplus, the cost shares x_i satisfy

$$0 < x_i < u_i(y^*) \quad \text{for all } i = 1, \dots, n.$$

Hint: for $x_i > 0$. Suppose $x_i = 0$ for some agent i. Note that $v(N) - v(N \setminus i) \le u_i(y^*)$ and deduce that $x_i = 0$ implies $v(N) - v(N \setminus i) = v(i) = u_i(y^*)$ (use the convexity of the stand alone game). Denote by y the stand alone level of public good for player i. Show that $v(N \setminus i) > u_{N \setminus i}(y)$ and deduce a contradiction.

Exercise 7.16. Universally Stable Families of Coalitions

(a) *Generalizing Example 7.2.* Consider a tree (connected undirected graph without cycles) of which the n nodes are occupied by the n players (one player per node). Call a set of nodes S *connected* if the path between any two nodes in S never visits a node outside S. Denote by $\mathcal{F}$ the set of all connected coalitions. To show that $\mathcal{F}$ is universally stable, we invoke the corollary of Theorem 7.3. A constructive proof uses a variant of the greedy algorithm.

Pick any node (player) i^* to be the root of the tree and denote by $S(i)$ the set of successors of node i, namely, the set of nodes j such that the path from i^* to j visits i. Note that $i \in S(i)$ and that $S(i^*) = N$.

Fix a superadditive value function v defined on $\mathcal{F}$ and consider the following allocation:

$$\pi_i = v(S(i)) - v^*(S(i) \setminus i) \quad \text{for all } i,$$

where v^* is defined by (5). Show that this allocation is in the core of (N, v^*). (*Hint:* Consider first the case of a linear tree as in Example 7.2.)

(b) *Applying the Corollary of Theorem 7.3.* The n players are located around a circle. The set $\mathscr{F}$ contains all singleton coalitions and all pairs of two *adjacent* players. Show that this family *is not* universally stable if n is odd. Show that it *is* universally stable if n is even.

Bibliography

Alkan, A. "Non Existence of Stable Threesome Matchings," *Mathematical Social Sciences*, 16, 207–209 (1988).

Alkan, A. "Monotonicity and Envyfree Assignments," *Economic Theory*, 4, 605–616 1994.

Alkan, A., G. Demange, and D. Gale. "Fair Allocation of Indivisible Goods and Criteria of Justice," *Econometrica*, 59, 1023–1039 (1991).

Aragones, E. "A Derivation of the Money Rawlsian Solution," *Social Choice and Welfare* (1992).

Arrow, K. and F. Hahn. *General Competitive Analysis*, San Francisco: Holden Day, 1971.

Atkinson, A. and J.E. Stiglitz *Lectures on Public Economics*. New York: McGraw-Hill, 1980.

Aumann, R.J. "Agreeing to Disagree," *The Annals of Statistics*, 4, 1236–1239 (1976).

Axelrod, R. *The Evolution of Cooperation*. New York: Basic Books, 1984.

Banks, J.S. "Singularity Theory and Core Existence in the Spatial Model," unpublished working paper, University of Rochester, 1994.

Barbera, S. and M. Jackson. "Strategy-Proof Exchange," *Econometrica* 63, 1, 51-84, 1995.

Barbera, S. and B. Peleg. "Strategyproof Voting Schemes with Continuous Preferences," *Social Choice and Welfare*, 7 (1), 31–38 (1980).

Barbera, S., F. Gül, and E. Stachetti. "Generalized Median Voter Schemes and Committees," *Journal of Economic Theory*, 61 (2), 262–289 (1993).

Bar-Hillel, M. and M. Yaari. "Judgements of Justice," in B. Mellers and J. Baron (eds.), *Psychological Perspectives on Justice*. Cambridge: Cambridge University Press, 55–84, 1993.

Baumol, W. *Superfairness*. Cambridge, MA: MIT Press, 1986.

Baumol, W., J. Panzar, and R. Willig. *Contestable Markets and the Theory of Industry Structure*. New York: Harcourt, Brace, Jovanovitch, 1982.

Becker, G.S. *A Treatise on the Family*, Cambridge, MA: Harvard University Press, 1981.

Berliant, M. "An Equilibrium Existence Result for an Economy with Land," *Journal of Mathematical Economics*, 14, 53–56 (1985).

Berliant, M., K. Dunz, and W. Thomson. "On the Fair Division of A Heterogeneous Commodity," *Journal of Mathematical Economics*, 21, 201–216 (1992).

Binmore, K. "Nash Bargaining Theory, 1," in K. Binmore and P Dasgupta (eds.), *The Economics of Bargaining*, Oxford: Blackwell Publisher, 1987.

Black, D. "On the Rationale of Group Decision Making," *Journal of Political Economy* 56, 23–34 (1948).

Bliss, C. and B. Nalebuff. "Dragon Slaying and Ballroom Dancing: The Private Supply of a Public Good," *Journal of Public Economics* 25, 1–12 (1984).

Bondareva, O.N. "Some Applications of Linear Programming Methods to the Theory of Cooperative Games" (in Russian), *Problemy Kibernetiki*, 10, 119–139 (1963).

Border, K.C. *Fixed Point Theorems with Applications to Economics and Game Theory*. New York: Cambridge University Press, 1985.

Brams, S. "Voting Procedures," in R.J. Aumann and S. Hart (eds.), *The Handbook of Game Theory with Economic Applications, Vol. 2.* Amsterdam: North-Holland, 1993.

Brams, S. and A. Taylor. "Fair Division: Procedures for Allocating Divisible and Indivisible Goods," forthcoming Cambridge University Press, 1995.

Bryant, J. "A Simple Rational Expectations Keynes-Type Model," *Quarterly Journal of Economics*, 98, 525–528 (1983).

Case, J.H. *Economics and the Competitive Process, Studies in Game Theory and Mathematical Economics*. New York: New York University Press, 1979.

Champsaur, P. and G. Laroque. "Fair Allocations in Large Economics," *Journal of Economic Theory*, 25, 269–282 (1981).

Chichilnisky, G. and W. Thomson. "The Walrasian Mechanism from Equal Division is not Monotonic with Respect to Variations in the Number of Consumers," *Journal of Public Economics*, 32, 119–124 (1987).

Ching, S. "An Alternative characterization of the Uniform Rule," *Social Choice and Welfare*, 10, 1–6 (1993).

Chun, Y. "A New Axiomatization of the Shapley Value," *Games and Economic Behaviour*, 1, 119–130 (1989).

Chun, Y. and W. Thomson. "Monotonicity Properties of Bargaining Solutions When Applied to Economics," *Mathematical Social Sciences*, 15, 11–27 (1988).

Coase, R.H. "The Problem of Social Cost," *Journal of Law and Economics*, 3, 1–44 (1960).

Cohen, G.A. "Self Ownership, World Ownership and Equality," *Social Philosophy and Policy*, 3, 77–96 (1986).

Condorcet, Marquis de. *Esquisse d'un Tableau Historique des Progrès de l'Esprit Humain (Ninth Epoch)*, Paris, 1793.

Cooper, R. and A. John. "Coordinating Coordination Failures in Keynesian Models," *Quarterly Journal of Economics*, 103, 441–463 (1988).

Cournot, A. *Recherches sur les Principes Mathématiques de la Theorie des Richesses*. Paris: Hachette, 1838.

Crawford, V.P. "A Procedure for Generating Pareto Efficient Egalitarian-Equivalent Allocations," *Econometrica*, 47, 49–60 (1979).

Crawford, V.P. "An Evolutionary Interpretation of Van Huyck, Battalio and Bell's Experimental Result on Coordination," *Games and Economic Behavior*, 3 (1), 25–59 (1991).

Debreu, G. *Theory of Value*. New York: Wiley, 1959.

Debreu, G. and H. Scarf. "A Limit Theorem on the Core of an Economy," *International Economic Review*, 4, 235–246 (1963).

Demange, G. "Implementing Efficient and Egalitarian Equivalent Allocations," *Econometrica*, 52, 5 (1984).

Demange, G. and D. Gale. "The Strategy Structure of Two-Sided Matching Markets," *Econometrica*, 53, 873–888 (1985).

Diamantaras, D. "On Equity With Public Goods," *Social Choice and Welfare*, 9, 141–157 (1992).

Diamantaras, D. and W. Thomson. "An Extension and Refinement of the No-Envy Concept," *Economics Letters*, 33, 217–222 (1990).

Diamantaras, D. and S. Wilkie. "A Generalization of Kaneko's Ratio Equilibrium for Economies with Private and Public Goods," *Journal of Economic Theory*, 62 (2), 499–512 (1994).

Dubey, P. "Price Quantity Strategic Market Games," *Econometrica*, 50 (1), 111–126 (1982).

Dubins, L. and E. Spanier. "How to Cut a Cake Fairly," *American Mathematical Monthly*, 68 (1), 17 (1961).

Dutta, B. and D. Ray. "A Concept of Egalitarianism under Participation Constraints," *Econometrica*, 57 (3), 615–636 (1989).

Dutta, B., A. Sen, and R. Vohra. "Nash Implementation through Elementary Mechanisms in Economic Environments," Economic Design (1993).

Dworkin, R. "What is Equality? Part 2: Equality of Resources," *Philosophy and Public Affairs*, 10, 283–345 (1981).

Elster, J. *Solomonic Judgments*. New York: Cambridge University Press, 1989.

Faulhaber, G. "Cross Subsidization: Pricing in Public Enterprises," *American Economic Review*, 65, 966–977 (1975).

Feldman, A. and A. Kirman. "Fairness and Envy," *American Economic Review*, 64, 995–1005 (1974).

Finley, M. *Studies in Land and Credit in Ancient Athens, 500–200 B.C.* New Brunswick: Rutgers University Press, 1951.

Fishburn, P.C. *The Theory of Social Choice*. Princeton: Princeton University Press, 1973.

Fleurbaey, M. "Three Solutions for the Compensation Problem," *Journal of Economic Theory* (1993).

Foley, D. "Resource Allocation and the Public Sector," *Yale Economic Essays*, 7 (1), 45–98 (1967).

Friedman, J.W. *Game Theory with Applications to Economics*. New York: Oxford University Press, 1986.

Fudenberg, D. and J. Tirole. *Game Theory*. Cambridge, MA: MIT Press, 1991.

Gale, D. and L.S. Shapley. "College Admission and the Stability of Marriage," *American Mathematical Monthly*, 69, 9–14 (1962).

Gibbard, A. "Manipulation of Voting Schemes: A General Result," *Econometrica*, 45, 665–682 (1973).

Granot, D. and G. Huberman. "Minimum Cost Spanning Tree Games," *Mathematical Programming*, 21, 1–18 (1981).

Greenberg, J. "Cores of Convex Games without Side Payments," *Mathematics of Operations Research*, 523–525 (1985).

Greenberg, J. *The Theory of Social Situations: An Alternative Game Theoretic Approach*. Cambridge: Cambridge University Press, 1990.

Groves, T. and J. Ledyard. "Incentive Compatibility Since 1972," in T. Groves, R. Radner, and S. Reiter (eds.), *Information, Incentives and Economic Mechanisms*. Minneapolis: University of Minnesota Press, 1987.

Harsanyi, J.C. "A Simplified Bargaining Model for the N-Persons Cooperative Games," *International Economic Review* 4, 194–220 (1963).

Hardin, G. "The Tragedy of the Commons," *Science*, 162, 1243–1248 (1968).

Hart, S. 1985, "An Axiomatization of Harsanyi's Non-Transferable Utility Solution," *Econometrica*, 53, 1295–1313 (1985).

Hart, S. and A. Mas-Colell. "Potential, Value and Consistency," *Econometrica*, 57, 589–614 (1989).

Hart, S. and A. Mas-Colell. "Bargaining and Value," unpublished working paper, The Hebrew University of Jerusalem, 1992.

Hayek, F. *The Mirage of Social Justice, Vol. 2.* Chicago: University of Chicago Press, 1976.

Hildenbrand, W. *Core and Equilibria of a Large Economy.* Princeton: Princeton University Press, 1974.

Hildenbrand, W. and A. Kirman. *Introduction to Equilibrium Analysis.* Amsterdam: North-Holland, 1974.

Homans, G.C. *Social Behavior: Its Elementry Forms.* New York: Harcourt, Brace, Jovanovitch, 1961.

Hurwicz, L. "On Informationally Decentralized Systems," in R. Radner and C.B. McGuire (eds.), *Decision and Organization.* Amsterdam: North-Holland, 1972.

Hylland, A. and R. Zeckhauser. "The Efficient Allocation of Individuals to Positions," *Journal of Political Economy*, 87, 293–314 (1979).

Ichiishi, T. "Super-Modularity: Applications to Convex Games and to the Greedy Algorithm for LP," *Journal of Economic Theory*, 25, 283–286 (1981).

Ichiishi, T. *Game Theory for Economic Analysis.* New York: Academic Press, 1983.

Ichiishi, T. "Comparative Cooperative Game Theory," *International Journal of Game Theory*, 19, 139–152 (1990).

Isaac, R.M., D. Mathieu, and E.E. Zajac. "Institutional Framing and Perceptions of Fairness," *Constitutional Political Economy*, 2 (3), 329–370 (1991).

Israelsen, D. "Collectives, Communes, and Incentives," *Journal of Comparative Economics*, 4, 99–124 (1980).

Kalai, E. and E. Zemel. "Totally Balanced Games and Games of Flow," *Mathematics of Operations Research*, 7, 476–478 (1982).

Kalai, E. and D. Samet. "Monotonic Solutions to General Cooperative Games," *Econometrica*, 53, 307–327 (1985).

Kaneko, M. "The Ratio Equilibrium and a Voting Game in a Public Good Economy," *Journal of Economic Theory*, 16, 2 (1977).

Kaneko, M. and M. Wooders. "Cores of Partitioning Games," *Mathematical Social Sciences*, 3, 313–327 (1982).

Kannai, Y. "Concavifiability and Constructions of Concave Utility Functions," *Journal of Mathematical Economics*, 4, 1–56 (1977).

Knuth, D.E. *Marriages Stables.* Montreal: Presses de l'Université de Montreal, 1976.

Kolm, S.C. *Justice et Equité*, Paris: Editions du CNRS, 1972.

Kolm, S.C. "Super-Equité," *Kyklos*, 26, 841–843 (1973).

Kolm, S.C. *Le Contrat Social Liberal*. Paris: Presses Universitaires de France, 1985.

Kreps, D. *A Course in Microeconomic Theory*. Princeton: Princeton University Press, 1990.

Le Breton, M., G. Owen, and S. Weber. "Strongly Balanced Cooperative Games," *International Journal of Game Theory*, 20, 419–427 (1992).

Leonard, H.B. "Elicitation of Honest Preferences for the Assignment of Individuals to Positions," *Journal of Political Economy*, 91, 461–479 (1983).

Lindahl, E. "Just Taxation—A Positive Solution" (translated from German), in R. Musgrave and A. Peacock, (eds.), *Classics in the Theory of Public Finance*. London: Macmillan, 1958 (1919).

Littlechild, S.C. and G. Owen. "A Simple Expression for the Shapley Value in a Special Case," *Management Science*, 20, 370–372 (1973).

Locke, J. "Second Treatise on Government," in *Of Civil Government*. London: Dent and Sons, 1924 (1690).

Lowry, S.T. *The Archeology of Economic Ideas*. Durham: Duke University Press, 1987.

Maniquet, F. "On Equity and Implementation in Economic Environments," Ph.D. thesis, Facultés Universitaires Notre-Dame de la Paix, Namur, 1994.

Maschler, M. "An Advantage of the Bargaining Set Over the Core," *Journal of Economic Theory*, 13, 184–192 (1976).

Mas-Collell, A. "Remarks on the Game Theoretic Analysis of a Simple Distribution of Surplus Problem," *International Journal of Game Theory*, 9, 125–140 (1980a).

Mas-Colell, A. (ed.), "Non Cooperative Approaches to the Theory of Perfect Competition," *Journal of Economic Theory*, 22, 2 (1980b).

Mas-Colell, A. *The Theory of General Economic Equilibrium: A Differentiable Approach*, Econometric Society Monograph 9. Cambridge: Cambridge University Press, 1985.

Mas-Colell, A. "Equilibrium Theory with Possibly Satiated Preferences," in M. Majumdar (ed.), *Equilibrium and Dynamics*. London: MacMillan Press, 1988, pp. 201–213.

Mas-Colell, A. and J. Silvestre. "Cost Share Equilibria: A Lindahlian Approach," *Journal of Economic Theory*, 47, 239–256 (1989).

Maskin, E. "The Theory of Implementation in Nash Equilibrium: A Survey," in L. Hurwicz, D. Schmeidler and H. Sonnenschein (eds.), *Social Goals and Social Organization*. Cambridge: Cambridge University Press, 1985.

McAfee, R.P. "A Dominant Strategy Double Auction," *mimeo, University of Texas, Austin* (1994).

McKelvey, R.D. "Intransitivities in Multi-Dimensional Voting Models and Some Implications for Agenda Control," *Journal of Economic Theory*, 12, 472–482 (1976).

McKelvey, R.D. and N. Schofield. "Generalized Symmetry Conditions at a Core Point," *Econometrica*, 55, 923–934 (1987).

Milgrom, P. and J. Roberts. "Adaptive and Sophisticated Learning in Normal Form Games," *Games and Economic Behavior*, 3 (1), 82–100 (1991).

Milgrom, P. and C. Shannon. "Monotonic Comparative Statics," *Econometrica*, 62, 157–180 (1994).

Miller, N. "Pluralism and Social Choice," *American Political Science Review*, 77, 734–745 (1983).

Moore, J.M. *Aristotle and Xenophon on Democracy and Oligarchy*, Berkeley: University of California Press, 1972.

Moore, J. "Implementation in Environments with Complete Information," in J.J. Laffont (ed.), *Advances in Economic Theory*. Cambridge, MA: Cambridge University Press, 1993.

Moulin, H. "On Strategy-Proofness and Single Peakedness," *Public Choice*, 35, 437–455 (1980).

Moulin, H. *Game Theory for the Social Sciences, 2nd Ed.* New York: New York University Press, 1986, p. 278.

Moulin, H. "A Core Selection for Pricing a Single Output Monopoly," *The Rand Journal of Economics*, Autumn, 18 (3), 397–407 (1987).

Moulin, H. *Axioms of Cooperative Decision Making*, Econometric Society Monograph 15. Cambridge, MA: Cambridge University Press, 1988.

Moulin, H. "Cores and Large Cores When Population Varies," *International Journal of Game Theory*, 19, 219–232 (1990a).

Moulin, H. "Uniform Externalities: Two Axioms for Fair Allocation," *Journal of Public Economics*, 43, 305–326 (1990b).

Moulin, H. "Welfare Bounds in the Fair Division Problem," *Journal of Economic Theory*, 54 (2), 321–337 (1991).

Moulin, H. "An Application of the Shapley Value to Fair Division with Money," *Econometrica*, 60 (6), 1331–1349 (1992).

Moulin, H. "Cost-Sharing Under Increasing Returns: A Comparison of Simple Mechanisms," *Games and Economic Behaviour* (1993).

Moulin, H. "Serial Cost Sharing of Excludable Public Goods," *Review of Economic Studies*, 61, 305–325 (1994).

Moulin, H. and S. Shenker. "Serial Cost Sharing," *Econometrica*, 60, 1009–1037 (1992).

Moulin, H. and W. Thomson. "Can Everyone Benefit from Growth? Two Difficulties," *Journal of Mathematical Economics*, 17, 339–345 (1988).

Moulin, H. and A. Watts. "Two Versions of the Tragedy of the Commons," unpublished working paper, Duke University, 1994.

Muto, S. "An Information Good Market with Symmetric Externalities," *Econometrica*, 54, 295–312 (1986).

Myerson, R.B. "Graphs and Cooperation in Games," *Mathematics of Operations Research*, 2, 225–229 (1977).

Myerson, R.B. *Game Theory: Analysis of Conflict*. Cambridge, MA: Harvard University Press, 1991.

Nozick, R. *Anarchy, State and Utopia*. New York: Basic Books, 1974.

Olson, M. *The Logic of Collective Action*. Cambridge, MA: Harvard University Press, 1965.

O'Neill, B. "A Problem of Right Arbitration in the Talmud," *Mathematical Social Sciences*, 2, 345–371 (1982).

Osborne, M.J. and A. Rubinstein. *Bargaining and Markets*. New York: Academic Press, 1990.

Ostrom, E. *Governing the Commons*. New York: Cambridge University Press, 1991.

Owen, G. *Game Theory, 2nd Ed.* New York: Academic Press, 1982.

Pazner, E. and D. Schmeidler. "A Difficulty in the Concept of Fairness," *Review of Economic Studies*, 41, 441–443 (1974).

Pazner, E. and D. Schmeidler. "Egalitarian-Equivalent Allocations: A New Concept of Economic Equity," *Quarterly Journal of Economics*, 92, 671–687 (1978).

Popper, K. *The Open Society and Its Enemies, Vol. 1*. Princeton: Princeton University Press, 1971.

Postlewaite, A. "Manipulation via Endowments," *Review of Economic Studies*, 46, 255–262 (1979).

Quinzii, M. "Core and Competitive Equilibria with Indivisibilities," *International Journal of Game Theory*, 13, 41–60 (1984).

Rawls, J. *A Theory of Justice*. Cambridge, MA: Harvard University Press, 1971.

Riker, W.H. *Liberalism Against Populism*. San Francisco: W.H. Freeman and Company, 1982.

Roemer, J. "Equality of Resources Implies Equality of Welfare," *Quarterly Journal of Economics*, 101, 751–784 (1986).

Roemer, J. "Theories of Distributive Justice," unpublished working paper, University of California, Davis, 1994.

Roemer, J. and J. Silvestre. "The Proportional Solution for Economies with Both Private and Public Ownership," *Journal of Economic Theory*, 59, 426–444 (1993).

Roth, A.E. (ed.). *The Shapley Value: Essays in Honor of Lloyd Shapley*. New York: Cambridge University Press, 1988.

Roth, A.E. "Laboratory Experimentation in Economics: A Methodological Overview," *Economic Journal*, 98, 974–1031 (1988).

Roth, A.E. and M. Sotomayor. *Two-Sided Matching*, Econometric Society Monograph 18. Cambridge: Cambridge University Press, 1990.

Rudin, W. *Real and Complex Analysis*. New York: McGraw-Hill, 1974.

Saijo, T. "Incentive Compatibility and Individual Rationality in Public Good Economics," *Journal of Economic Theory*, 55 (1), 203–212 (1991).

Samuelson, P. *Foundations of Economic Analysis*. Cambridge, MA: Harvard University Press, 1947.

Satterthwaite, M. "Strategy-Proofness and Arrow's Conditions: Existence and Correspondence Theorems for Voting Procedures and Social Choice Functions," *Journal of Economic Theory*, 10, 187–217 (1975).

Scarf, H. "The Core of a *N* Person Game," *Econometrica*, 35, 50–69 (1967).

Scarf, H. "Notes on the Core of a Productive Economy," in W. Hildenbrand and A. Mas-Colell (eds.), *Contributions to Mathematical Economics in Honor of Gerard Debreu*. Amsterdam: North-Holland, 1986.

Schelling, T.C. *The Strategy of Conflict*. Cambridge, MA: Harvard University Press, 1971.

Schmeidler, D. and K. Vind. "Fair Net Trades," *Econometrica*, 40, 637–664 (1972).

Schofield, N. *Social Choice and Democracy*. Berlin: Springer-Verlag, 1985.

Sebenius, J. *Negotiating the Law of the Sea*. Cambridge, MA: Harvard University Press, 1984.

Sen, A.K. "Labour Allocation in a Cooperative Enterprise," *Review of Economic Studies*, 33, 361–371 (1966).

Sen, A.K. *Commodities and Capabilities*. Amsterdam: North-Holland, 1985.

Shafer, W. "Equilibrium in Economies without Ordered Preferences or Free Disposal," *Journal of Mathematical Economics*, 3, 135–137 (1976).

Shapley, L.S. "A Value for *N*-person Games," in H.W. Kuhn and W. Tucker (eds.), *Contributions to the Theory of Games II*, Annals of Mathematical Studies, 28. Princeton, NJ: Princeton University Press, 1953.

Shapley, L.S. "Utility Comparisons and the Theory of Games," in *La Decision*. Paris: Editions du CNRS, 1969, pp. 251–263.

Shapley, L.S. "Cores of Convex Games," *International Journal of Game Theory*, 1, 11–26 (1971).

Shapley, L.S. and H. Scarf. "On Cores and Indivisibility," *Journal of Mathematical Economics*, 1, 23–28 (1974).

Shapley, L.S. and M. Shubik. "A Method of Evaluating Power in a Committee System," *American Political Science Review*, 48, 787–792 (1954).

Shapley, L.S. and M. Shubik. "Pure Competition, Coalitional Power and Fair Division," *International Economic Review*, 10 (3), 337–362 (1969a).

Shapley, L.S. and M. Shubik. "On the Core of an Economic System with Externalities," *American Economic Review*, 59, 678–684 (1969b).

Shapley, L.S. and M. Shubik. "The Assignment Game 1: The Core," *International Journal of Game Theory*, 1, 111–130 (1972).

Shapley, L.S. and M. Shubik. "Trade Using One Commodity as a Means of Payment," *Journal of Political Economy*, 85, 937–968 (1977).

Sharkey, W. "Cooperative Games with Large Cores," *International Journal of Game Theory*, 11, 175–182 (1982a).

Sharkey, W. *The Theory of Natural Monopoly*. Cambridge: Cambridge University Press, 1982b.

Sharkey, W. "Network Models in Economics," in *Handbook of Operations Research and Management Science*, 1993.

Shubik, M. *Game Theory in the Social Sciences: Concepts and Solutions*. Cambridge, MA: MIT Press, 1984.

Smith, A. *An Inquiry Into the Nature and Causes of the Wealth of Nations*. London, 1784.

Sprumont, Y. "The Division Problem with Single-Peaked Preferences: A Characterization of the Uniform Allocation Rule," *Econometrica*, 59 (2), 509–520 (1991).

Starrett, D. *Foundations of Public Economics*. Cambridge: Cambridge University Press, 1988.

Steinhaus, H. "The Problem of Fair Division," *Econometrica*, 16, 101–104 (1948).

Tamir, A. "On the Core of a Traveling Salesman Cost Allocation Game," *Operation Research Letters*, 8, 31–34 (1989).

Telser, L.G. *Theories of Competition*. Amsterdam: North-Holland, 1988.

Thomson, W. "The Fair Division of a Fixed Supply Among a Growing Population," *Mathematics of Operations Research*, 8, 319–326 (1983).

Thomson, W. and H. Varian. "Theories of Justice Based on Symmetry," in L. Hurwicz, D. Schmeidler, and H. Sonnenschein (eds.), *Social Goals and Social Organization*. Cambridge, MA: Cambridge University Press, 1985.

Tocqueville, A. de. *Democracy in America* (translated from French). New York: Anchor Books, 1969 (1860).

Tullock, G. "Efficient Rent Seeking," in J. Buchanan, R. Tollison, and G. Tullock (eds.), *Toward a Theory of the Rent Seeking Society*. College Station, TX: Texas A & M Press, 1980.

Varian, H. "Equity, Envy and Efficiency," *Journal of Economic Theory*, 29 (2), 217–244 (1974).

Varian, H. "Two Problems in the Theory of Fairness," *Journal of Public Economics*, 5, 249–260 (1976).

Vilkov, V.V. "Convex Games without Side Payments" (in Russian), *Vestnik Leningrad Skiva Universitata*, 7, 21–24 (1977).

Vohra, R. "Equity and Efficiency in Non-Convex Economies," *Social Choice and Welfare*, 9, 185–202 (1992).

Von Neumann, J. and O. Morgenstern. *Theory of Games and Economic Behavior, 2nd Ed.* Princeton: Princeton University Press, 1947.

Wako, J. "Some Properties of Weak Domination in an Exchange Market with Indivisible Goods," *The Economics Studies Quarterly*, 42 (4), 303–314 (1991).

Watts, A. "Provision of a Public Good: A Comparison of Lower and Upper Bounds," unpublished working paper, Duke University, 1991.

Watts, A. "On the Uniqueness of the Equilibrium in Cournot Oligopoly and Other Games," *Games and Economic Behavior* (1995).

Weber, R.J. "Probabilistic Values for Games," in A. Roth (ed.), *The Shapley Value*. New York: Cambridge University Press, 1988.

Weitzman, M. "Free Access vs Private Ownership as Alternative Systems for Managing Common Property," *Journal of Economic Theory*, 8, 225–234 (1974).

Weller, D. "Fair Division of a Measurable Space," *Journal of Mathematical Economics*, 14, 5–17 (1989).

Wicksell, K. "A New Principle of Just Taxation" (translated from German), in R. Musgrave and A. Peacock (eds.), *Classics in the Theory of Public Finance*. London: Macmillan, 1958 (1896).

Yaari, M. and M. Bar-Hillel. "On Dividing Justly," *Social Choice and Welfare*, 1, 1–14 (1984).

Young, H.P. (ed.). *Cost Allocation: Methods, Principles, Applications*. Amsterdam: North-Holland, 1985a.

Young, H.P. "Monotonic Solution of Cooperative Games," *International Journal of Game Theory*, 14 (2), 65–72 (1985b).

Young, H.P. "The Evolution of Conventions," *Econometrica*, 61, 57–84 (1993).

Young, H.P. *Equity, in Theory and Practice*. Princeton: Princeton University Press, 1994.

Zajac, E.E. "Perceived Economic Justice: The Example of Public Utility Regulation," in H.P. Young (ed.), *Cost Allocation: Methods, Principles, Applications*. Amsterdam: North-Holland, 1985.

Zhou, L. "Inefficiency of Strategyproof Allocation Mechanisms in Pure Exchange Economies," *Social Choice and Welfare*, 8, 247–254 (1991).

Zhou, L. "Strictly Fair Allocations and Walrasian Equilibria in Large Exchange Economies," *Journal of Economic Theory*, 57, 158–175 (1992).

Index